Computability in Combinatory Spaces

Mathematics and Its Applications (*East European Series*)

Volume 55

Computability in Combinatory Spaces

An Algebraic Generalization of Abstract First Order Computability

by

Dimiter G. Skordev

Department of Mathematics and Computer Science,
Sofia University,
Sofia, Bulgaria

SPRINGER SCIENCE+BUSINESS MEDIA, B.V.

Library of Congress Cataloging-in-Publication Data

Skordev, Dimitŭr Genchev.
Computability in combinatory spaces : an algebraic generalization of abstract first order computability / by Dimiter G. Skordev.
p. cm. -- (Mathematics and its applications (East European series) ; v.55)
Completely new version of: Kombinatorny postranstva i rekursivnost n nikh. 1980.
Includes bibliographical references (p.) and indexes.
ISBN 978-94-010-5165-1 ISBN 978-94-011-2636-6 (eBook)
DOI 10.1007/978-94-011-2636-6
1. Recursion theory. 2. Combinatory logic. I. Skordev, Dimitŭr Genchev. Kombinatorny postranstva i rekursivnost'v nikh. II. Title. III. Series: Mathematics and its applications (Kluwer Academic Publishers). East European series ; v. 55.
QA9.6.S558 1992
511.3'5--dc20 91-44535

ISBN 978-94-010-5165-1

This is a completely new version of the book
Combinatory Spaces and Recursiveness in them
published in Russian by the Bulgarian Academy of Sciences in 1980.

SERIES EDITOR'S PREFACE

'Et moi, ..., si j'avait su comment en revenir, je n'y serais point allé.'
Jules Verne

The series is divergent; therefore we may be able to do something with it.
O. Heaviside

One service mathematics has rendered the human race. It has put common sense back where it belongs, on the topmost shelf next to the dusty canister labelled 'discarded nonsense'.
Eric T. Bell

Mathematics is a tool for thought. A highly necessary tool in a world where both feedback and nonlinearities abound. Similarly, all kinds of parts of mathematics serve as tools for other parts and for other sciences.

Applying a simple rewriting rule to the quote on the right above one finds such statements as: 'One service topology has rendered mathematical physics ...'; 'One service logic has rendered computer science ...'; 'One service category theory has rendered mathematics ...'. All arguably true. And all statements obtainable this way form part of the raison d'être of this series.

This series, *Mathematics and Its Applications*, started in 1977. Now that over one hundred volumes have appeared it seems opportune to reexamine its scope. At the time I wrote

> "Growing specialization and diversification have brought a host of monographs and textbooks on increasingly specialized topics. However, the 'tree' of knowledge of mathematics and related fields does not grow only by putting forth new branches. It also happens, quite often in fact, that branches which were thought to be completely disparate are suddenly seen to be related. Further, the kind and level of sophistication of mathematics applied in various sciences has changed drastically in recent years: measure theory is used (non-trivially) in regional and theoretical economics; algebraic geometry interacts with physics; the Minkowsky lemma, coding theory and the structure of water meet one another in packing and covering theory; quantum fields, crystal defects and mathematical programming profit from homotopy theory; Lie algebras are relevant to filtering; and prediction and electrical engineering can use Stein spaces. And in addition to this there are such new emerging subdisciplines as 'experimental mathematics', 'CFD', 'completely integrable systems', 'chaos, synergetics and large-scale order', which are almost impossible to fit into the existing classification schemes. They draw upon widely different sections of mathematics."

By and large, all this still applies today. It is still true that at first sight mathematics seems rather fragmented and that to find, see, and exploit the deeper underlying interrelations more effort is needed and so are books that can help mathematicians and scientists do so. Accordingly MIA will continue to try to make such books available.

If anything, the description I gave in 1977 is now an understatement. To the examples of interaction areas one should add string theory where Riemann surfaces, algebraic geometry, modular functions, knots, quantum field theory, Kac-Moody algebras, monstrous moonshine (and more) all come together. And to the examples of things which can be usefully applied let me add the topic 'finite geometry'; a combination of words which sounds like it might not even exist, let alone be applicable. And yet it is being applied: to statistics via designs, to radar/sonar detection arrays (via finite projective planes), and to bus connections of VLSI chips (via difference sets). There seems to be no part of (so-called pure) mathematics that is not in immediate danger of being applied. And, accordingly, the applied mathematician needs to be aware of much more. Besides analysis and numerics, the traditional workhorses, he may need all kinds of combinatorics, algebra, probability, and so on.

In addition, the applied scientist needs to cope increasingly with the nonlinear world and the extra mathematical sophistication that this requires. For that is where the rewards are. Linear models are honest and a bit sad and depressing: proportional efforts and results. It is in the nonlinear world that infinitesimal inputs may result in macroscopic outputs (or vice versa). To appreciate what I am hinting at: if electronics were linear we would have no fun with transistors and computers; we would have no TV; in fact you would not be reading these lines.

There is also no safety in ignoring such outlandish things as nonstandard analysis, superspace and anticommuting integration, p-adic and ultrametric space. All three have applications in both electrical engineering and physics. Once, complex numbers were equally outlandish, but they frequently proved the shortest path between 'real' results. Similarly, the first two topics named have already provided a number of 'wormhole' paths. There is no telling where all this is leading - fortunately.

Thus the original scope of the series, which for various (sound) reasons now comprises five subseries: white (Japan), yellow (China), red (USSR), blue (Eastern Europe), and green (everything else), still applies. It has been enlarged a bit to include books treating of the tools from one subdiscipline which are used in others. Thus the series still aims at books dealing with:

- a central concept which plays an important role in several different mathematical and/or scientific specialization areas;
- new applications of the results and ideas from one area of scientific endeavour into another;
- influences which the results, problems and concepts of one field of enquiry have, and have had, on the development of another.

One of the relative newcomers to the family of mathematical subjects is the theory of computation and computability. Just what it means for an object to be computable is the central question and it is far from answered. Recursiveness is important and the theory of general recursive (partial) functions constitutes an early success.

The author of the present volume originated in 1974 an algebraic approach to capture the idea of more general recursive objects which has led to something of a school in the topic. Central in this are his '(iterative) combinatory spaces'.

The theory of these objects by now has a substantial number of deep results which generalize and at times simplify (as good generalizations tend to do) corresponding results in recursive function theory.

Here is the comprehensive account of this approach to 'recursiveness' and 'computability' by the author who started it almost 20 years ago.

The shortest path between two truths in the real domain passes through the complex domain.

J. Hadamard

La physique ne nous donne pas seulement l'occasion de résoudre des problèmes ... elle nous fait pressentir la solution.

H. Poincaré

Never lend books, for no one ever returns them; the only books I have in my library are books that other folk have lent me.

Anatole France

The function of an expert is not to be more right than other people, but to be wrong for more sophisticated reasons.

David Butler

Bussum, 9 February 1992

Michiel Hazewinkel

TO THE MEMORY OF MY PARENTS

CONTENTS

PREFACE

About thirty years ago, logicians and computer scientists, more or less independently, began to study systematically computability and computations on abstract algebraic structures. The papers Fraïssé [1961], Lacombe [1964, 1964a], Kreisel [1965] are among the earliest in this direction, which are written by logicians, and the papers Ershov [1958] and McCarthy [1962, 1963] are among the first ones, which approached the same problems from the other side. A series of other essential contributions to the field have been produced, and a theory of computability arose, generalizing the ordinary recursive function theory (the monographs Fenstad [1980] and Fitting [1981] give a comprehensive account of certain directions of this general theory).

The present book is a monograph on a further generalization of a part of the general theory in question, namely of the theory of abstract first order computability from the paper Moschovakis [1969]. When reading that paper, the present author was, in particular, intrigued by the use made of multiple-valued functions. In 1974, he observed certain useful algebraic properties of the operations on such functions. This step enabled him to make a further one, namely to generalize certain notions and results of the theory of the abstract first order computability in such a way that also the role of functions can be played by objects of a more general nature. It has been proposed to use as such function-like objects the elements of appropriate partially ordered algebraic structures, called *combinatory spaces*. Among all possible combinatory spaces, the so-called *iterative* ones turned out to be suitable for the development of a theory of computability. Various examples of such structures have been found, and the intuitive idea about most of these examples can be described as follows.

It is usual to study data processing devices (in particular, computational procedures) by means of some mathematical models which describe their behaviour. The most frequently used model is the input-output relation showing which are the possible output data corresponding to any concrete instance of the input data (in the case of a deterministic device, this relation is the graph of a partial function). One could supplement this model with information about those instances of the input data at which some unpleasant effects during the work of the device are possible, such as, for example, unproductive termination, the presence of a failure or entering into an infinite loop. A further model arises if, in conjunction with this, one neglects information about the possible output data corresponding to such unsafe instances of the input data and removes this information from the model. In the case of a probabilistic device, the appropriate mathematical model would be not a relation, but a real-valued function showing, for each concrete instance of the input data, which are the probabilities of the various possible output data. Many other kinds of mathematical models can turn out to be appropriate in different situations. An important point is that there are some widely used ways of combining data processing devices, and to these ways usually some operations on the mathematical models of the devices correspond. For example, one can usually define an associative operation of composition of the mathematical models, which corresponds to sequential composition (piping) of the devices. In most examples of combinatory spaces, the elements of the space can be regarded as possible mathematical models (of some fixed kind) corresponding to concrete data processing devices, and the algebraic operations on these elements will correspond to some ways of combining the devices.

The above intuitive idea gives an explanation of our choice of the term "combinatory space". The adjective "combinatory" is connected with combining the devices, and there is no close connection between our theory and Combinatory Logic (although similarities in certain respects can be observed). A combinatory space can be regarded as a system for functional programming, dealing with some kind of functionlike objects, and the FP-systems from Backus [1978] are a particular case of this.

A mathematical theory of a general nature must be justified by the existence of sufficient examples. But this is not enough. The theory must also contain serious mathematical results. We hope that such is the case of the theory of combinatory spaces. At this moment we cannot enter into details, and we shall only mention that our generalization of the recursive function theory contains (among others) such results as the First Recursion Theorem, Normal Form Theorems, Theorem on Existence of Universal Computable Elements with Second Recursion Theorem. Moreover, the corresponding results from the ordinary recursive function theory and from Moschovakis' theory of the abstract first order computability can be obtained as particular instances. We note also that the notions of prime and search computability from the latter theory obtain characterizations, which are simpler and easier to use than the definitions in the paper Moschovakis [1969].

As already mentioned, the theory of combinatory spaces arose about 1974. Before 1980, parts of it have been presented in the papers Skordev [1975, 1976a–e, 1977, 1978, 1979] and Ivanov [1978] (in the paper Petrov and Skordev [1979] an attempt is made to use categories instead of semi-groups, but the success is only partial). The state of this theory at the beginning of 1978 is presented in the monograph Skordev [1980]. During the time when that book was in print, and later, many additions and improvements have been done (in particular, one of the conditions in the definition of the notion of combinatory space has been considerably weakened). A number of colleagues and students have taken an active part in the development of the theory and in its application since 1975. Here are the names of these people, listed in alphabetical order: A. Ditchev, N. Georgieva, O. Ignatov, L. Ivanov, R. Lukanova, S. Nikolova, E. Pazova, V. Petrov, I. Soskov, A. Soskova, M. Tabakov. Ideas from the theory have been used also in papers of G. Gargov, S. Passy, A. Radensky, T. Tinchev, D. Vakarelov and J. Zashev.

The present book is a completely new version of the above-mentioned monograph and aims at giving an account of the current state of the theory of combinatory spaces and of some of its applications. When starting the work on the manuscript, the author had rather ambitious plans. He planned to present not only his own results, but also those of his colleagues, and also to give a short account of the very interesting other algebraic generalizations of the theory of computability, obtained by L. Ivanov and J. Zashev. Unfortunately, the manuscript became too large. Therefore the author had to restrict his plans. The contributions of other authors are presented only partially in this book. As to Ivanov's and Zashev's generalizations (and especially Zashev's one), they are only mentioned with corresponding references (one of these references is the monograph Ivanov [1986], which contains also much information about combinatory spaces).

Some other reductions have been also made. Many results are given in the form of exercises (some of them are accompanied by hints). The details from Skordev [1980], concerning the possibility of using intuitionistic logic in most of the proofs, are omitted now. In the preface of that book, there was a brief survey of earlier publications to which the theory of combinatory spaces is more or less related. Now this survey is omitted, but the main list of references is supplemented with an additional bibliography, in whose first part those publications are listed (the second part is a list of publications later than 1980, which have a certain relationship to the theory of combinatory spaces, but are not mentioned in the text).

The author has tried to avoid most references to his previous publications (especially to the previous version of this book). However, in some places such references are given to indicate additional sources of information or for making possible a discussion about the course of the development of the studies.

The main text of this book consists of three chapters. All sections except those in the appendices, are accompanied by exercises. The exposition is organized in such a way, that results from the exercises are usually not logically needed in the other part of the text. An exception is made in Section III.8, where results from some exercises in Section II.1 are used. In Section 8 of the Appendix, some exercises from Sections II.4 and III.2 are needed not directly, but in the role of hints. The material from any of the sections III.6-III.9, as well as from any of the sections 3, 5, 7-10 of the Appendix, is not used in the further exposition.

For the reading of some places of the book, a knowledge of mathematical logic or recursive function theory is needed.

The help of many people and of some institutions was very essential for writing this book. Discussions with A. Ditchev, L. Ivanov, R. Lukanova, S. Nikolova, I. Soskov, A. Soskova, J. Zashev have been extremely useful for the work. I. Soskov and S. Nikolova helped also by reading some parts of the manuscript. L. Ivanov, I. Soskov and T. Tinchev introduced me to the text-editing by computer and gave me many practical advices in this respect. P. Petkov organized the printing of the text, A. Zinoviev did most of the printing, and V. Goranko provided me promptly with xerox copies. During this last month of the work, D. Vakarelov took my official duties upon himself in order to give me more free time. Throughout several months, my wife and my sons had to endure to see me at home only for staying overnight, and all this time they did everything in their power to assist me. The work of the manuscript has been partially supported by the Bulgarian Ministry of Science and Higher Education under Contract no. 247/1987, 1990. Kluwer Academic Publishers have beeen patient enough to wait for the manuscript much longer than it was initially planned.

I am deeply grateful to all these people and institutions.

September, 1991

Dimiter Skordev

CHAPTER I

COMPUTATIONAL STRUCTURES AND COMPUTABILITY ON THEM

1. Computational structures

As noted in the preface, we shall consider computability not only of functions, but also of a large variety of other kinds of function-like objects. However, we feel it would be not wise to start with the general case from the very beginning. Therefore we shall first consider a certain notion of computability concerning ordinary functions, as well as its natural generalization for the case of multiple-valued functions (cf. Sections 2 and 5 of this chapter). Although this notion can be reduced to other ones which are well-known[1], it provides a class of examples which we consider useful for the better understanding of the general theory and for the demonstration of its applicability.

The above mentioned notion of computability will concern unary functions in so-called computational structures. These will be a certain kind of algebraic structures (possibly partial).

Definition 1. A *computational structure* is a 7-tuple $\langle M, J, L, R, T, F, H\rangle$, where M is an infinite set, J is an injective mapping of M^2 into M, L and R are partial mappings of M into M, T and F are total mappings of M into M, H is a partial predicate on M and the following equalities are satisfied for all s, t, u in M:

$$L(J(s, t)) = s, \quad R(J(s, t)) = t,$$

$$H(T(u)) = \text{true}, \quad H(F(u)) = \text{false}.$$

If $\langle M, J, L, R, T, F, H\rangle$ is a computational structure,

---

[1]For a typical case of the considered situation, such a reduction can be found in Soskov [1985].

then the mapping J can be used for coding ordered pairs of elements of M by elements of M, and L, R provide us with the corresponding means for decoding ($\operatorname{dom} L$ and $\operatorname{dom} R$ must include the set $\operatorname{rng} J$, without necessarily being equal to it). As to T, F, H, we could regard all values of T and F as codes of the logical values true and false, respectively, and the partial predicate H can be regarded as a means for the corresponding decoding (of course, this predicate will transform into logical values all elements from its domain, although some of these elements could belong neither to $\operatorname{rng} T$ nor to $\operatorname{rng} F$).

Three examples of computational structures $\langle M, J, L, R, T, F, H\rangle$ follow.

Example 1. Let M be the set $\mathbb{N}=\{0, 1, 2, 3, \ldots\}$ of the natural numbers, and let

$$J=\lambda st.\, 2^s . 3^t,\quad L=\lambda u.\,(u)_0,\quad R=\lambda u.\,(u)_1$$

(where $(u)_0$ and $(u)_1$ denote the exponent of 2 and of 3, respectively, in the prime decomposition of u if $u>0$, and they denote 0 if $u=0$). Let $T=\lambda u.\,1$, $F=\lambda u.\,0$, and

$$H(u)=\begin{cases}\text{true} & \text{if } u>0,\\ \text{false} & \text{if } u=0.\end{cases}$$

Example 2 (cf. Moschovakis [1969]). Let B be some set, B° be the set $B\cup\{O\}$, where O is some object not in B, and M be the least set containing B° and closed under formation of ordered pairs (in the mentioned paper, this set is denoted by B^*). The ordered pair operation used in the construction of M is assumed to be chosen in such a way that no element of B° is an ordered pair. Let J be the mapping $\lambda st.\,\langle s, t\rangle$, and the mappings L, R be defined by the conditions

$$L(O)=R(O)=O,$$

$$L(u)=R(u)=\langle O, O\rangle \text{ for all } u \text{ in } B,$$

$$L(\langle s, t\rangle)=s,\quad R(\langle s, t\rangle)=t \text{ for all } s, t \text{ in } M$$

(in Moschovakis [1969], these mappings are denoted by π and δ, respectively). Let $T=\lambda u.\,\langle O, O\rangle$, $F=\lambda u.\,O$, and

$$H(u)=\begin{cases}\text{true} & \text{if } u\notin B^\circ,\\ \text{false} & \text{if } u\in B^\circ.\end{cases}$$

The computational structure obtained in this way will be called *the Moschovakis computational structure based on* B and will be denoted by $\mathfrak{M}_B$ (the element O is not indicated in the name and in the denotation, since the concrete choice of O can be usually considered immaterial).

Example 3. Let an FP-system in the sense of Backus [1978] be given. Denote by M the set of all its objects with the exception of the special atom $\perp$.[2] Let J be the mapping $\lambda st.\langle s, t\rangle$, and L, R be the functions having as their domain the set of all non-empty finite sequences of elements of M and selecting from each such sequence its first and its last member, respectively[3]. Let T and F be the constant functions whose values at all elements of M are equal, respectively, to the atoms meaning truth and falsity. Let $\mathrm{dom}H$ be the subset of M containing exactly these two atoms, and let H have the value **true** for the first of them and the value **false** for the second one.

Note that in all three above examples T and F are constant functions, and this could be considered to be a typical case.

We should like to add several further examples of computational structures. In some of them, T and F will not be constant functions.

Example 4. Let M be the set of all infinite sequences of real numbers, and J, L, R, T, F, H be defined as follows:

$$J(\langle s_0, s_1, s_2, \ldots\rangle, \langle t_0, t_1, t_2, \ldots\rangle) = \langle s_0, t_0, s_1, t_1, \ldots\rangle,$$

$$L(\langle u_0, u_1, u_2, \ldots\rangle) = \langle u_0, u_2, u_4, \ldots\rangle,$$

$$R(\langle u_0, u_1, u_2, \ldots\rangle) = \langle u_1, u_3, u_5, \ldots\rangle,$$

$$T(\langle u_0, u_1, u_2, \ldots\rangle) = \langle 1, u_0, u_1, u_2, \ldots\rangle,$$

$$F(\langle u_0, u_1, u_2, \ldots\rangle) = \langle -1, u_0, u_1, u_2, \ldots\rangle,$$

---

[2] The objects of the FP-system can be described as follows. One starts from some objects called *atoms*, among them one meaning truth and another one meaning falsity (Backus denotes them by T and F, respectively). Also a special atom $\perp$ is provided whose meaning is "not defined". Then the set of the objects of the FP-system consists of all atoms, of the empty sequence $\emptyset$ and of all non-empty finite sequences $\langle u_1, \ldots, u_n\rangle$ whose members u_i are already constructed objects of the system distinct from $\perp$.

[3] L and R could be identified with the functions which Backus denotes by 1 and 1r.

$$H(\langle u_0, u_1, u_2, \ldots \rangle) = \begin{cases} \text{true if } u_0 > 0, \\ \text{false if } u_0 < 0 \end{cases}$$

($H(\langle u_0, u_1, u_2, \ldots \rangle)$ is not defined if $u_0 = 0$).

Example 5. Let $\langle \mathcal{F}, I, \Pi_*, L_*, R_* \rangle$ be an operative space in the sense of Ivanov [1986] (cf. Section II.2 of the present book for the definition). Let M be the set of the elements of the semigroup $\mathcal{F}$, and let H be defined as follows:

$$\mathrm{dom}\, H = \{L_*, R_*\}, \quad H(u) = \begin{cases} \text{true} & \text{if } u = L_*, \\ \text{false} & \text{if } u = R_*. \end{cases}$$

Then $\langle M, \Pi, \lambda u. L_* u, \lambda u. R_* u, \lambda u. L_*, \lambda u. R_*, H \rangle$ is a computational structure.

Example 6. Let $\langle M, J, L, R, T, F, H \rangle$ be an arbitrary computational structure. Then the following three 7-tuples are also computational structures:

$$\langle M, \lambda ts. J(s, t), R, L, T, F, H \rangle,$$

$$\langle M, J, L, R, F, T, \mathrm{not}\, H \rangle,$$

$$\langle M, \lambda ts. J(s, t), R, L, F, T, \mathrm{not}\, H \rangle.$$

If L', R' are the restrictions of L and R, respectively, to the set $\mathrm{rng}\, J$, and H' is the restriction of H to the union of $\mathrm{rng}\, T$ and $\mathrm{rng}\, F$, then $\langle M, J, L', R', T, F, H \rangle$ and $\langle M, J, L, R, T, F, H' \rangle$, too, are computational structures.

Example 7. Let B be an arbitrary non-empty set whose elements are not ordered pairs, and let M be the least set containing B and closed under formation of ordered pairs. Let $J = \lambda st. \langle s, t \rangle$, and the mappings L, R be defined by the conditions

$$\mathrm{dom}\, L = \mathrm{dom}\, R = M \setminus B,$$

$$L(\langle s, t \rangle) = s, \quad R(\langle s, t \rangle) = t \text{ for all } s, t \text{ in } M.$$

Let $T = \lambda u. \langle u, u \rangle$, and F be defined by induction as follows: $F(u) = u$ for all u in B, and $F(\langle s, t \rangle) = F(s)$ for all s, t in M. Let

$$H(u) = \begin{cases} \text{true} & \text{if } u \notin B, \\ \text{false} & \text{if } u \in B. \end{cases}$$

Then $\langle M, J, L, R, T, F, H \rangle$ is a computational structure. This structure (with an exchange between the codes of truth and falsity) has been introduced and used, in essence, in the thesis Soskova [1979].

Example 8. Let M, J, L, R be defined in the same way as in the previous example, but for the particular case of

$B=\mathbb{N}$. Let $T=\lambda u.\langle 0, 0\rangle$, $F=\lambda u.0$, and

$$H(u)=\begin{cases}\text{true} & \text{if } u\neq 0,\\ \text{false} & \text{if } u=0.\end{cases}$$

Then $\langle M, J, L, R, T, F, H\rangle$ is again a computational structure. It has been used, in essence, in the thesis Ignatov [1979].

Exercise. Show that we get an equivalent form of the definition of computational structure if we make the following two modifications (or one of them) in it: (i) omitting the requirement J to be injective; (ii) replacing the requirement M to be infinite by the requirement M to be non-empty.

2. Computability of partial functions with respect to a given computational structure

For each set M, let $\mathcal{F}_p(M)$ denote the set of all partial mappings of M into M. If φ and ψ belong to this set then their *composition* $\varphi\psi$ (denoted also by $\varphi\circ\psi$) is the element θ of $\mathcal{F}_p(M)$ determined by the condition that

$$\theta(u)\simeq\varphi(\psi(u))$$

for all u in M. *The identity mapping* $\lambda u.u$ of M onto M will be denoted by I_M.

Suppose now a computational structure $\mathfrak{U}=\langle M, J, L, R, T, F, H\rangle$ is given. Then we shall define two other binary operations in $\mathcal{F}_p(M)$. The first one will be called $\mathfrak{U}$-*combination*, and it will be denoted by Π. By definition, for arbitrary φ, ψ in $\mathcal{F}_p(M)$, $\Pi(\varphi, \psi)$ is the element θ of $\mathcal{F}_p(M)$ determined by the condition that

$$\theta(u)\simeq J(\varphi(u), \psi(u))$$

for all u in M. The second one will be called $\mathfrak{U}$-*iteration* (*iteration*, for short). The result of its application to the elements σ and χ of $\mathcal{F}_p(M)$ will be denoted[4] by $[\sigma, \chi]$, and it will be named *the iteration of* σ *controlled by* χ. By definition, $[\sigma, \chi]$ is the function θ determined by the following condition: $\theta(u)=w$ iff there is a finite sequence $v_0, v_1, \ldots, v_m$ of elements of M such that

[4] Compare with the denotations used in Buchberger [1974].

$$v_0 = u \ \& \ v_m = w \ \& \ \forall_{j<m} j\, (H(\chi(v_j)) = \text{true} \ \& \ v_{j+1} = \sigma(v_j)) \ \& \ H(\chi(v_m)) = \text{false}$$

(in Pascal-like denotations, the function θ can be represented by means of the declaration

```
function θ(u: M): M;
  var v: M;
begin
  v: = u;
  while H(χ(v)) do v: = σ(v);
  θ: = v
end; ).
```

It is useful to note that

$$(1) \qquad [\sigma, \chi](u) \simeq \begin{cases} [\sigma, \chi](\sigma(u)) & \text{if } H(\chi(u)) = \text{true}, \\ u & \text{if } H(\chi(u)) = \text{false}. \end{cases}$$

One more operation looks very natural, namely an operation of definition by cases, which will be called $\mathfrak{A}$-*branching*. This is a ternary operation in $\mathcal{F}_p(M)$ which will be denoted by Σ. By definition, for arbitrary χ, φ, ψ in $\mathcal{F}_p(M)$, $\Sigma(\chi, \varphi, \psi)$ is the element θ of $\mathcal{F}_p(M)$ determined by the condition that $\theta(u) = w$ iff

$$H(\chi(u)) = \text{true} \ \& \ \varphi(u) = w \ \vee \ H(\chi(u)) = \text{false} \ \& \ \psi(u) = w.$$

In some issues, however, this operation could be not taken into account since it can be expressed by means of the preceding ones (see Exercises 1 and 2).

Using the operation Σ, we can formulate the following characterization of the iteration as a least fixed point: for arbitrary σ, χ in $\mathcal{F}_p(M)$, the equality

$$(2) \qquad [\sigma, \chi] = \Sigma(\chi, [\sigma, \chi]\sigma, I_M)$$

holds, and $[\sigma, \chi]$ is a subfunction of each $\tau \in \mathcal{F}_p(M)$ which satisfies the condition that $\Sigma(\chi, \tau\sigma, I_M)$ is a subfunction of τ (compare, for example, with Mazurkiewicz [1971] or Scott [1971, Section 7]). Of course, (2) is an easy consequence of (1) (the only additional thing needed for deriving (2) is the fact that $H(\chi(u))$ is defined for all u in $\text{dom}[\sigma, \chi]$). As to the second part of the statement (the minimality of the iteration), we prefer not to give its proof here, since in a further section of this chapter the more general (and a little more complicated) case of the iteration of multiple-valued functions will be considered (cf. also Exercises 3 and 8 after this section).

Now we shall define a notion of relative computability of elements of $\mathcal{F}_p(M)$ with respect to the given computational structure $\mathfrak{A}$.

Definition 1. Let $\mathcal{B}$ be some subset of $\mathcal{F}_p(M)$. The elements of $\mathcal{F}_p(M)$ *$\mathfrak{A}$-computable in $\mathcal{B}$* (*computable in $\mathcal{B}$*, for short) are those elements of $\mathcal{F}_p(M)$ which can be generated from elements of $\{L, R, T, F\}\cup\mathcal{B}$ by means of composition, $\mathfrak{A}$-combination and $\mathfrak{A}$-iteration.

Of course, if each element of $\mathcal{B}$ is computable in $\mathcal{B}'$ (in particular, if $\mathcal{B}$ is contained in $\mathcal{B}'$), then each element computable in $\mathcal{B}$ is computable in $\mathcal{B}'$. Exercise 1 below shows that I_M and the function, whose domain is empty, are $\mathfrak{A}$-computable in $\emptyset$. From Exercise 2 and the computability of I_M, it is seen that $\Sigma(\chi, \varphi, \psi)$ is always $\mathfrak{A}$-computable in $\{\chi, \varphi, \psi\}$; hence including Σ as an additional generating operation in the above definition of computability would not enlarge the set of the elements of $\mathcal{F}_p(M)$ which are $\mathfrak{A}$-computable in $\mathcal{B}$. However, one could ask whether there are not other reasonable effective constructions in $\mathcal{F}_p(M)$ which could enlarge this set. In this book it will be shown that in some sense such other constructions do not exist. In particular, functions computable by means of a large class of recursive programs will be found to be computable in our sense.

By considering an uniform variant of the introduced computability notion, we could also define $\mathfrak{A}$-computability for operators in $\mathcal{F}_p(M)$. Here is the corresponding definition.

Definition 2. Let $\mathcal{B}\subseteq\mathcal{F}_p(M)$, and let Γ be a mapping of $(\mathcal{F}_p(M))^l$ into $\mathcal{F}_p(M)$, where l is some positive integer. Then Γ is called (*an operator*) *$\mathfrak{A}$-computable in $\mathcal{B}$* (*computable in $\mathcal{B}$*, for short) iff, for arbitrary $\psi_1, \ldots, \psi_l$ in $\mathcal{F}_p(M)$, there is an explicit expression for $\Gamma(\psi_1, \ldots, \psi_l)$ through $L, R, T, F, \psi_1, \ldots, \psi_l$ and elements of $\mathcal{B}$ by means of composition, $\mathfrak{A}$-combination and $\mathfrak{A}$-iteration, the form of the expression not depending on the concrete choice of $\psi_1, \ldots, \psi_l$.

Remark 1. The above definition can be formulated more precisely using induction. We could, for example, adopt the following formulation:

(i) For each $\alpha\in\{L, R, T, F\}\cup\mathcal{B}$, the mapping $\lambda\psi_1\ldots\psi_l.\alpha$ is *$\mathfrak{A}$-computable in $\mathcal{B}$*.

(ii) The mappings $\lambda\psi_1 \ldots \psi_l . \psi_i$, $i=1, \ldots, l$, are $\mathfrak{A}$*-computable in* $\mathcal{B}$.

(iii) If Γ_1 and Γ_2 are mappings of $(\mathcal{F}_p(M))^l$ into $\mathcal{F}_p(M)$, which are $\mathfrak{A}$-computable in $\mathcal{B}$, then *so* are

$$\lambda\psi_1 \ldots \psi_l . \Gamma_1((\psi_1, \ldots, \psi_l))\Gamma_2((\psi_1, \ldots, \psi_l)),$$
$$\lambda\psi_1 \ldots \psi_l . \Pi(\Gamma_1(\psi_1, \ldots, \psi_l), \Gamma_2(\psi_1, \ldots, \psi_l)),$$
$$\lambda\psi_1 \ldots \psi_l . [\Gamma_1(\psi_1, \ldots, \psi_l), \Gamma_2(\psi_1, \ldots, \psi_l)].$$

Example 1. Exercises 1 and 2 below show that Σ is an operator computable in $\emptyset$.

In the next two sections, the problem will be studied which are the functions $\mathfrak{A}$-computable in certain sets $\mathcal{B}$ for computational structures $\mathfrak{A}$ as in Examples 1.1 and 1.3 (in the case of Example 1.1, also the computable operators will be considered). For the computational structures described in Example 1.2, the same problem will be studied a little later, as a part of the more general problem concerning multiple-valued functions.

Exercises

(In all these exercises, a computational structure $\mathfrak{A} = \langle M, J, L, R, T, F, H\rangle$ is supposed to be given)

1. Prove that $[\sigma, T] = \emptyset$,[5] and $[\sigma, F] = I_M$ for all $\sigma \in \mathcal{F}_p(M)$.

2. Let $L_* = \Pi(T, I_M)$, $R_* = \Pi(F, I_M)$. For arbitrary χ, φ, ψ in $\mathcal{F}_p(M)$, prove the equalities

$$\Sigma(L, \varphi R, \psi R)L_* = \varphi, \quad \Sigma(L, \varphi R, \psi R)R_* = \psi,$$
$$\Sigma(\chi, \varphi, \psi) = R[R_*\psi R, L]R[R_*^2\varphi R^2, L]\Pi(\chi, L_*)$$

(compare the last one with the equality

$$\Delta(\alpha, a, b) = \underline{\alpha}(KKaTT)K(KbT)K$$

in Böhm and Jacopini [1966, p. 369], taking into account the difference in denotations, the absence of our operation Π there and the fact that iteration there has the semantics of **while not**).

3. Let σ, χ, ρ be arbitrary elements of $\mathcal{F}_p(M)$, and let $\tau_0 = \rho[\sigma, \chi]$. Prove that τ_0 is a solution of the equality

---

[5] We identify functions with their graphs, hence $\emptyset$ is the function whose domain is empty.

$\tau = \Sigma(\chi, \tau\sigma, \rho)$, and τ_0 is a subfunction of each $\tau \in \mathcal{F}_p(M)$ which satisfies the condition that $\Sigma(\chi, \tau\sigma, \rho)$ is a subfunction of τ.

4. Suppose T and F are constant mappings. Let $T(u) = a$, $F(u) = b$ for all u in M. Define elements $0^{\#}$, $1^{\#}$, $2^{\#}$, ... of M as follows[6]:

$$k^{\#} = \begin{cases} J(b, b) & \text{if } k = 0, \\ J(a, (k-1)^{\#}) & \text{otherwise.} \end{cases}$$

Prove the existence of an element φ of $\mathcal{F}_p(M)$ which is $\mathfrak{A}$-computable in $\emptyset$ and satisfies the condition $\varphi(n^{\#}) = (2n)^{\#}$ for all $n \in \mathbb{N}$.

Hint. First construct an element σ of $\mathcal{F}_p(M)$ which is $\mathfrak{A}$-computable in $\emptyset$ and satisfies the equality

$$\sigma(J(k^{\#}, l^{\#})) = J((k-1)^{\#}, (l+1)^{\#})$$

for all non-zero k in $\mathbb{N}$ and all $l \in \mathbb{N}$.

5. Prove the conclusion of Exercise 4 without the assumption that T and F are constant mappings, and using the following weaker assumption about a, b: a and b are fixed elements of M such that $H(a) = \text{true}$, $H(b) = \text{false}$.

6. Let *the dual* (or *while not*) *iteration* $[\sigma, \chi]_d$ *of* σ *controlled by* χ be defined be exchanging true and false in the definition of $[\sigma, \chi]$ (i. e. $[\sigma, \chi]_d$ is the $\mathfrak{A}'$-iteration of σ controlled by χ, where $\mathfrak{A}' = \langle M, J, L, R, F, T, \text{not}\, H\rangle$). Prove that the mapping $\lambda\sigma\chi.[\sigma, \chi]_d$ is $\mathfrak{A}$-computable in $\emptyset$.[7]

Hint. Show that $\text{not}\, H(\chi(u)) = H(\Sigma(\chi, F, T)(u))$ for all χ in $\mathcal{F}_p(M)$ and all u in M.

---

[6] This generalizes a representation of natural numbers from Moschovakis [1969] (cf. Section 7 of his chapter).

[7] Consequently, $\lambda\sigma\chi.[\sigma, \chi]$ is $\mathfrak{A}'$-computable in $\emptyset$ (since $[\sigma, \chi]$ is the dual iteration in $\mathfrak{A}'$). Note that iteration in the examples, considered in the papers Skordev [1975, 1976a, 1976b, 1976c] and in Chapter I of the book Skordev [1980], is the same as in the examples which will be given in the next sections, since it was defined, roughly speaking, as the dual iteration in the corresponding structures $\mathfrak{A}$.

7. Let $\mathcal{B} \subseteq \mathcal{F}_p(M)$, and let $\mathfrak{U}'$ be some of the three other computational structures related to $\mathfrak{U}$, which were mentioned at the end of Section 1. Prove that the elements of $\mathcal{F}_p(M)$, $\mathfrak{U}'$-computable in $\mathcal{B}$, are the same as its elements $\mathfrak{U}$-computable in $\mathcal{B}$.

8. Let σ, χ, ρ be arbitrary elements of $\mathcal{F}_p(M)$, and let $\tau_0 = \rho[\sigma, \chi]$. Suppose K is a subset of M such that $\sigma(u) \in K$ for all $u \in K \cap \mathrm{dom}\,\sigma$, and τ is an element of $\mathcal{F}_p(M)$ such that the restriction of $\Sigma(\chi, \tau\sigma, \rho)$ to K is a subfunction of τ. Prove that the restriction of τ_0 to K is a also a subfunction of τ.

3. On a procedure for generating the unary partial recursive functions

In this section, a characterization will be given of the unary partial recursive functions as the elements of $\mathcal{F}_p(\mathbb{N})$ $\mathfrak{U}$-computable in $\mathcal{B}$, where $\mathfrak{U}$ is a certain computational structure whose carrier is $\mathbb{N}$, and $\mathcal{B}$ is a certain finite subset of $\mathcal{F}_p(\mathbb{N})$. By giving such a characterization we aim to show the place of the theory of partial recursive functions as a special case of the general theory developed further in this book. As a suitable computational structure $\mathfrak{U}$, the one from Example 1.1 can be taken.

Let T, F, H be the same as in Example 1.1, i. e. $T = \lambda u.1$, $F = \lambda u.0$, and

$$H(u) = \begin{cases} \text{true} & \text{if } u > 0, \\ \text{false} & \text{if } u = 0. \end{cases}$$

Let $S = \lambda u.u+1$, $P = \lambda u.u \dot{-} 1$. It will be said that $\mathfrak{U}$ is *a standard computational structure on the natural numbers* iff $\mathfrak{U} = \langle \mathbb{N}, J, L, R, T, F, H \rangle$, where J is a recursive function of two variables, L, R are partial recursive functions of one variable, and the equalities $L(J(s, t)) = s$, $R(J(s, t)) = t$ hold for all s, t in $\mathbb{N}$ (as an example of such a structure the computational structure from the above mentioned example can be taken).

Theorem 1. Let $\mathfrak{U}$ be a standard computational structure on the natural numbers. Then the unary partial recursive functions are exactly those elements of $\mathcal{F}_p(\mathbb{N})$ which are $\mathfrak{U}$-computable in $\{S, P\}$.

Remark 1. If $\mathfrak{U} = \langle \mathbb{N}, J, L, R, T, F, H \rangle$ satisfies the above assumption, and $\mathcal{B}$ is a subset of $\mathcal{F}_p(\mathbb{N})$ containing

$\{S, P\}$, then T and F can be omitted as initial elements in the definition of $\mathfrak{U}$-computability in $\mathfrak{B}$. This follows from the equalities $I_{\mathbb{N}} = PS$, $F = [P, I_{\mathbb{N}}]$, $T = SF$. So the theorem states that the unary partial recursive functions are exactly those elements of $\mathcal{F}_p(\mathbb{N})$ which can be generated from L, R, S, P by means of finitely many applications of composition, $\mathfrak{U}$-combination and $\mathfrak{U}$-iteration. We think this result must be considered well-known, but we are not able to give a relevant bibliographical reference.

Proof. Let $\mathfrak{U} = \langle \mathbb{N}, J, L, R, T, F, H \rangle$. We set $\mathcal{F} = \mathcal{F}_p(\mathbb{N})$, $\mathfrak{B} = \{S, P\}$ for short. Let $\mathfrak{G}_1$ be the set of all elements of $\mathcal{F}$ which are $\mathfrak{U}$-computable in $\mathfrak{B}$. Since all elements of $\mathfrak{G}_1$ are partial recursive, we have only to show that, conversely, all unary partial recursive functions belong to $\mathfrak{G}_1$.

For each integer n greater than 1, we set

$$J_n(s_1, \dots, s_n) = J(s_1, J(s_2, \dots, J(s_{n-1}, s_n)\dots)),$$

and then we define $\mathfrak{G}_n$ to be the set of all functions having the form $\lambda s_1 \dots s_n . \psi(J_n(s_1, \dots, s_n))$, where $\psi \in \mathfrak{G}_1$. The theorem will be proved if we succeed to show that the union of the sets $\mathfrak{G}_1, \mathfrak{G}_2, \mathfrak{G}_3, \dots$ contains the initial partial recursive functions S, F and $\lambda s_1 \dots s_n . s_i$ for $n = 1, 2, \dots$, $i = 1, 2, \dots, n$, and this union is closed under substitution, primitive recursion and μ-operation.

Of course, $S, F \in \mathfrak{G}_1$, and also $\lambda s_1 . s_1 = I_{\mathbb{N}}$ belongs to $\mathfrak{G}_1$. If $n > 1$ then

$$s_i = LR^{i-1}(J_n(s_1, \dots, s_n))$$

for $i = 1, 2, \dots, n-1$, and

$$s_n = R^{n-1}(J_n(s_1, \dots, s_n));$$

hence $\lambda s_1 \dots s_n . s_i \in \mathfrak{G}_n$ for $i = 1, 2, \dots, n$.

For showing that $\bigcup_{n=1}^{\infty} \mathfrak{G}_n$ is closed under substitution, it is sufficient to prove the following two statements:

(i) If $f \in \mathfrak{G}_n$ $(n \geq 1)$, and $\varphi \in \mathfrak{G}_1$ then

$$\lambda s_1 \dots s_n . \varphi(f(s_1, \dots, s_n)) \in \mathfrak{G}_n.$$

(ii) If $f_1, f_2, \dots, f_m \in \mathfrak{G}_n$ $(m > 1,\ n \geq 1)$ then

$$\lambda s_1 \dots s_n . J_m(f_1(s_1, \dots, s_n), \dots, f_n(s_1, \dots, s_n)) \in \mathfrak{G}_n.$$

The truth of (i) is obvious in the case when $n = 1$. If $n > 1$ then (i) follows from the fact that

$$f(s_1, \ldots, s_n) \simeq \psi(J_n(s_1, \ldots, s_n))$$

implies

$$\varphi(f(s_1, \ldots, s_n)) \simeq \varphi\psi(J_n(s_1, \ldots, s_n)),$$

and $\varphi\psi \in \mathfrak{F}_1$, whenever $\varphi, \psi \in \mathfrak{F}_1$.

For the proof of (ii), it is sufficient to show that $\lambda s_1 \ldots s_n . J(f_1(s_1, \ldots, s_n), f_2(s_1, \ldots, s_n)) \in \mathfrak{F}_n$, whenever $f_1, f_2 \in \mathfrak{F}_n$ $(n \geq 1)$. The last is obvious in the case when $n = 1$. If $n > 1$ then we use that

$$f_i(s_1, \ldots, s_n) \simeq \psi_i(J_n(s_1, \ldots, s_n)), \quad i = 1, 2,$$

implies

$$J(f_1(s_1, \ldots, s_n), f_2(s_1, \ldots, s_n)) \simeq \Pi(\psi_1, \psi_2)(J_n(s_1, \ldots, s_n)),$$

and $\Pi(\psi_1, \psi_2) \in \mathfrak{F}_1$, whenever $\psi_1, \psi_2 \in \mathfrak{F}_1$.

Let $f \in \mathfrak{F}_1$, $g \in \mathfrak{F}_3$, and h be the two-argument function defined by

$$h(0, u) \simeq f(u), \quad h(j+1, u) \simeq g(h(j, u), j, u).$$

We shall prove that $h \in \mathfrak{F}_2$.[8] From an intuitive point of view, the proof will be based on a functional-style translation[9] of the following Pascal-like function declaration which represents h:

```
function h(t, u: ℕ): ℕ;
  var s, v, j: ℕ;
begin
  s: = t;  v: = f(u);  j: = 0;
  while s > 0 do
    begin s: = s-1;  v: = g(v, j, u);  j: = j+1  end;
  h: = v
end; .
```

---

[8] The more general case of primitive recursion, when $f \in \mathfrak{F}_n$, $g \in \mathfrak{F}_{n+2}$, and h is $n+1$-ary, can be easily reduced to the case considered now. Roughly speaking, we have only to substitute $J_n(u_1, \ldots, u_n)$ for u.

[9] Compare with Backus [1978]. Our $\mathfrak{U}$-combination and $\mathfrak{U}$-iteration correspond to the binary case of his operation called construction and to his **while**-operation, respectively (note, however, that no explicit use of list objects is made in our case, and natural numbers are the only objects).

Namely, we first construct $\alpha, \beta \in \mathfrak{F}_1$ such that

$$\alpha(J(t, u)) \simeq J_4(t, f(u), 0, u),$$

$$\beta(J_4(s, v, j, u)) \simeq J_4(s \dot{-} 1, g(v, j, u), j+1, u)$$

for all t, u, v, j in $\mathbb{N}$. For that purpose, we take

$$\alpha = \Pi(L, \Pi(fR, \Pi(F, R))),$$

$$\beta = \Pi(PL, \Pi(\psi R, \Pi(SLR^2, R^3))),$$

where $\psi \in \mathfrak{F}_1$ and $g = \lambda vjs.\, \psi(J_3(v, j, s))$. Having such α, β at our disposal, we prove that

$$[\beta, L]\alpha(J(t, u)) \simeq J_4(0, h(t, u), t, u)$$

for all t, u in $\mathbb{N}$ (to do this, we could, for example, use induction on j as well as the first case in (1) for proving that

$$[\beta, L]\alpha(J(t, u)) \simeq J_4(t-j, h(j, u), j, u)$$

for all j, t, u in $\mathbb{N}$ satisfying $j \leq t$; then we could take $j = t$ and use the second case in (1)). From the established equality, we get

$$h(t, u) \simeq LR[\beta, L]\alpha(J(t, u)),$$

and we have only to note that $LR[\beta, L]\alpha \in \mathfrak{F}_1$.

We are now going to the case of μ-operation. Suppose $f \in \mathfrak{F}_2$, and g is the unary function defined by

$$g(u) \simeq \mu j[f(j, u) = 0].$$

We shall prove that $g \in \mathfrak{F}_1$.[10] Again a Pascal-like function declaration will be written for the intuitive explanation of the proof, namely the following one which represents g:

```
function g(u: N): N;
  var j: N;
begin
  j: = 0;
  while f(j, u) > 0 do j: = j+1;
  g: = j
end; .
```

The corresponding functional-style translation needs functions $\alpha, \beta \in \mathfrak{F}_1$ such that

$$\alpha(u) \simeq J(0, u), \quad \beta(J(j, u)) \simeq J(j+1, u),$$

and such functions are $\alpha = \Pi(F, I)$, $\beta = \Pi(SL, R)$. Taking

---

[10]The more general case, when $f \in \mathfrak{F}_{n+1}$ and g is n-ary, can be reduced to this case.

$\psi \in \mathcal{G}_1$ such that $f = \lambda ju.\psi(J(j, u))$, one proves that

$$[\beta, \psi]\alpha(u) \simeq J(g(u), u)$$

for all u in $\mathbb{N}$. (To do this, one could express what the equality $[\beta, \psi]\alpha(u) = w$ means by the definition of iteration, and then see that necessarily $v_j = J(j, u)$, $j = 0, 1, \ldots, m$, in the corresponding finite sequence $v_0, v_1, \ldots, v_m$. So $[\beta, \psi]\alpha(u) = w$ turns out to be equivalent to the existence of a natural number m such that

$$J(m, u) = w \ \& \ \forall_{j<m} j(\psi(J(j, u)) > 0) \ \& \ \psi(J(m, u)) = 0,$$

i.e.

$$J(m, u) = w \ \& \ \forall_{j<m} j(f(j, u) > 0) \ \& \ f(m, u) = 0.$$

Obviously, this condition is equivalent to the condition $J(g(u), u) = w$). From the proven equality, it follows that

$$g(u) \simeq L[\beta, \psi]\alpha(u),$$

and the only thing left is to note that $L[\beta, \psi]\alpha \in \mathcal{G}_1$. ∎

If $\psi_1, \ldots, \psi_l \in \mathcal{F}_p(\mathbb{N})$ then *the functions μ-recursive in* $\psi_1, \ldots, \psi_l$ are, by definition, those partial functions which can be generated from the initial partial recursive functions and the functions $\psi_1, \ldots, \psi_l$ by means of substitution, primitive recursion and μ-operation.[11] Using almost the same proof as above, one can prove

Theorem 2. Let $\mathfrak{A}$ be a standard computational structure on the natural numbers. If $\psi_1, \ldots, \psi_l \in \mathcal{F}_p(\mathbb{N})$ then the unary functions μ-recursive in $\psi_1, \ldots, \psi_l$ are exactly those elements of $\mathcal{F}_p(\mathbb{N})$ which are $\mathfrak{A}$-computable in the set $\{S, P, \psi_1, \ldots, \psi_l\}$.

Also a uniform version of the above theorem is valid. A mapping Γ of $(\mathcal{F}_p(\mathbb{N}))^l$ into $\mathcal{F}_p(\mathbb{N})$ is called a *μ-recursive operator* iff there is an explicit expression for

---

[11] Since $\psi_1, \ldots, \psi_l$ are not necessarily total, the notion of μ-recursiveness in $\psi_1, \ldots, \psi_l$ has, in general, a narrower scope than the notion of partial recursiveness in $\psi_1, \ldots, \psi_l$ (cf. Myhill [1961], Skordev [1963], Rogers [1967, Ch. 13, Theorem XIX, and also the footnote on p. 362 of the Russian translation] or Sasso [1975]).

$\Gamma(\psi_1, \ldots, \psi_l)$ through the initial partial recursive functions and $\psi_1, \ldots, \psi_l$ by means of substitution, primitive recursion and μ-operation, and the form of this expression does not depend on the concrete choice of $\psi_1, \ldots, \psi_l$.[12] Then a simple analysis of the proof of Theorem 2 shows that, whenever $\mathfrak{U}$ is a standard computational structure on the natural numbers, the μ-recursiveness of Γ is equivalent to its $\mathfrak{U}$-computability in $\{S, P\}$.

Exercise. Prove that Theorem 1 remains valid if we replace $\{S, P\}$ in its formulation by $\{P\}$. (This result is essentially contained in Soskov [1985, pp. 9-10]).

Hint. Prove that S is $\mathfrak{U}$-computable in $\{P\}$. To do this, first establish the existence of an element δ of $\mathcal{F}_p(\mathbb{N})$ such that $\delta(J(s, t)) = s \dot{-} t$ for all s, t in $\mathbb{N}$, and δ is $\mathfrak{U}$-computable in $\{P\}$. Then construct an element θ of $\mathcal{F}_p(\mathbb{N})$ which is also $\mathfrak{U}$-computable in $\{P\}$ and satisfies the condition $\theta(u) > u$ for all u in $\mathbb{N}$ (you could, for example, take $\theta(u)$ to be the first one greater than u among the numbers $J(0, 0)$, $J(1, J(0, 0))$, $J(1, J(1, J(0, 0)))$, ...).

4. On the interconnection between programmability in a FP-system and $\mathfrak{U}$-computability

In this section, we suppose that an FP-system in the sense of Backus [1978] is given. Let $\mathfrak{U} = \langle M, J, L, R, T, F, H\rangle$ be the corresponding computational structure described in Example 1.3. We aim to characterize the programmability in the given FP-system by means of $\mathfrak{U}$-computability in a certain subset $\mathcal{B}$ of $\mathcal{F}_p(M)$. Programmability in the FP-system, as defined by Backus, concerns strict total functions in $M \cup \{\perp\}$, i.e. total mappings of this set into itself which transform $\perp$ into $\perp$. These functions are in one-to-one correspondence with the elements of $\mathcal{F}_p(M)$, each φ from $\mathcal{F}_p(M)$ corresponding to its natural extension to $M \cup \{\perp\}$ obtained by assigning the value $\perp$ to all elements

---

[12] If (as it is the case here) partial functions are allowed as arguments of the operators then the class of the μ-recursive operators is narrower than the class of all recursive operators (in the sense of Rogers [1967, Section 9.8]). Cf., for example, Skordev [1963, 1976], Bird [1975] or Sasso [1975].

of $M\cup\{\perp\}$ not belonging to $\mathrm{dom}\varphi$. For our purposes, it is convenient to modify inessentially Backus' notion of programmability by replacing the strict total functions in $M\cup\{\perp\}$ by the corresponding elements of $\mathcal{F}_p(M)$. After this modification, the notion can be briefly described as follows.

Some elements of $\mathcal{F}_p(M)$ are chosen as primitive functions, among them the functions L and R from Example 3.1 and the functions **null**, tl and **apndl** defined by the equalities

$$\mathbf{null}(s)=\begin{cases}\mathrm{t} & \text{if } s=\emptyset,\\ \mathrm{f} & \text{otherwise,}\end{cases}$$

$$\mathrm{tl}(\langle s\rangle)=\emptyset,\quad \mathbf{apndl}(\langle s,\emptyset\rangle)=\langle s\rangle,$$

$$\mathrm{tl}(\langle s,t_1,\ldots,t_k\rangle)=\langle t_1,\ldots,t_k\rangle,$$

$$\mathbf{apndl}(\langle s,\langle t_1,\ldots,t_k\rangle\rangle)=\langle s,t_1,\ldots,t_k\rangle,$$

where t and f are the atoms meaning truth and falsity, respectively, and tl and **apndl** are defined only for such types of objects which are considered in the left-hand sides of the corresponding equalities. Starting from primitive functions, new ones are constructed using so-called functional forms and recursion. At this stage of our exposition, we restrict ourselves to the construction by means of functional forms[13]. The corresponding notion of programmable function can be described by the following inductive definition:

(i) all primitive functions are *programmable*;

(ii) for each s in M, the constant function $\bar{s}$, assigning the value s to all elements of M, is *programmable*;

(iii) the composition of every two programmable elements of $\mathcal{F}_p(M)$ is *programmable*;

(iv) if $\varphi_1,\ldots,\varphi_n$ $(n\geq 1)$ are programmable elements of $\mathcal{F}_p(M)$, then so is the function $(\varphi_1,\ldots,\varphi_n)$ defined by the condition that

$$(\varphi_1,\ldots,\varphi_n)(t)\simeq\langle\varphi_1(t),\ldots,\varphi_n(t)\rangle$$

---

[13] It will be shown later in this book (namely in Subsection (II) of Section III.5) that recursion does not enlarge the class of the programmable functions (cf. also Skordev [1982a]).

for all t in M;[14]

(v) if χ, φ, ψ are programmable elements of $\mathcal{F}_p(M)$, then so is $\Sigma(\chi, \varphi, \psi)$, where Σ is the branching operation in $\mathfrak{U}$ defined in Section 2;[15]

(vi) if σ and χ are programmable elements of $\mathcal{F}_p(M)$, then so is $[\sigma, \chi]$ (the $\mathfrak{U}$-iteration of σ controlled by χ);[16]

(vii) for each s in M, if φ is a programmable element of $\mathcal{F}_p(M)$, then so is the function $\mathrm{bu}\,\varphi\, s$ defined by

$$\mathrm{bu}\,\varphi\, s(t) \simeq \varphi(\langle s, t\rangle);$$

(viii) if φ is a programmable element of $\mathcal{F}_p(M)$, then so is the function $\mathrm{a}\varphi$ defined by

$$\mathrm{a}\varphi(\emptyset) = \emptyset, \quad \mathrm{a}\varphi(\langle t_1, \ldots, t_k\rangle) \simeq \langle \varphi(t_1), \ldots, \varphi(t_k)\rangle;$$ [17]

(ix) if φ is a programmable element of $\mathcal{F}_p(M)$, then so is the function $/\varphi$ defined by

$$/\varphi(\langle t_1, \ldots, t_k\rangle) \simeq \varphi(\langle t_1, \varphi(\langle t_2, \ldots \varphi(\langle t_{k-1}, t_k\rangle)\ldots\rangle\rangle)$$

(dom$(/\varphi)$ consists only of non-empty sequences of elements of M)[18].

Let $\mathcal{B}_0$ consist of all primitive functions of the given FP-system and of all functions of the form $\bar{s}$, where $s \in M$ (cf. clause (ii) in the above inductive definition). If

---

[14] Backus denotes this function by $[\varphi_1, \ldots, \varphi_n]$. We do not use this denotation due to the conflict with our denotation for iteration in the case of $n = 2$.

[15] Backus uses the denotation $(\chi \rightarrow \varphi, \psi)$ for the element $\Sigma(\chi, \varphi, \psi)$.

[16] Backus' denotation for the function $[\sigma, \chi]$ is **while** $\chi\,\sigma$.

[17] Backus' denotation for $\mathrm{a}\varphi$ is $\alpha\varphi$.

[18] Another variant is also considered by Backus, where $\emptyset$ also belongs to dom$(/\varphi)$. Then the function $/\varphi$ is defined by the equalities $/\varphi(\emptyset) = u$ and

$$/\varphi(\langle t_1, \ldots, t_k\rangle) \simeq \varphi(\langle t_1, \varphi(\langle t_2, \ldots \varphi(\langle t_k, u\rangle)\ldots\rangle\rangle),$$

where u is some fixed element of M. Including this variant in the inductive definition formulated now would cause no essential modification in our exposition.

$\mathcal{B}=\mathcal{B}_0$ then the following holds:

Theorem 1. The programmable elements of $\mathcal{F}_p(M)$ are exactly those elements of $\mathcal{F}_p(M)$ which are $\mathfrak{U}$-computable in $\mathcal{B}$.

The proof of this theorem is based on 7 lemmas.

Lemma 1. $\mathfrak{U}$-computability in $\mathcal{B}_0$ implies programmability.

Proof. All functions from the set $\{L, R, T, F\}\cup\mathcal{B}_0$ are programmable according to clauses (i) and (ii) above. On the other hand, clauses (iii), (iv) and (vi) assure that programmability is preserved by composition, $\mathfrak{U}$-combination and $\mathfrak{U}$-iteration. ∎

Lemma 2. All primitive functions and all functions $\bar{s}$, where $s\in M$, are $\mathfrak{U}$-computable in $\mathcal{B}_0$.

Proof. All such functions belong to $\mathcal{B}_0$. ∎

Lemma 3. The operations considered in clauses (iii), (v) and (vi) preserve $\mathfrak{U}$-computability in $\mathcal{B}_0$.

Proof. We use the definition of $\mathfrak{U}$-computability in $\mathcal{B}_0$ and Exercises 2.1, 2.2. ∎

Lemma 4. The operation considered in clause (iv) preserves $\mathfrak{U}$-computability in $\mathcal{B}_0$.

Proof. The following equalities hold for all $\varphi, \varphi_1, \dots, \varphi_n$ in $\mathcal{F}_p(M)$, where $n>1$:

$$(\varphi)=\mathrm{apndl}\circ\Pi(\varphi, \bar{\emptyset}),$$

$$(\varphi_1, \dots, \varphi_n)=\mathrm{apndl}\circ\Pi(\varphi_1, (\varphi_2, \dots, \varphi_n)).$$ ∎

Lemma 5. For each function φ, which is $\mathfrak{U}$-computable in $\mathcal{B}_0$, and each element s of M, the function $\mathrm{bu}\,\varphi\, s$ is also $\mathfrak{U}$-computable in $\mathcal{B}_0$.

Proof. We use the equality $\mathrm{bu}\,\varphi\, s=\varphi\circ\Pi(\bar{s}, I_M)$ and Exercise 2.1. ∎

Lemma 6. Whenever φ is a function $\mathfrak{U}$-computable in $\mathcal{B}_0$, then $\mathrm{a}\varphi$ is also $\mathfrak{U}$-computable in $\mathcal{B}_0$.

For the proof of this lemma, see Exercise 3 of the present section. An easier proof will be given later in the book, after proving an algebraic generalization of the First Recursion Theorem.

Lemma 7. Whenever φ is a function $\mathfrak{U}$-computable in $\mathcal{B}_0$, then $/\varphi$ is also $\mathfrak{U}$-computable in $\mathcal{B}_0$.

For the proof, see Exercise 4 of the present section. An easier proof could be given after Section III.4.

Of course, the non-trivial part of the proven theorem consists in the converse statement of Lemma 1, i.e. in the statement that programmability implies $\mathfrak{A}$-computability in the set $\mathcal{B}$ used in the theorem (this part was proven by means of the remaining 6 lemmas). Therefore a reduction of the set $\mathcal{B}$ will increase the value of the theorem. When defining the set $\mathcal{B}$ used in this theorem (namely the set $\mathcal{B}_0$), we have put all primitive functions of the FP-system in it. Now we should like to mention that the theorem remains valid after leaving only a small number of the primitive functions in $\mathcal{B}$. In Skordev [1982a], a variant of the theorem was proven, where only 8 among the primitive functions are taken as elements of $\mathcal{B}$, namely the four arithmetical operations $+, -, *, \div$ and the functions **tl**, **apndl**, **atom**, **eq**.[19] The corresponding proof uses the generalization of the First Recursion Theorem mentioned above and can be given later in the book. However, some simple parts of that proof are included in exercises after the present section (for the rest, cf. Exercises III.5.1 and III.5.2).

Exercises

(In all these exercises, $\mathfrak{A}$ is the computational structure considered in this section, and $\mathcal{B}$ is some subset of $\mathcal{F}_p(M)$ containing the functions **tl**, **apndl**, **eq** and $\bar{\emptyset}$)

1. Prove the $\mathfrak{A}$-computability of the function **null** in $\mathcal{B}$.

2. Let **reverse** be the element of $\mathcal{F}_p(M)$ determined by the condition that dom(**reverse**) consists of all finite sequences of elements of M and by the equalities

$$\mathbf{reverse}(\emptyset) = \emptyset,$$

---

[19] The last two of these functions have the following definitions:

$$\mathbf{atom}(s) = \begin{cases} \mathrm{t} & \text{if } s \text{ is an atom,} \\ \mathrm{f} & \text{otherwise,} \end{cases}$$

$$\mathbf{eq}(\langle s, t\rangle) = \begin{cases} \mathrm{t} & \text{if } s = t, \\ \mathrm{f} & \text{otherwise,} \end{cases}$$

where dom(**eq**) consists only of two-element sequences from M. We note the following small difference between the computational structure $\mathfrak{A}$ used here and the one implicitly used in the quoted paper: the function R used in that paper is defined only for sequences from M having more than one member, and the value of R on such a sequence is equal to its second member.

$$\text{reverse}(\langle t_1, \ldots, t_k\rangle) = \langle t_k, \ldots, t_1\rangle.$$

Prove the $\mathfrak{A}$-computability of this function in $\mathcal{B}$.

Hint. Prove the equality

$$\text{reverse} = L \circ [\sigma, \text{null} \circ R]_{\text{d}} \circ \Pi(\bar{\emptyset}, I_M),$$

where $\sigma = \Pi(\text{apndl} \circ \Pi(LR, L), \text{tl} \circ R)$ (for the meaning of $[\sigma, \chi]_{\text{d}}$, cf. Exercise 2.6).

3. Let φ be a function $\mathfrak{A}$-computable in $\mathcal{B}$. Prove the $\mathfrak{A}$-computability of the function $\text{a}\varphi$ in $\mathcal{B}$.

Hint. Prove the equality

$$\text{a}\varphi = L \circ [\sigma, \text{null} \circ R]_{\text{d}} \circ \Pi(\bar{\emptyset}, \text{reverse}),$$

where $\sigma = \Pi(\text{apndl} \circ \Pi(\varphi LR, L), \text{tl} \circ R)$.

4. Let φ be a function $\mathfrak{A}$-computable in $\mathcal{B}$. Prove the $\mathfrak{A}$-computability of the function $/\varphi$ in $\mathcal{B}$.

Hint. Prove the equality

$$/\varphi = L \circ [\sigma, \text{null} \circ R]_{\text{d}} \circ \Pi(L, \text{tl}) \circ \text{reverse},$$

where $\sigma = \Pi(\varphi\Pi(LR, L), \text{tl} \circ R)$.

5. For each non-zero natural number i, let $i\text{l}$ and $i\text{r}$ be the elements of $\mathcal{F}_{\text{p}}(M)$ determined by the condition that $\text{dom}(i\text{l})$ and $\text{dom}(i\text{r})$ consist of all finite sequences of at least i elements of M, and, for each such sequence s, the values $i\text{l}(s)$ and $i\text{r}(s)$ are equal, respectively, to the i-th member of s from the left and to its i-th member from the right.[20] Prove the $\mathfrak{A}$-computability of the functions $i\text{l}$ and $i\text{r}$ in $\mathcal{B}$.

6. Let tlr and apndr be the functions determined by the equalities

$$\text{tlr}(\langle s\rangle) = \emptyset, \quad \text{apndr}(\langle \emptyset, s\rangle) = \langle s\rangle,$$

$$\text{tlr}(\langle t_1, \ldots, t_k, s\rangle) = \langle t_1, \ldots, t_k\rangle,$$

$$\text{apndr}(\langle\langle t_1, \ldots, t_k\rangle, s\rangle) = \langle t_1, \ldots, t_k, s\rangle$$

and by the condition that tlr and apndr are defined only for such types of objects which are indicated in the left-hand sides of the corresponding equalities. Prove the $\mathfrak{A}$-computability of the functions tlr and apndr in $\mathcal{B}$.

7. Let the functions rotl and rotr be determined by the equalities

$$\text{rotl}(\emptyset) = \text{rotr}(\emptyset) = \emptyset,$$

---

[20] Backus denotes the function $i\text{l}$ simply by i.

$$\mathrm{rotl}\,(\langle t_1, \ldots, t_k\rangle) = \langle t_2, \ldots, t_k, t_1\rangle,$$
$$\mathrm{rotr}\,(\langle t_1, \ldots, t_k\rangle) = \langle t_k, t_1, \ldots, t_{k-1}\rangle$$

and by the condition that dom(rotl) and dom(rotr) consist of all finite sequences of elements of M. Prove the $\mathfrak{U}$-computability of the functions rotl and rotr in $\mathcal{B}$.

Hint. Prove the equalities

$$\mathrm{rotl} = \Sigma(\mathrm{null}, \bar{\emptyset}, \mathrm{apndr} \circ \Pi(\mathrm{tl}, L)),$$
$$\mathrm{rotr} = \Sigma(\mathrm{null}, \bar{\emptyset}, \mathrm{apndl} \circ \Pi(R, \mathrm{tlr})).$$

8. Let not be the element of $\mathcal{F}_{\mathrm{p}}(M)$ determined by the equalities $\mathrm{dom}(\mathrm{not}) = \{\mathrm{t}, \mathrm{f}\}$, $\mathrm{not}(\mathrm{t}) = \mathrm{f}$, $\mathrm{not}(\mathrm{f}) = \mathrm{t}$. Prove the $\mathfrak{U}$-computability of the function not in $\mathcal{B}$.

Hint. Prove the equality $\mathrm{not} = \Sigma(I_M, F, T)$.

9. Let and and or be the elements of $\mathcal{F}_{\mathrm{p}}(M)$ determined by the equalities $\mathrm{dom}(\mathrm{and}) = \mathrm{dom}(\mathrm{or}) = \{\mathrm{t}, \mathrm{f}\}^2$, $\mathrm{and}(\langle \mathrm{t}, \mathrm{t}\rangle) = \mathrm{t}$, $\mathrm{and}(\langle \mathrm{t}, \mathrm{f}\rangle) = \mathrm{and}(\langle \mathrm{f}, \mathrm{t}\rangle) = \mathrm{and}(\langle \mathrm{f}, \mathrm{f}\rangle) = \mathrm{f}$, $\mathrm{or}(\langle \mathrm{f}, \mathrm{f}\rangle) = \mathrm{f}$, $\mathrm{or}(\langle \mathrm{f}, \mathrm{t}\rangle) = \mathrm{or}(\langle \mathrm{t}, \mathrm{f}\rangle) = \mathrm{or}(\langle \mathrm{t}, \mathrm{t}\rangle) = \mathrm{t}$. Prove the $\mathfrak{U}$-computability of the functions and and or in $\mathcal{B}$.

Hint. Prove the equalities

$$\mathrm{and} = \Sigma(\mathrm{null} \circ \mathrm{tl} \circ \mathrm{tl}, \varphi, o),$$
$$\mathrm{or} = \Sigma(\mathrm{null} \circ \mathrm{tl} \circ \mathrm{tl}, \psi, o),$$

where

$$\varphi = \Sigma(L, \Sigma(R, T, F), \Sigma(R, F, F)),$$
$$\psi = \Sigma(L, \Sigma(R, T, T), \Sigma(R, T, F)),$$

and o is the element of $\mathcal{F}_{\mathrm{p}}(M)$ whose domain is empty.

5. Computability of multiple-valued functions with respect to a given computational structure

Given a set M, we shall denote by $\mathcal{F}_{\mathrm{m}}(M)$ the set of all binary relations in M, i.e. the set of all subsets of M^2. The elements of $\mathcal{F}_{\mathrm{m}}(M)$ will be regarded as unary multiple-valued functions in the following sense: if $\varphi \in \mathcal{F}_{\mathrm{m}}(M)$ and $u \in M$ then *the values of* φ *at* u will be, by definition, those v in M which satisfy the condition $\langle u, v\rangle \in \varphi$. Identifying the unary partial functions in M with their graphs (i.e. adopting the equality

$$\varphi = \{\langle u, \varphi(u)\rangle : u \in \mathrm{dom}\,\varphi\}$$

for $\varphi\in\mathcal{F}_p(M)$), we shall regard them as elements of the above defined set $\mathcal{F}_m(M)$. Obviously, the above definition of values is in agreement with this identification. Of course, the operation of *composition* in $\mathcal{F}_m(M)$ which naturally extends the corresponding operation in $\mathcal{F}_p(M)$, will be defined as follows: if $\varphi, \psi\in\mathcal{F}_m(M)$ then

$$\varphi\psi=\{\langle u, w\rangle:\ \exists v(\langle u, v\rangle\in\psi\ \&\ \langle v, w\rangle\in\varphi)\}$$

(i.e. $\varphi\psi$ is the usual composition of the binary relations ψ and φ),

Suppose now a computational structure $\mathfrak{U}=\langle M, J, L, R, T, F, H\rangle$ is given. Then natural extensions of the $\mathfrak{U}$-*combination* and the $\mathfrak{U}$-*iteration* from $\mathcal{F}_p(M)$ on $\mathcal{F}_m(M)$ are the operations defined as follows: for all $\varphi, \psi, \sigma, \chi$ in $\mathcal{F}_m(M)$,

$$\Pi(\varphi, \psi)=\{\langle u, w\rangle:\ \exists s\exists t(\langle u, s\rangle\in\varphi\ \&\ \langle u, t\rangle\in\psi\ \&\ J(s, t)=w)\},$$

and $\langle u, w\rangle\in[\sigma, \chi]$ iff there is a finite sequence $v_0, v_1, \ldots, v_m$ of elements of M such that

(1) $$v_0=u\ \&\ v_m=w\ \&\ \forall_{j<m} j(\langle v_j, \mathrm{true}\rangle\in H\chi\ \&\ \langle v_j, v_{j+1}\rangle\in\sigma)\ \&\ \langle v_m, \mathrm{false}\rangle\in H\chi,$$

where $H\chi=\{\langle u, p\rangle:\ \exists v(\langle u, v\rangle\in\chi\ \&\ H(v)=p)\}$. Also $\mathfrak{U}$-*branching* will be defined in $\mathcal{F}_m(M)$. This will be the ternary operation Σ defined as follows:

$$\Sigma(\chi, \varphi, \psi)=\{\langle u, w\rangle:\ \langle u, \mathrm{true}\rangle\in H\chi\ \&\ \langle u, w\rangle\in\varphi\ \vee\ \langle u, \mathrm{false}\rangle\in H\chi\ \&\ \langle u, w\rangle\in\psi\}.$$

A least-fixed-point characterization of iteration in $\mathcal{F}_m(M)$ is contained in the following

Proposition 1. Let $\sigma, \chi\in\mathcal{F}_m(M)$. Then the equality

$$[\sigma, \chi]=\Sigma(\chi, [\sigma, \chi]\sigma, I_M)$$

holds. More generally, for each ρ in $\mathcal{F}_m(M)$ the equality

$$\rho[\sigma, \chi]=\Sigma(\chi, \rho[\sigma, \chi]\sigma, \rho)$$

holds, and $\rho[\sigma, \chi]$ is the least element τ of $\mathcal{F}_m(M)$ satisfying

(2) $$\tau\geq\Sigma(\chi, \tau\sigma, \rho).$$

Proof. Let

$$\alpha=\{\langle u, v\rangle:\ \langle u, \mathrm{true}\rangle\in H\chi\ \&\ \langle u, v\rangle\in\sigma\},$$

$$\beta=\{\langle u, v\rangle:\ \langle u, \mathrm{false}\rangle\in H\chi\ \&\ \langle u, v\rangle\in\rho\}.$$

Then $\Sigma(\chi, \tau\sigma, \rho) = \tau\alpha \cup \beta$ for all τ in $\mathcal{F}_m(M)$. It is known (cf., for example, Blikle [1971]) that the element

$$\beta\alpha^* = \beta \bigcup_{m=0}^{\infty} \alpha^m = \bigcup_{m=0}^{\infty} \beta\alpha^m$$

of $\mathcal{F}_m(M)$ is the least solution of the equation $\tau = \tau\alpha \cup \beta$ and of the inequality $\tau \supseteq \tau\alpha \cup \beta$. Thus $\beta\alpha^*$ is the least solution of the equation $\tau = \Sigma(\chi, \tau\sigma, \rho)$ and of the inequality (2). On the other hand, $\langle u, v\rangle \in \beta\alpha^m$ iff there are an element w of M and a sequence $v_0, v_1, \ldots, v_m$ of elements of M such that

(3) $\quad v_0 = u \;\&\; \forall_{j<m} j(\langle v_j, v_{j+1}\rangle \in \alpha) \;\&\; v_m = w \;\&\; \langle w, v\rangle \in \beta.$

After taking into account the definitions of α and β, we see that (3) is equivalent to the conjunction of (1) and $\langle w, v\rangle \in \rho$. Hence $\beta\alpha^* = \rho[\sigma, \chi]$. ∎

Remark 1. The given definitions of iteration and branching use only the components M and H of $\mathfrak{U}$. Thus we could consider such operations in every situation when a set M and a partial predicate H on it are given. Obviously, the above proof and hence the proven proposition remain valid in such a more general case.

Now we shall define the notion of relative computability of elements of $\mathcal{F}_m(M)$ with respect to the given computational structure $\mathfrak{U}$. The definition will be quite similar to the corresponding definition for elements of $\mathcal{F}_p(M)$.

Definition 1. Let $\mathcal{B}$ be some subset of $\mathcal{F}_m(M)$. The elements of $\mathcal{F}_m(M)$ *$\mathfrak{U}$-computable in* $\mathcal{B}$ (*computable in* $\mathcal{B}$, for short) are those elements of $\mathcal{F}_m(M)$ which can be generated from elements of $\{L, R, T, F\} \cup \mathcal{B}$ by means of composition, $\mathfrak{U}$-combination and $\mathfrak{U}$-iteration.

As before, if φ is computable in $\mathcal{B}$, and each element of $\mathcal{B}$ is computable in $\mathcal{B}'$ then φ is computable in $\mathcal{B}'$. In the case when $\mathcal{B}$ is a subset of $\mathcal{F}_p(M)$, the elements of $\mathcal{F}_m(M)$ computable in $\mathcal{B}$ are exactly the same as the elements of $\mathcal{F}_p(M)$ computable in $\mathcal{B}$. In particular, I_M and the empty relation are elements of $\mathcal{F}_m(M)$ which are $\mathfrak{U}$-computable in $\emptyset$. As before (cf. Exercise 1 below), the element $\Sigma(\chi, \varphi, \psi)$ is always $\mathfrak{U}$-computable in $\{\chi, \varphi, \psi\}$; hence including Σ as an additional generating operation in the above definition of computability would not enlarge the set of the elements of $\mathcal{F}_m(M)$ which are $\mathfrak{U}$-computable in

$\mathcal{B}$. Again, one could ask whether some other reasonable effective constructions in $\mathcal{F}_{\mathrm{m}}(M)$ could, however, enlarge this set. And again, it will be shown in this book that in some sense such other constructions do not exist.

We could also define $\mathfrak{U}$-computability for operators in $\mathcal{F}_{\mathrm{p}}(M)$.

Definition 2. Let $\mathcal{B} \subseteq \mathcal{F}_{\mathrm{m}}(M)$, and let Γ be a mapping of $(\mathcal{F}_{\mathrm{m}}(M))^{l}$ into $\mathcal{F}_{\mathrm{m}}(M)$, where l is some positive integer. Then Γ is called *(an operator) $\mathfrak{U}$-computable in $\mathcal{B}$ (computable in $\mathcal{B}$*, for short) iff, for arbitrary $\psi_1, \ldots, \psi_l$ in $\mathcal{F}_{\mathrm{m}}(M)$, there is an explicit expression for $\Gamma(\psi_1, \ldots, \psi_l)$ through $L, R, T, F, \psi_1, \ldots, \psi_l$ and elements of $\mathcal{B}$ by means of composition, $\mathfrak{U}$-combination and $\mathfrak{U}$-iteration, the form of the expression not depending on the concrete choice of $\psi_1, \ldots, \psi_l$.

Of course, the above definition can be formulated more precisely using induction. We omit the corresponding formulation.

Exercises

(In all these exercises, a computational structure $\mathfrak{U} = \langle M, J, L, R, T, F, H\rangle$ is supposed to be given, the corresponding set $\mathcal{F}_{\mathrm{m}}(M)$ is denoted by $\mathcal{F}$, and the functional relation I_M is denoted by I)

1. Prove the statements of Exercises 2.2, 2.6 and 2.7 for the case when $\mathcal{F}_{\mathrm{p}}(M)$ is replaced by $\mathcal{F} = \mathcal{F}_{\mathrm{m}}(M)$.

2. Prove that $\mathcal{F}$ is a semigroup with respect to composition, and I is a unit of this semigroup (this means that $\mathcal{F}$ is closed under composition, composition is associative in $\mathcal{F}$, and the equalities $I\theta = \theta I = \theta$ hold for all θ in $\mathcal{F}$).

3. Prove that composition, $\mathfrak{U}$-combination, $\mathfrak{U}$-branching and $\mathfrak{U}$-iteration are monotonically increasing operations in $\mathcal{F}$ with respect to the partial ordering of $\mathcal{F}$ by inclusion.

4. For all $\varphi, \psi, \theta, \chi$ in $\mathcal{F}$, prove the equalities

$$\Sigma(T, \varphi, \psi) = \varphi, \quad \Sigma(F, \varphi, \psi) = \psi,$$
$$\theta\Sigma(\chi, \varphi, \psi) = \Sigma(\chi, \theta\varphi, \theta\psi)$$

(compare with McCarthy [1963]).

5. For each s in M, let $\bar{s}$ be the constant function, assigning the value s to all elements of M. Let $\mathcal{C}$ be

the set of all elements of $\mathcal{F}$ having the form $\bar{s}$, where $s \in M$. Suppose φ and ψ are such elements of $\mathcal{F}$ that $\varphi x \geq \psi x$ for all x in $\mathcal{C}$. Prove that $\varphi \geq \psi$.

6. Let $\mathcal{C}$ be the same as in Exercise 5. For all x and y in $\mathcal{C}$, prove that $\Pi(x, y)$ also belongs to $\mathcal{C}$, and the equalities $L\Pi(x, y) = x$, $R\Pi(x, y) = y$ hold.

7. Let $\mathcal{C}$ be the same as in Exercise 5. For all $\varphi, \psi, \chi, \theta$ in $\mathcal{F}$ and all x in $\mathcal{C}$, prove the equalities

$$\Pi(\varphi, \psi)x = \Pi(\varphi x, \psi x), \quad \Sigma(\chi, \varphi, \psi)x = \Sigma(\chi x, \varphi x, \psi x),$$

$$\Pi(\varphi x, I)\theta = \Pi(\varphi x, \theta), \quad \Pi(I, \psi x)\theta = \Pi(\theta, \psi x)$$

$$\Sigma(I, \varphi x, \psi x)\theta = \Sigma(\theta, \varphi x, \psi x).$$

8. Let $\theta \in \mathcal{F}$. Prove that $\Pi(I, I)\theta = \Pi(\theta, \theta)$ iff $\theta \in \mathcal{F}_p(M)$.

9. Let σ, χ, ρ be arbitrary elements of $\mathcal{F}_m(M)$, and let $\tau_0 = \rho[\sigma, \chi]$. Suppose K is a subset of M such that $v \in K$ for all $\langle u, v \rangle$ in σ with $u \in K$, and τ is an element of $\mathcal{F}_m(M)$ such that $\tau \geq \Sigma(\chi, \tau\sigma, \rho) \cap (K \times M)$. Prove that $\tau \geq \tau_0 \cap (K \times M)$.

6. The recursively enumerable binary relations considered as multiple-valued functions

In this section, an application of the notion of computability in $\mathcal{F}_m(M)$ will be made, which will be similar to the application of the notion of computability in $\mathcal{F}_p(M)$, made in Section 3. Namely, a characterization will be given of the recursively enumerable binary relations as the elements of $\mathcal{F}_m(\mathbb{N})$ $\mathfrak{U}$-computable in $\mathcal{B}$, where $\mathfrak{U}$ is a certain computational structure whose carrier is $\mathbb{N}$, and $\mathcal{B}$ is a certain finite subset of $\mathcal{F}_m(\mathbb{N})$.

Let $\mathfrak{U} = \langle \mathbb{N}, J, L, R, T, F, H \rangle$ be a standard computational structure on the natural numbers in the sense of Section 3 (in particular, $\mathfrak{U}$ could be the computational structure from Example 1.1). Let the denotations S and P have the same meaning as in Section 3 (namely, $S = \lambda u.\, u+1$, $P = \lambda u.\, u \dot{-} 1$). Let $I = I_{\mathbb{N}}$. Before formulating and proving analogues of Theorems 3.1 and 3.2, we shall indicate a way for reducing some problems about multiple-valued functions to problems concerning partial functions.

Lemma 1. For arbitrary $\theta \in \mathcal{F}$, let $\Delta(\theta)$ be the restriction of the function R to the set

$$\{u \in \operatorname{dom} L \cap \operatorname{dom} R:\ \langle L(u),\ R(u)\rangle \in \theta\}.$$

Then each of the elements θ and $\Delta(\theta)$ of $\mathcal{F}$ can be generated from the other one and L, R, S, P, $\mathbb{N}^2$ by means of finitely many applications of composition, J-combination and iteration.

Proof. It is easy to check that the following two equalities hold:

$$\theta = \Delta(\theta)\Pi(I,\ \mathbb{N}^2),\quad \Delta(\theta) = \Delta(I)\Pi(\theta L,\ R).$$

Since $\Delta(I)$ is a partial recursive function, Theorem 3.1 and the remark after that theorem imply that $\Delta(I)$ can be generated from L, R, S, P by means of finitely many applications of composition, J-combination and iteration.■

Now we shall formulate the analog of Theorem 3.1.

Theorem 1. The recursively enumerable binary relations in $\mathbb{N}$ are exactly those elements of $\mathcal{F}_m(\mathbb{N})$, which are $\mathfrak{A}$-computable in the set $\{S,\ P,\ \mathbb{N}^2\}$.

Proof. Clearly, all elements of $\mathcal{F}_m(\mathbb{N})$, $\mathfrak{A}$-computable in $\{S,\ P,\ \mathbb{N}^2\}$, are recursively enumerable relations in $\mathbb{N}$. For proving the converse statement, consider an arbitrary recursively enumerable binary relation φ in $\mathbb{N}$. Let $\Delta(\varphi)$ be the corresponding partial function defined as in Lemma 1. Since $\Delta(\varphi)$ is partial recursive, Theorem 3.1 shows that $\Delta(\varphi)$ is $\mathfrak{A}$-computable in the set $\{S,\ P\}$. Now it is sufficient to apply the lemma.■

Remark 1. The exercise after Section 3 shows that we could formulate the above theorem with $\{P,\ \mathbb{N}^2\}$ instead of $\{S,\ P,\ \mathbb{N}^2\}$.

Remark 2. An inspection of the proof of the above theorem shows that each element of $\mathcal{F}_m(\mathbb{N})$ $\mathfrak{A}$-computable in the set $\{S,\ P,\ \mathbb{N}^2\}$ can be represented in the form $\delta\Pi(I,\ \mathbb{N}^2)$, where δ is some element $\mathfrak{A}$-computable in the set $\{S,\ P\}$.

A certain generalization of the above theorem will be the analog of Theorem 3.2. The generalization concerns enumeration reducibility of binary relations in $\mathbb{N}$ (for the definition of this notion, cf. Rogers [1967, Section 9.7], where the enumeration reducibility of a subset of $\mathbb{N}$ to another one is considered, and it is obvious how the definition could be generalized in order to consider reducibility to several relations).

Theorem 2. Let $\psi_1,\ \ldots,\ \psi_l \in \mathcal{F}_m(\mathbb{N})$. Then the binary relations in $\mathbb{N}$ enumeration reducible to $\psi_1,\ \ldots,\ \psi_l$ are exactly those elements of $\mathcal{F}_m(\mathbb{N})$ which are $\mathfrak{U}$-computable in the set $\{S, P, \mathbb{N}^2, \psi_1, \ldots, \psi_l\}$.

Proof. Again only that part of the proof needs to be exposed, where one has to show that each binary relation enumeration reducible to $\psi_1, \ldots, \psi_l$ is $\mathfrak{U}$-computable in the above set. Suppose $\varphi \in \mathcal{F}_m(\mathbb{N})$, and φ is enumeration reducible to $\psi_1, \ldots, \psi_l$. It is seen from Lemma 1 that θ and $\Delta(\theta)$ are mutually enumeration reducible for each $\theta \in \mathcal{F}$. Since enumeration reducibility is transitive, it follows that $\Delta(\varphi)$ is enumeration reducible in $\Delta(\psi_1), \ldots, \Delta(\psi_l)$. But $\Delta(\varphi)$ and $\Delta(\psi_1), \ldots, \Delta(\psi_l)$ are partial functions. Therefore $\Delta(\varphi)$ is partial recursive in $\Delta(\psi_1), \ldots, \Delta(\psi_l)$. From here, the existence of a two-argument function h follows which is μ-recursive in $\Delta(\psi_1), \ldots, \Delta(\psi_l)$ and satisfies the condition

$$\forall u \forall v (\Delta(\varphi)(u) = v \iff \exists t (h(u, t) = v)).$$

In the case when $l = 1$, this follows from Lemma 5 in Skordev [1973, pp. 164-165], and the general case can be considered using the corresponding straight-forward generalization of the mentioned lemma. Consider now the unary function

$$\chi = \lambda u . h(L(u), R(u)).$$

It is also μ-recursive in $\Delta(\psi_1), \ldots, \Delta(\psi_l)$. Therefore (by Theorem 3.2) χ is $\mathfrak{U}$-computable in the set $\{S, P, \Delta(\psi_1), \ldots, \Delta(\psi_l)\}$. Since $h(u, t) \simeq \chi(J(u, t))$ for all u, t in $\mathbb{N}$, the equality $\Delta(\varphi) = \chi \Pi(I, \mathbb{N}^2)$ holds. This, together with Lemma 1, completes the proof. ∎

A uniform version of the above theorem is also true. It concerns enumeration operators (for the definition of this notion, cf. Uspensky [1955], where such operators are called computable operations, or Rogers [1967, Section 9.7]). Let a mapping Γ of $(\mathcal{F}_m(\mathbb{N}))^l$ into $\mathcal{F}_m(\mathbb{N})$ be given. Then Γ is an enumeration operator iff Γ is $\mathfrak{U}$-computable in the set $\{S, P, \mathbb{N}^2\}$.

Exercise. Prove that $\mathbb{N}^2$ in Theorems 1 and 2 can be replaced by $\mathbb{N} \times \{0, 1\}$.

Hint. Prove the equality $\mathbb{N}^2 = [S, \mathbb{N} \times \{0, 1\}] F$.

7. On the notions of prime and search computability

(I) **A recollection of definitions.** Moschovakis [1969] introduced the notions of prime and search computability which generalize the notion of computability for the case of an arbitrary object domain. We shall recall here (up to unessential technical details) the formulations of some definitions from that paper.

Let B, O and B° be such as in Example 1.2, i.e. B is some set, O is some object not in B, and B° is the set $B\cup\{O\}$. A set B^{*} is defined by the following inductive clauses:

(i) if $u\in B^{\circ}$ then $u\in B^{*}$;

(ii) if $s, t\in B^{*}$ then $\langle s, t\rangle\in B^{*}$,

where the definition of ordered pair is chosen in such a way that no element of B° is an ordered pair (i.e. B^{*} is the set M from the example in question). If $A\subseteq B^{*}$ then A^{*} is, by definition, the least subset X of B^{*} containing $A\cup\{O\}$ and satisfying the condition

$$\forall s\forall t(\langle s, t\rangle\in X \Longleftrightarrow s\in X \ \&\ t\in X).^{21}$$

The natural numbers $0, 1, 2, 3, \ldots$ are identified with the elements O, $\langle O, O\rangle$, $\langle\langle O, O\rangle, O\rangle$, $\langle\langle\langle O, O\rangle, O\rangle, O\rangle, \ldots$ of B^{*}, respectively.

Mappings L and R of B^{*} into itself are defined as in the example mentioned above, i.e.by the conditions

$$L(O) = R(O) = 0,$$

$$L(u) = R(u) = 1 \text{ for all } u \text{ in } B,$$

$$L(\langle s, t\rangle) = s, \quad R(\langle s, t\rangle) = t \text{ for all } s, t \text{ in } B^{*}$$

(as we pointed, Moschovakis denotes these mappings by π and δ, respectively).

If $u_1, \ldots, u_m\in B^{*}$ then $\langle\langle u_1, \ldots, u_m\rangle\rangle$ is an abbreviation

---

[21] The definition of A^{*} given in Moschovakis [1969] sounds somewhat differently, but it is equivalent to the present one. In the special case when $A = B$, this least subset is obviously B^{*}, hence the above definition does not cause inconsistency in the denotations.

for $\langle m, \langle u_1, \langle u_2, \ldots, \langle u_m, 0\rangle \ldots \rangle\rangle\rangle$ (in Moschovakis [1969], this element is denoted by $\langle u_1, \ldots, u_m\rangle$, but in the denotation of ordered pair, round brackets are used in that paper).

Suppose now some partial multiple-valued functions $\psi_1, \ldots, \psi_l$ in B^* are given, ψ_j being n_j-ary for $j=1, \ldots, l$.[22] We are going to describe now an index construction used by Moschovakis for defining the notions of prime and search computability. For the definition of prime computability, a partial multiple-valued operation $\{e\}(q_1, \ldots, q_n)$ from elements $e, q_1, \ldots, q_n$ of B^* into B^* is defined by means of the following recursive definition (here and further in this section, the letters e, g, h, q, r, s, t denote elements of B^*, and the letters j, k, m, n denote natural numbers):

0) if $1 \le j \le l$ then

$$\{\langle\langle 0, n_j+m, j\rangle\rangle\}(s_1, \ldots, s_{n_j}, t_1, \ldots, t_m) = \psi_j(s_1, \ldots, s_{n_j});$$

1) $\{\langle\langle 1, n, r\rangle\rangle\}(q_1, \ldots, q_n) = r$;

2) $\{\langle\langle 2, m+1\rangle\rangle\}(s, t_1, \ldots, t_m) = s$;

3) $\{\langle\langle 3, m+2\rangle\rangle\}(s_1, s_2, t_1, \ldots, t_m) = \langle s_1, s_2\rangle$;

4_0) $\{\langle\langle 4, m+1, 0\rangle\rangle\}(s, t_1, \ldots, t_m) = L(s)$;

4_1) $\{\langle\langle 4, m+1, 1\rangle\rangle\}(s, t_1, \ldots, t_m) = R(s)$;

---

[22] We treat an n-ary partial multiple-valued function ψ in B^* as a subset of $(B^*)^{n+1}$, the values of ψ at $\langle s_1, \ldots, s_n\rangle$ being all objects r satisfying the condition that $\langle s_1, \ldots, s_k, r\rangle$ belongs to ψ. This point of view is only formally different from the one in Moschovakis [1969], where an n-ary partial multiple-valued function ψ in B^* is a mapping of $(B^*)^n$ into the set of all subsets of B^*. (The set of all values of ψ at $\langle s_1, \ldots, s_n\rangle$ will be denoted by $\psi(s_1, \ldots, s_n)$, and in the special case when this set consists of a single element, that element will be also denoted by $\psi(s_1, \ldots, s_n)$). Let us mention that denotations $\varphi_1, \ldots, \varphi_l$ instead of $\psi_1, \ldots, \psi_l$ are used in Moschovakis [1969].

5) $\{\langle\langle 5, m, g, h\rangle\rangle\}(t_1, \dots, t_m) =$

$$\{g\}(\{h\}(t_1, \dots, t_m), t_1, \dots, t_m);$$

6_0) if $s \in B^\circ$ then

$$\{\langle\langle 6, m+1, g, h\rangle\rangle\}(s, t_1, \dots, t_m) = \{g\}(s, t_1, \dots, t_m);$$

6_1) $\{\langle\langle 6, m+1, g, h\rangle\rangle\}(\langle s_1, s_2\rangle, t_1, \dots, t_m) =$

$$\{h\}(\{\langle\langle 6, m+1, g, h\rangle\rangle\}(s_1, t_1, \dots, t_m), \{\langle\langle 6, m+1, g, h\rangle\rangle\}(s_2, t_1, \dots, t_m), s_1, s_2, t_1, \dots, t_m);$$

7) whenever $k < n$, then

$$\{\langle\langle 7, n, k, g\rangle\rangle\}(q_1, \dots, q_k, q_{k+1}, q_{k+2}, \dots, q_n) = \{g\}(q_{k+1}, q_1, \dots, q_k, q_{k+2}, \dots, q_n);$$

8) $\{\langle\langle 8, k+m+1, k\rangle\rangle\}(e, s_1, \dots, s_k, t_1, \dots, t_m) =$

$$\{e\}(s_1, \dots, s_k).$$

For the definition of search computability, an operation $\{e\}_\nu(q_1, \dots, q_n)$ is defined recursively through replacing of $\}$ by $\}_\nu$ in the above definition and appending the following additional clause:

9) $\{\langle\langle 9, n, g\rangle\rangle\}_\nu(q_1, \dots, q_n) = \{r\colon \{g\}_\nu(r, q_1, \dots, q_n) \ni 0\}$.

Let φ be a n-ary partial multiple-valued function in B^*, and let A be some subset of B^*. The function φ is called *prime computable from* A *in* $\psi_1, \dots, \psi_l$ iff there is some e in A^* such that

$$\varphi(q_1, \dots, q_n) = \{e\}(q_1, \dots, q_n)$$

for all $q_1, \dots, q_n$ in B^*. It is called *search computable from* A *in* $\psi_1, \dots, \psi_l$ iff there is some e in A^* such that

$$\varphi(q_1, \dots, q_n) = \{e\}_\nu(q_1, \dots, q_n)$$

for all $q_1, \dots, q_n$ in B^*. The set of all partial multiple-valued functions in B^* which are prime computable from A in $\psi_1, \dots, \psi_l$ and the set of all ones which are search computable from A in $\psi_1, \dots, \psi_l$ are denoted by $PC(A, \psi_1, \dots, \psi_l)$ and $SC(A, \psi_1, \dots, \psi_l)$, respectively. The elements of the sets $PC(\emptyset, \psi_1, \dots, \psi_l)$ and $SC(\emptyset, \psi_1, \dots, \psi_l)$ are called, respectively, *absolutely prime computable in* $\psi_1, \dots, \psi_l$ and *absolutely search computable in* $\psi_1, \dots, \psi_l$.

(II) Prime and search computability of one-argument functions and $\mathfrak{M}_B$-computability. In the sequel, we shall assume that $\psi_1, \ldots, \psi_l$ belong to $\mathcal{F}_m(B^*)$, i.e. they are one-argument partial multiple-valued functions in B^*. This is not an essential restriction since, according to Lemmas 22 and 32 in Moschovakis [1969], it is always possible to replace a system $\psi_1, \ldots, \psi_l$ of arbitrary partial multiple-valued functions in B^* by some $\psi_1', \ldots, \psi_l'$ from $\mathcal{F}_m(B^*)$ satisfying the conditions

$$\mathrm{PC}(A, \psi_1, \ldots, \psi_l) = \mathrm{PC}(A, \psi_1', \ldots, \psi_l'),$$
$$\mathrm{SC}(A, \psi_1, \ldots, \psi_l) = \mathrm{SC}(A, \psi_1', \ldots, \psi_l').$$

Let $\langle B^*, J, L, R, T, F, H\rangle$ be the computational structure $\mathfrak{M}_B$ from Example 1.2, i.e. $J = \lambda st.\langle s, t\rangle$, the mappings L, R are the same as in Subsection (I), $T = \lambda u.1$, $F = \lambda u.0$, and

$$H(u) = \begin{cases} \text{true} & \text{if } u \notin B^\circ, \\ \text{false} & \text{if } u \in B^\circ. \end{cases}$$

Let $A \subseteq B^*$, and let $\mathcal{C}_A$ consist of all constant single-valued functions whose domain is B^* and whose values belong to A. We shall prove now the following two propositions:

Proposition 1. All elements of $\mathcal{F}_m(B^*)$, which are $\mathfrak{M}_B$-computable in $\mathcal{C}_A \cup \{\psi_1, \ldots, \psi_l\}$, belong to $\mathrm{PC}(A, \psi_1, \ldots, \psi_l)$.

Proposition 2. All elements of $\mathcal{F}_m(B^*)$, which are $\mathfrak{M}_B$-computable in $\mathcal{C}_A \cup \{\psi_1, \ldots, \psi_l, (B^*)^2\}$, belong to $\mathrm{SC}(A, \psi_1, \ldots, \psi_l)$.

Proof (of both propositions). We shall use some of the notations whose meaning is explained in the upper part of page 430 in Moschovakis [1969]. For short, let us set

$$\mathcal{G}_1 = \mathrm{PC}(A, \psi_1, \ldots, \psi_l), \quad \mathcal{G}_2 = \mathrm{SC}(A, \psi_1, \ldots, \psi_l),$$
$$\mathcal{H}_i = \mathcal{F}_m(B^*) \cap \mathcal{G}_i, \quad i = 1, 2.$$

Obviously,

$$\{L, R, T, F\} \cup \mathcal{C}_A \cup \{\psi_1, \ldots, \psi_l\} \subseteq \mathcal{H}_1,$$
$$\{L, R, T, F\} \cup \mathcal{C}_A \cup \{\psi_1, \ldots, \psi_l, (B^*)^2\} \subseteq \mathcal{H}_2.$$

By means of Lemmas 2, 17 and 26 in Moschovakis [1969] it is easily shown that $\mathcal{H}_1$ and $\mathcal{H}_2$ are closed under composition and $\mathfrak{M}_B$-combinatiion. The proof will be completed if we succeed to show that $\mathcal{H}_1$ and $\mathcal{H}_2$ are closed also under $\mathfrak{M}_B$-iteration.

For uniformity in the denotations, we set

$$\Phi_1(e, q_1, \dots, q_n) = \{r: \{e\}(q_1, \dots, q_n) \ni r\},$$
$$\Phi_2(e, q_1, \dots, q_n) = \{r: \{e\}_\nu(q_1, \dots, q_n) \ni r\}.$$

Let σ and χ belong to $\mathcal{H}_i$, where $i=1$ or $i=2$. We shall prove that $[\sigma, \chi]$ also belongs to $\mathcal{H}_i$. Denote by α the function $\lambda v. \langle L(v), \langle R(v), 0\rangle\rangle$ (having the property that

$$\alpha(\langle v, k\rangle) = \langle v, k+1\rangle$$

for all v in B^* and all natural numbers k), and consider the partial multiple-valued function of three variables γ which is defined by the equality

$$\gamma(s, e, v) = \begin{cases} \langle v, 0\rangle & \text{if } s \in B^\circ, \\ \alpha(\Phi_i(e, \sigma(v))) & \text{otherwise.} \end{cases}$$

It is easy to see that $\gamma \in \mathcal{G}_i$ (to do this, we make use of the equality

$$\gamma(\langle s_1, s_2\rangle, e, v) = \beta(\gamma(s_1, e, v), \gamma(\langle s_1, e, v), t_1, t_2, e, v),$$

where β is the partial multiple-valued function of six variables which is defined by means of the equality

$$\beta(s_1, s_2, t_1, t_2, e, v) = \alpha(\Phi_i(e, \sigma(v)))).$$

Then the partial multiple-valued function $\lambda ev. \gamma(\chi(v), e, v)$ also belongs to $\mathcal{G}_i$. This allows an application of the recursion theorem from Moschovakis [1969] given by Lemma 21 in the case when $i=1$ or by Lemma 29 in the case when $i=2$. Its application provides us with an element e_0 of A^* such that

$$\Phi_i(e_0, v) = \gamma(\chi(v), e_0, v)$$

for all v in B^*. We shall now show the equality

$$[\sigma, \chi](u) = L(\Phi_i(e_0, u)),$$

and this will complete the proof.

Suppose $w \in [\sigma, \chi](u)$. By the definition of $\mathfrak{M}_B$-iteration, there is a finite sequence $v_0, v_1, \dots, v_m$ of elements of M such that

$$v_0 = u \ \& \ v_m = w \ \& \ \forall_{j<m} j \, (\langle v_j, \text{true}\rangle \in H\chi \ \& \ \langle v_j, v_{j+1}\rangle \in \sigma) \\ \& \ \langle v_m, \text{false}\rangle \in H\chi.$$

According to the definition of H, the above condition is

equivalent to the following one:

$$v_0 = u \ \& \ v_m = w \ \& \ \forall_{j<m} j\,(\chi(v_j)\setminus B^\circ \neq \emptyset \ \& \ \langle v_j, v_{j+1}\rangle \in \sigma) \ \& \ \chi(v_m)\cap B^\circ \neq \emptyset.$$

From here, using the definition of γ, we conclude that

$$\gamma(\chi(v_j), e_0, v_j) \supseteq \alpha(\Phi_i(e_0, \sigma(v_j))) \supseteq \alpha(\Phi_i(e_0, v_{j+1}))$$

for $j = 0, 1, 2, \ldots, m-1$, and

$$\gamma(\chi(v_m), e_0, v) \ni \langle v_m, 0\rangle.$$

Hence

$$\Phi_i(e_0, v_j) \supseteq \alpha(\Phi_i(e_0, v_{j+1})), \quad j = 0, 1, 2, \ldots, m-1,$$

$$\Phi_i(e_0, v_m) \ni \langle v_m, 0\rangle.$$

Now an easy induction shows that

$$\Phi_i(e_0, v_{m-k}) \ni \langle v_m, k\rangle, \quad k = 0, 1, 2, \ldots, m.$$

Namely, if $k < m$ and $\langle v_m, k\rangle$ belongs to $\Phi_i(e_0, v_{m-k})$, then

$$\langle v_m, k+1\rangle = \alpha(\langle v_m, k\rangle) \in \alpha(\Phi_i(e_0, v_{m-k})) \subseteq \Phi_i(e_0, v_{m-k-1}).$$

In particular,

$$\Phi_i(e_0, v_0) \ni \langle v_m, m\rangle.$$

Consequently,

$$w = v_m \in L(\Phi_i(e_0, v_0)) = L(\Phi_i(e_0, u)).$$

Conversely, we have to prove that

$$L(\Phi_i(e_0, u)) \subseteq [\sigma, \chi](u)$$

for all u in B^*. For each element r of B^*, let us define a natural number $\|r\|$ (the complexity of r) in the following way: $\|r\| = 0$ for all r in B°, and $\|\langle r_1, r_2\rangle\| = \|r_1\| + \|r_2\| + 1$ for all r_1, r_2 in B^*.

Our goal will be reached if we succeed to prove the following statement: whenever $r \in \Phi_i(e_0, u)$, then $L(r) \in [\sigma, \chi](u)$. This statement will be proven by induction on the value of $\|r\|$. Suppose $r \in \Phi_i(e_0, u)$. Then $r \in \gamma(\chi(u), e_0, u)$. Hence $r \in \gamma(h, e_0, u)$ for some h belonging to $\chi(u)$. If $h \in B^\circ$, then $r = \langle u, 0\rangle$ and $H\chi \ni \langle u, \text{false}\rangle$; consequently,

$$\Sigma(\chi, [\sigma, \chi]\sigma, I_{B^*}) \ni \langle u, u\rangle$$

and therefore, by Proposition 5.1, $[\sigma, \chi] \ni \langle u, u\rangle$, i.e. $L(r) = u \in [\sigma, \chi](u)$. Consider now the case when $h \notin B^\circ$. Then $H\chi \ni \langle u, \text{true}\rangle$ and $r \in \alpha(\Phi_i(e_0, \sigma(u)))$, i.e. $r = \alpha(\bar{r})$,

$\bar{r}\in\Phi_i(e_0, \bar{u}))$ and $\bar{u}\in\sigma(u)$ for some $\bar{r}$ and $\bar{u}$ in B^*. From the definition of α, it follows that $\|r\| > \|\bar{r}\|$, and hence we may assume that $L(\bar{r})\in[\sigma, \chi](\bar{u})$. But then we see that $L(r) = L(\bar{r})\in[\sigma, \chi]\sigma(u)$, and consequently

$$\Sigma(\chi, [\sigma, \chi]\sigma, I_{B^*})\ni\langle u, L(r)\rangle.$$

By Proposition 5.1, it follows that $[\sigma, \chi]\ni\langle u, L(r)\rangle$, i.e. $L(r)\in[\sigma, \chi](u)$ again.■

Remark 1. In the above proof we used Proposition 5.1 only partially, since the inclusion

$$[\sigma, \chi]\supseteq\Sigma(\chi, [\sigma, \chi]\sigma, I_{B^*})$$

is sufficient for the application of the proposition in the proof. On the other hand, that proposition gives a least-fixed-point characterization of iteration, and such a characterization suggests another way of proving Propositions 1 and 2 above, namely by application of the First Recursion Theorems for prime and for search computable functions. For the case of search computability, the First Recursion Theorem is formulated as Theorem 2 in Moschovakis [1969], but its assumptions there include the superfluous one that $\psi_1, \ldots, \psi_l$ are totally defined and single-valued. As to the case of prime computability, the validity of the First Recursion Theorem is noted in Remark 11 of the same paper (without explicit listing of the assumptions needed for the proof).

Of course, the converse statements of Propositions 1 and 2 are not true, since $\mathrm{PC}(A, \psi_1, \ldots, \psi_l)$ and $\mathrm{SC}(A, \psi_1, \ldots, \psi_l)$ contain functions of arbitrary number of arguments. However, if we replace $\mathrm{PC}(A, \psi_1, \ldots, \psi_l)$ and $\mathrm{SC}(A, \psi_1, \ldots, \psi_l)$ by $\mathcal{F}_m(B^*)\cap\mathrm{PC}(A, \psi_1, \ldots, \psi_l)$ and $\mathcal{F}_m(B^*)\cap\mathrm{SC}(A, \psi_1, \ldots, \psi_l)$, respectively, then also the converse statements of Propositions 1 and 2 are valid. This fact will be proved in a natural way in Subsection (III) of Section III.5 on the basis of results from the general theory which we are going to present. We note that a direct proof of the same fact is given in the previous version Skordev [1980] of this book, but that proof is quite a long one (more than eleven pages).

The above mentioned conversion of Propositions 1 and 2 will give a characterization of the classes $\mathcal{F}_m(B^*)\cap\mathrm{PC}(A, \psi_1, \ldots, \psi_l)$ and $\mathcal{F}_m(B^*)\cap\mathrm{SC}(A, \psi_1, \ldots, \psi_l)$ (for the case when $\psi_1, \ldots, \psi_l$ are unary), which is much more simpler than their characterization in Moschovakis [1969]. Since prime and search computability in/of functions of several variables are easily reducible to prime and search computability in/of corresponding one-argument functions, the details about the number of arguments could be considered not very essential, and we could just say that a considerable simplification of the definitions of prime and search computability will be reached by means of the mentioned conversion.

Remark 2. The presented proof of Propositions 1 and 2 essentially uses (via the recursion theorem) the eight clause of the definitions of $\{e\}(q_1, \ldots, q_n)$ and $\{e\}_\nu(q_1, \ldots, q_n)$. On the other hand, as noted in Remark 8 of Moschovakis [1969], that clause is superfluous in the definition of $\{e\}_\nu(q_1, \ldots, q_n)$, at least in the case when $\psi_1, \ldots, \psi_l$ are single-valued and total. Therefore a proof of Proposition 2 not using that clause is desirable. For such a proof, cf. the exercise below. Note that similar things can be done also in the case of Proposition 1, but in this case one must compensate the removing of the eight clause by a clause concerning μ-operation (cf. Remark 10 in Moschovakis [1969]).

Exercise. Prove Proposition 2 in the case of a definition of $\{e\}_\nu(q_1, \ldots, q_n)$ not including the eight clause.

Hint. To show that $\mathcal{H}_2$ is closed under $\mathfrak{M}_B$-iteration, suppose σ and χ belonging to $\mathcal{H}_2$ are given, and show the existence of functions θ_0 and θ_1 in $SC(A, \psi_1, \ldots, \psi_l)$, with the following properties:

$$\theta_0(s, u) = u \text{ for all } s \text{ in } B^\circ,$$

$$\theta_0(\langle s_1, s_2\rangle, u) = \sigma(\theta_0(s_1, u)),$$

$$\theta_1(s, r) = \begin{cases} r & \text{if } s \in B^\circ, \\ \emptyset & \text{otherwise.} \end{cases}$$

Then prove the equality

$$[\sigma, \chi](u) = \theta_3(\theta_0(\theta_2(u), u)),$$

where $\theta_2 = (B^*)^2$ and $\theta_3 = \lambda r.\, \theta_1(\chi(r), r)$.

8. Computability in the case of unproductive termination taken into account

The intuitive idea behind our considerations up to now was connected with characterizing of computational procedures by their input-output relations. Suppose a set M is given. Then, given a computational procedure transforming elements of M into elements of M, its input-output relation consists of all ordered pairs $\langle u, v\rangle$ such that u can be transformed into v by means of the given procedure. This relation surely belongs to $\mathcal{F}_m(M)$, and it belongs to $\mathcal{F}_p(M)$ in the case, when the given computational procedure is a deterministic one.

When computational procedures are characterized by their input-output relations, then no distinction is made between cases when the computational process never terminates and ones when this process terminates without yielding a result. However, the difference between these cases is an important one from the point of view of practice. Therefore it is natural to look for some more detailed mathematical characterization of computational procedures, which takes also this difference into account. Such a characterization will be considered now. The characterization will be based on considering a set E, whose elements can be regarded as error messages, and on the convention that unproductive termination of the application of the procedure to the element u transforms u into some element of E.[29]

Suppose E is some fixed set having no common elements with the set M. Then we shall consider sets $\mathcal{F}_p(M, E)$ and $\mathcal{F}_m(M, E)$, defined as follows: $\mathcal{F}_p(M, E)$ is the set of all partial functions φ such that $\mathrm{dom}\varphi \subseteq M$ and $\mathrm{rng}\varphi \subseteq M \cup E$,

---

[23] There is also another intuitive interpretation of the elements of E, namely as sorts of failures which may arise during computation. When using this interpretation, we may adopt that the rise of a failure during the application of the procedure to an element u of M transforms u into the element of E corresponding to the concrete failure. It is not obligatory to assume that the rise of a failure necessarily causes unproductive termination – the computation could sometimes go on and lead to some (possibly incorrect) result (also more than one failure could arise during the course of a certain application of the procedure). This intuitive interpretation of the elements of E is suggested by an idea of S. Nikolova arisen in joint work with I. Soskov and expressed and used by her in 1988. The idea is to characterize a computational procedure by the ordered pair consisting of the corresponding input-output relation and the set of those elements of M which, taken as input values, are safe with respect to rise of failures during the application of the procedure (for the respective technical details, cf. Exercise 3 after this section).

and $\mathcal{F}_{m}(M, E)$ is the set of all subsets of the Cartesian product $M\times(M\cup E)$. Clearly, $\mathcal{F}_{p}(M, E)$ is a subset of $\mathcal{F}_{m}(M, E)$ (note also that $\mathcal{F}_{p}(M)=\mathcal{F}_{p}(M, \emptyset)\subseteq \mathcal{F}_{p}(M, E)$ and $\mathcal{F}_{m}(M)=\mathcal{F}_{m}(M, \emptyset)\subseteq\mathcal{F}_{m}(M, E)$). The *composition* $\varphi\psi$ of two elements φ and ψ of $\mathcal{F}_{p}(M, E)$ will be defined as the union of their usual composition and the relation $\psi\cap(M\times E)$, i.e.

$$\varphi\psi=\{<u, w>: \exists v(<u, v>\in\psi \mathbin{\&} <v, w>\in\varphi) \vee <u, w>\in\psi \mathbin{\&} w\in E\}.$$ [24]

Obviously, $\varphi\psi\in\mathcal{F}_{p}(M, E)$ whenever φ and ψ belong to $\mathcal{F}_{p}(M, E)$.

Suppose now a computational structure $\mathfrak{U}=<M, J, L, R, T, F, H>$ is given, where, for the sake of simplicity, the predicate H is assumed to be total. Then we shall define also $\mathfrak{U}$*-combination*, $\mathfrak{U}$*-branching* and $\mathfrak{U}$*-iteration* in $\mathcal{F}_{m}(M, E)$. The following definitions can be intuitively motivated in the spirit of the footnote concerning the definition of composition in $\mathcal{F}_{m}(M, E)$:

$$\Pi(\varphi, \psi)=\{<u, w>: \exists s\in M(<u, s>\in\varphi \mathbin{\&} (\exists t\in M(<u, t>\in\psi \mathbin{\&} J(s, t)=w) \vee <u, w>\in\psi \mathbin{\&} w\in E) \vee <u, w>\in\varphi \mathbin{\&} w\in E)\},$$

$$\Sigma(\chi, \varphi, \psi)=\{<u, w>: <u, \text{true}>\in H\chi \mathbin{\&} <u, w>\in\varphi \vee <u, \text{false}>\in H\chi \mathbin{\&} <u, w>\in\psi \vee <u, w>\in\chi \mathbin{\&} w\in E\},$$

and $<u, w>\in[\sigma, \chi]$ iff there is a finite sequence $v_0, v_1, \ldots, v_m$ of elements of M such that

(1) $$v_0=u \mathbin{\&} \forall_{j<m} j(<v_j, \text{true}>\in H\chi \mathbin{\&} <v_j, v_{j+1}>\in\sigma) \mathbin{\&} (<v_m, \text{false}>\in H\chi \mathbin{\&} v_m=w \vee (<v_m, w>\in\chi \vee <v_m, \text{true}>\in H\chi \mathbin{\&} <v_m, w>\in\sigma) \mathbin{\&} w\in E),$$

where $H\chi=\{<u, p>: \exists v\in M(<u, v>\in\chi \mathbin{\&} H(v)=p)\}$. Again it is easy to see that $\mathcal{F}_{p}(M, E)$ is closed under the introduced operations.

---

[24] The intuitive motivation for including $\psi\cap(M\times E)$ into $\varphi\psi$ is the following: $\varphi\psi$ must characterize the procedure consisting in consecutive execution of the procedures characterized by ψ and φ, but unproductive termination of the execution of the first of these two procedures implies that unproductive termination of the consecutive execution of both is present, and the error message is the same one.

From a purely mathematical point of view, the above definition of iteration looks somewhat messy. However, the iteration again has a least-fixed-point-characterization as in the previous situations we considered.

Proposition 1. Let $\sigma, \chi \in \mathcal{F}_{\mathrm{m}}(M, E)$. Then the equality

$$[\sigma, \chi] = \Sigma(\chi, [\sigma, \chi]\sigma, I_M)$$

holds. More generally, for each ρ in $\mathcal{F}_{\mathrm{m}}(M, E)$ the equality

(2) $$\rho[\sigma, \chi] = \Sigma(\chi, \rho[\sigma, \chi]\sigma, \rho)$$

holds, and $\rho[\sigma, \chi]$ is the least element τ of $\mathcal{F}_{\mathrm{m}}(M, E)$ satisfying

(3) $$\tau \geq \Sigma(\chi, \tau\sigma, \rho).$$

Proof. For each τ in $\mathcal{F}_{\mathrm{m}}(M, E)$, let τ' be the element $\tau \cup I_E$ of $\mathcal{F}_{\mathrm{m}}(M \cup E)$. Then the mapping $\lambda\tau.\tau'$ is injective, and $\tau_1' \geq \tau_2'$ always implies $\tau_1 \geq \tau_2$. For all σ, χ, τ in $\mathcal{F}_{\mathrm{m}}(M, E)$, the equality

$$(\Sigma(\chi, \tau\sigma, \rho))' = \tau'\alpha \cup \beta$$

holds, where

$$\alpha = \{\langle u, v\rangle : \langle u, \mathrm{true}\rangle \in H\chi \ \&\ \langle u, v\rangle \in \sigma\},$$

$$\beta = \{\langle u, v\rangle : \langle u, \mathrm{false}\rangle \in H\chi \ \&\ \langle u, v\rangle \in \rho \ \vee \ \langle u, v\rangle \in \chi' \ \&\ v \in E\},$$

and $\tau'\alpha$ is the ordinary composition of τ' and α as elements of $\mathcal{F}_{\mathrm{m}}(M \cup E)$. Consider now the equation $\tau' = \tau'\alpha \cup \beta$ and the inequality $\tau' \geq \tau'\alpha \cup \beta$ with τ' ranging over $\mathcal{F}_{\mathrm{m}}(M \cup E)$. Their least solution is the element $\beta\alpha^*$, which is the union of the elements $\beta\alpha^m$, $m = 0, 1, 2, \ldots$ (cf. the analogous proof in Section 5). On the other hand, it is easy to verify that $\beta\alpha^* = (\rho[\sigma, \chi])'$. From here, using the properties of $\lambda\tau.\tau'$ mentioned at the beginning, we conclude that the equality (2) holds, and $\rho[\sigma, \chi]$ is contained in each τ satisfying (3). ∎

Remark 1. If τ' is defined as in the above proof then, for all φ and ψ in $\mathcal{F}_{\mathrm{m}}(M, E)$, the equality $(\varphi\psi)' = \varphi'\psi'$ holds, where $\varphi'\psi'$ is the ordinary composition of φ' and ψ' in $\mathcal{F}_{\mathrm{m}}(M \cup E)$. This suggests another way of treatment of the subject of this section: to consider

$$\{\varphi \in \mathcal{F}_{\mathrm{m}}(M \cup E) : \varphi \restriction E = I_E\}$$

with the ordinary composition in it instead of $\mathcal{F}_m(M, E)$ with the unusual composition which we introduced. Roughly speaking, this is the way used in Example 2 of Skordev [1980 a] for the case of E consisting of a single element. A similar situation is present also in Example 2 of Lukanova [1986] (given previously in Lukanova [1978]), where, in essence, the case corresponding to $\mathcal{F}_p(M, E)$ is considered for such an E. Note however that only composition turns into the ordinary one when using this other way, while combination, branching and iteration remain unusual (i.e. not exactly of the type considered in the previous sections).

Having composition, combination and iteration in $\mathcal{F}_p(M, E)$ and in $\mathcal{F}_m(M, E)$ at our disposal, we can define relative computability in $\mathcal{F}_p(M, E)$ and in $\mathcal{F}_m(M, E)$ in a similar way as in the previous sections.

Definition 1. Let $\mathcal{F}$ denote $\mathcal{F}_p(M, E)$ or $\mathcal{F}_m(M, E)$, and let $\mathcal{B}$ be some subset of $\mathcal{F}$. The elements of $\mathcal{F}$ *$\mathfrak{U}$-computable in $\mathcal{B}$* are those elements of $\mathcal{F}$ which can be generated from elements of $\{L, R, T, F\}\cup\mathcal{B}$ by means of composition, $\mathfrak{U}$-combination and $\mathfrak{U}$-iteration.

At the present moment, we shall not comment the introduced computability notion in the general case, but we shall demonstrate how things look in a natural special case. Namely, we shall consider the case when E consists of a single element, and $\mathfrak{U}$ is a standard computational structure over the natural numbers in the sense of Section 3, i.e. $M=\mathbb{N}$, J is a recursive function, L, R are partial recursive functions, $T=\lambda u.1$, $F=\lambda u.0$,

$$H(u)=\begin{cases}\text{true} & \text{if } u>0,\\ \text{false} & \text{if } u=0.\end{cases}$$

Theorem 1. Let E consist of a single element $\bullet$, and let $\mathfrak{U}$ be a standard computational structure on the natural numbers. Let $\mathcal{F}=\mathcal{F}_p(\mathbb{N}, \{\bullet\})$, and let S and $P^{\bullet}$ be the elements of $\mathcal{F}$ defined as follows: $S=\lambda u.u+1$,

$$P^{\bullet}(u)=\begin{cases}u-1 & \text{if } u>0,\\ \bullet & \text{if } u=0.\end{cases}$$

An element φ of $\mathcal{F}$ is $\mathfrak{U}$-computable in $\{S, P^{\bullet}\}$ iff $\varphi\cap\mathbb{N}^2$ is a partial recursive function and the set $\{u: \langle u, \bullet\rangle\in\varphi\}$ is recursively enumerable.

Proof. Let $\mathcal{H}$ be the set of all φ in $\mathcal{F}$ such that $\varphi\cap\mathbb{N}^2$ is partial recursive and $\{u: \langle u, \bullet\rangle\in\varphi\}$ is recursively enumerable. Obviously, $\{L, R, T, F, S, P^{\bullet}\}\subseteq\mathcal{H}$, and it is easy to verify that $\mathcal{H}$ is closed under composition, $\mathfrak{U}$-

combination and $\mathfrak{A}$-iteration. Consequently, all elements of $\mathcal{F}$ $\mathfrak{A}$-computable in $\{S, P^{\bullet}\}$ belong to $\mathcal{H}$. So it remains to prove that all elements of $\mathcal{H}$ are $\mathfrak{A}$-computable in $\{S, P^{\bullet}\}$.

Let φ be an arbitrary element of $\mathcal{H}$. Denote by ψ the element of $\mathcal{F}$ defined as follows:

$$\psi(u) \simeq \begin{cases} 0 & \text{if } \varphi(u) = \bullet, \\ \varphi(u) + 1 & \text{otherwise.} \end{cases}$$

Then ψ is a partial recursive function. By Theorem 3.1, ψ is $\mathfrak{A}$-computable in $\{S, P\}$, where $P = \lambda u.\, u \dot{-} 1$. Since composition, $\mathfrak{A}$-combination and $\mathfrak{A}$-iteration in $\mathcal{F}_p(\mathbb{N})$ are restrictions of the corresponding operations in $\mathcal{F}$, it follows that ψ is $\mathfrak{A}$-computable in $\{S, P\}$ also as an element of $\mathcal{F}$. On the other hand, again in $\mathcal{F}$, we have the equality $\varphi = P^{\bullet}\psi$. Hence φ is an element of $\mathcal{F}$ $\mathfrak{A}$-computable in $\{S, P, P^{\bullet}\}$. Thus it is sufficient to prove that P is $\mathfrak{A}$-computable in $\{S, P^{\bullet}\}$, and this can be done by means of the equality $P = L[\Pi(PL, F), R]\Pi(I_{\mathbb{N}}, I_{\mathbb{N}})$. ∎

Exercises

(in all these exercises, a computational structure $\mathfrak{A} = \langle M, J, L, R, T, F, H\rangle$, where H is total, and a set E with $M \cap E = \varnothing$ are supposed to be given)

1. Prove the statements of Exercises 2.1, 2.2, 2.6 and 2.7 for the case when $\mathcal{F}_p(M)$ is replaced by $\mathcal{F}_m(M, E)$.

2. Show that composition, $\mathfrak{A}$-combination, $\mathfrak{A}$-branching and $\mathfrak{A}$-iteration in $\mathcal{F}_m(M, E)$ are extensions of the corresponding operations in $\mathcal{F}_m(M)$. Prove that the statements of Exercises 5.2-5.6 remain valid with $\mathcal{F} = \mathcal{F}_m(M, E)$. Prove the same for the statement of Exercise 5.7 with the equality $\Pi(x, I)\theta = \Pi(x, \theta)$ instead of $\Pi(\varphi x, I)\theta = \Pi(\varphi x, \theta)$. Under the assumption that E is non-empty, construct a counter-example to the equality $\Pi(\varphi x, I)\theta = \Pi(\varphi x, \theta)$. Show that such a counter-example is not possible if the set E has only one element and the requirement $\operatorname{dom}\varphi = \operatorname{dom}\theta = M$ is imposed (but $\{\tau \in \mathcal{F}_m(M, E) : \operatorname{dom}\tau = M\}$ is not closed under iteration).

3. Let $E = \{\bullet\}$, and let $\mathcal{F}$ be the set of all ordered pairs $\langle f, A\rangle$, where $f \in \mathcal{F}_m(M)$ and $A \subseteq M$. Let composition, combination and branching in $\mathcal{F}$ be defined in the following way:

$$\langle f, A\rangle\langle g, B\rangle = \langle fg, \{u \in B :\ \forall v(\langle u, v\rangle \in g \Rightarrow v \in A)\}\rangle,$$

$$\Pi(\langle f, A\rangle, \langle g, B\rangle) = \langle \Pi(f, g), \{u \in A :\ \exists v(\langle u, v\rangle \in f) \Rightarrow u \in B\}\rangle,$$

$$\Sigma(\langle h, C\rangle, \langle f, A\rangle, \langle g, B\rangle) = \langle \Sigma(h, f, g), \{u \in C: (\langle u, \text{true}\rangle \in Hh \Rightarrow u \in A) \,\&\, (\langle u, \text{false}\rangle \in Hh \Rightarrow u \in B)\}\rangle,$$

where fg, $\Pi(f, g)$, $\Sigma(h, f, g)$ and Hh are understood in the sense of Section 5. Let iteration in $\mathcal{F}$ be defined by the equality

$$[\langle f, A\rangle, \langle h, C\rangle] = \langle [f, h], D\rangle,$$

where $[f, h]$ is understood again in the sense of Section 5, and D is the set of all elements u of M satisfying the following condition: whenever $v_0, v_1, \ldots, v_m$ is a finite sequence of elements of M with the property that

$$v_0 = u \,\&\, \forall_{j<m} j\,(\langle v_j, \text{true}\rangle \in Hh \,\&\, \langle v_j, v_{j+1}\rangle \in f),$$

then $v_m \in C \,\&\, (\langle v_m, \text{true}\rangle \in Hh \Rightarrow v_m \in A)$.[25] Let partial ordering in $\mathcal{F}$ be defined by the equivalence

$$\langle f, A\rangle \geq \langle g, B\rangle \Longleftrightarrow f \geq g \,\&\, A \subseteq B.$$

Now define a mapping Φ of $\mathcal{F}$ into $\mathcal{F}_{\mathrm{m}}(M, E)$ as follows:

$$\Phi(\langle f, A\rangle) = f \cup ((M \setminus A) \times E)$$

for all elements $\langle f, A\rangle$ of $\mathcal{F}$. Prove that Φ is an one-to-one correspondence between $\mathcal{F}$ and $\mathcal{F}_{\mathrm{m}}(M, E)$, and this correspondence is an isomorphism with respect to composition, combination, branching, iteration and partial ordering (i.e.

$$\Phi(\varphi\psi) = \Phi(\varphi)\Phi(\psi), \quad \Phi(\Pi(\varphi, \psi)) = \Pi(\Phi(\varphi), \Phi(\psi)),$$

$$\Phi(\Sigma(\chi, \varphi, \psi)) = \Sigma(\Phi(\chi), \Phi(\varphi), \Phi(\psi)),$$

---

[25] The above definitions of composition, combination, branching and iteration in $\mathcal{F}$ are in essential definitions given by S. Nikolova in 1988. The definitions of composition and branching replay, up to unessential details, the corresponding definitions adopted in Example 4 of Skordev [1976b] for the elements of a certain subset of $\mathcal{F}$, and the operation Π defined above is an extension of the operation Π from the mentioned example (see also Exercise II.4.13 in the present book). That example, however, corresponds to a quite different intuitive interpretation of the second members of the ordered pairs belonging to the subset in question, and the corresponding iteration (studied in Skordev [1980, Chapter III, Section 3.2, Example 11]; cf. also Exercises II.4.17 and II.4.18 in this book) turns out to be quite different from Nikolova's iteration.

$$\Phi([\varphi, \chi]) = [\Phi(\varphi), \Phi(\chi)], \quad \varphi \geq \psi \iff \Phi(\varphi) \geq \Phi(\psi),$$

for all φ, ψ, χ in $\mathcal{F}$).

4. Let the set $\mathcal{F}$ and the operations composition, combination, branching and iteration in it be defined as in the previous exercise. Let I be the element $\langle I_M, M\rangle$ of $\mathcal{F}$. Give direct proofs (not using Exercises 2 and 3 above) of the statements of Exercises 5.2–5.4. After changing the definition of $\mathcal{C}$ in Exercise 2.5 by setting $\mathcal{C}$ to consist of all pairs $\langle \bar{s}, M\rangle$ with $s \in M$, give such direct proofs also of the statements of Exercises 5.5, 5.6 and of the statement of Exercise 5.7 with $\Pi(x, I)\theta = \Pi(x, \theta)$ instead of $\Pi(\varphi x, I)\theta = \Pi(\varphi x, \theta)$. Give also a direct proof of Proposition 1 with $\mathcal{F}$ and $\geq$ instead of $\mathcal{F}_m(M, E)$ and $\geq$.

5. Let $E = \{\bullet\}$, and let $\mathfrak{U}$ be a standard computational structure on the natural numbers. Let $\mathcal{F} = \mathcal{F}_m(\mathbb{N}, \{\bullet\})$, and let the function $P^{\bullet}$ from $\mathcal{F}$ be defined as in Theorem 1. Prove that an element φ of $\mathcal{F}$ is $\mathfrak{U}$-computable in the set $\{S, P^{\bullet}, \mathbb{N}^2\}$ iff both the relation $\varphi \cap \mathbb{N}^2$ and the set $\{u\colon \langle u, \bullet\rangle \in \varphi\}$ are recursively enumerable.

6. Let $\bar{\bullet}$ be the function, determined by the condition that $\mathrm{dom}\,\bar{\bullet} = \mathbb{N}$ and $\bar{\bullet}(u) = \bullet$ for all u in $\mathbb{N}$. Prove that Theorem 1 remains valid after replacing $\{S, P^{\bullet}\}$ by $\{S, P, \bar{\bullet}\}$, and the statement of Exercise 5 remains valid after replacing $\{S, P^{\bullet}, \mathbb{N}^2\}$ by $\{S, P, \bar{\bullet}, \mathbb{N}^2\}$, where $P = \lambda u.\, u \dot{-} 1$ (from the point of view of the correspondence Φ mentioned in Exercise 3, this is equivalent to some results obtained by S. Nikolova in 1988).

7. Let $E = \{\bullet\}$, and let $\mathfrak{U}$ be a standard computational structure on the natural numbers. Let $\mathcal{F} = \mathcal{F}_m(\mathbb{N}, \{\bullet\})$, and let $P^{?} = P \cup \{\langle 0, \bullet\rangle\}$, where $P = \lambda u.\, u \dot{-} 1$. Prove that an element φ of $\mathcal{F}$ is $\mathfrak{U}$-computable in $\{S, P^{?}\}$ iff $\varphi \cap \mathbb{N}^2$ is a partial recursive function, the set $\{u\colon \langle u, \bullet\rangle \in \varphi\}$ is recursively enumerable, and there is a partial recursive function χ which satisfies the following conditions:

$$\mathrm{dom}\,\chi = \mathrm{dom}(\varphi \cap \mathbb{N}^2), \quad \forall u \in \mathrm{dom}\,\chi\, (\chi(u) = 0 \iff \langle u, \bullet\rangle \in \varphi).$$

Hint. If $\varphi \in \mathcal{F}$, $\varphi \cap \mathbb{N}^2$ is a function, and χ is a partial function in $\mathbb{N}$ satisfying the above conditions, then

$$\varphi = (\varphi \cap \mathbb{N}^2) R \Pi(P^{?} \chi_1, I_{\mathbb{N}}),$$

where $\chi_1 = \chi \cup \{\langle u, 0\rangle\colon \langle u, \bullet\rangle \in \varphi\}$.

8. Let σ, χ, ρ be arbitrary elements of $\mathcal{F}_m(M, E)$, and let $\tau_0 = \rho[\sigma, \chi]$. Suppose K is a subset of M such that $v \in K \cup E$ for all $\langle u, v\rangle$ in σ with $u \in K$, and τ is an el-

ement of $\mathcal{F}_m(M, E)$ such that $\tau \geqq \Sigma(\chi, \tau\sigma, \rho) \cap (K \times (M \cup E))$. Prove that $\tau \geqq \tau_0 \cap (K \times (M \cup E))$.

CHAPTER II

COMBINATORY SPACES

1. The notion of combinatory space

In the previous chapter, a number of situations were described, where a set $\mathcal{F}$ of functions or function-like objects is fixed and a notion of relative computability for the elements of $\mathcal{F}$ can be considered. A common feature of these situations is that $\mathcal{F}$ contains an identity element I and is supplied with a composition operation, an operation Π of combination, a branching operation Σ, an operation of iteration and a partial ordering such that iteration has a least-fixed-point characterization in terms of branching, composition and I. In addition, elements L and R of $\mathcal{F}$ are fixed having a certain connection with the operation of combination, as well as elements T and F of $\mathcal{F}$ having a certain connection with branching. Now we shall give an abstract axiomatic treatment of such kind of situations. For the first time, we shall leave aside the operation of iteration (having in mind its characterizability by means of the other operations). Of course, a given number of concrete situations can be captured by a general notion in infinitely many different ways. However, we aim to introduce a notion capturing not only the considered concrete situations, but also other interesting ones, and giving the possibility to develop a sufficiently rich theory about it. These requirements leave not so much room for arbitrariness, and it is even not clear whether such a goal can be reached. As we hope, a positive answer of the last question will be seen from this book (another solution of the above problem is given by the notion of iterative operative space studied in Ivanov [1986]).

The definition, which we shall give now, makes use of the notion of partially ordered semigroup. We think this notion is well-known to the reader, but, for the sake of completeness, we shall recall its definition. Namely, a *par-*

tially ordered semigroup is a non-empty set supplied with a partial ordering and an associative monotonically increasing binary operation. The partial ordering will be considered reflexive[26]. The corresponding denotations will be $\geq$ and $\leq$. The binary operation mentioned above will be denoted as multiplication. Thus, if the semigroup is $\mathcal{F}$, then the following conditions must be satisfied for all φ, ψ, θ in $\mathcal{F}$:

$$(\varphi\psi)\theta = \varphi(\psi\theta), \quad \varphi \geq \varphi, \quad \varphi \geq \psi \; \& \; \psi \geq \theta \Longrightarrow \varphi \geq \theta,$$

$$\varphi \geq \psi \; \& \; \psi \geq \varphi \Longrightarrow \varphi = \psi, \quad \varphi \geq \psi \Longrightarrow \varphi\theta \geq \psi\theta \; \& \; \theta\varphi \geq \theta\psi.$$

An element I of $\mathcal{F}$ is called an *identity of* $\mathcal{F}$ iff

$$I\theta = \theta I = \theta$$

for all θ in $\mathcal{F}$.

Now we proceed to the main definition of this chapter – the definition of the notion of combinatory space.

Definition 1. A *combinatory space* is a 9-tuple

$$\mathfrak{S} = \langle \mathcal{F}, I, \mathcal{C}, \Pi, L, R, \Sigma, T, F \rangle,$$

where $\mathcal{F}$ is a partially ordered semigroup, I is an identity of $\mathcal{F}$, $\mathcal{C}$ is a subset of $\mathcal{F}$, Π and Σ are a binary and a ternary operation in $\mathcal{F}$, respectively, L, R, T, F are elements of $\mathcal{F}$, and the following sixteen conditions are identically satisfied, when $\varphi, \psi, \theta, \chi$ range over $\mathcal{F}$, x, y range over $\mathcal{C}$, and $\Pi(\varphi, \psi)$, $\Sigma(\chi, \varphi, \psi)$ are denoted by (φ, ψ) and $(\chi \rightarrow \varphi, \psi)$, respectively[27]:

(1) $$\forall x(\varphi x \geq \psi x) \Longrightarrow \varphi \geq \psi,$$

(2) $$(x, y) \in \mathcal{C},$$

(3) $$L(x, y) = x,$$

(4) $$R(x, y) = y,$$

(5) $$(\varphi, \psi)x = (\varphi x, \psi x),$$

(6) $$(I, \psi x)\theta = (\theta, \psi x),$$

(7) $$(x, I)\theta = (x, \theta),$$

---

[26] In other words, a partial ordering in a set will be any reflexive, transitive and anti-symmetric binary relation in it (the anti-symmetry means that inequalities in both directions between two given elements of the set always imply equality of these elements).

[27] These denotations will be systematically used not only in the present definition, but also in the further exposition (we should like to mention that we used $\supset$ instead of $\rightarrow$ in the previous publications on combinatory spaces).

(8) $$T \neq F,$$

(9) $$Tx \in \mathscr{C},$$

(10) $$Fx \in \mathscr{C},$$

(11) $$(T \rightarrow \varphi, \psi) = \varphi,$$

(12) $$(F \rightarrow \varphi, \psi) = \psi,$$

(13) $$\theta(\chi \rightarrow \varphi, \psi) = (\chi \rightarrow \theta\varphi, \theta\psi),$$

(14) $$(\chi \rightarrow \varphi, \psi)x = (\chi x \rightarrow \varphi x, \psi x),$$

(15) $$(I \rightarrow \varphi x, \psi x)\theta = (\theta \rightarrow \varphi x, \psi x),$$

(16) $$\varphi \geq \psi \ \& \ \theta \geq \chi \implies (I \rightarrow \varphi, \theta) \geq (I \rightarrow \psi, \chi).$$

The semigroup multiplication in $\mathscr{F}$, and the operations Π, Σ will be called *composition*, *combination* and *branching in* $\mathfrak{S}$, respectively. The combinatory space $\mathfrak{S}$ is called *symmetric* iff

(7^*) $$(\varphi x, I)\theta = (\varphi x, \theta)$$

for all φ, θ in $\mathscr{F}$ and all x in $\mathscr{C}$ (obviously, (7) is a particular instance of (7^*)).

Remark 1. The equalities (11)-(13) correspond to some well-known equivalences from the paper McCarthy [1963]. In connections with counterparts of some other equivalences from that paper, cf. Exercises 21, 23, 44, 45 after this section.

Remark 2. Our first publications, where a definition of the notion of combinatory space appears, are the papers Skordev [1975, 1976b].[28] There are three things in that first definition, which make it different from the present one, namely: (i) instead of (9) and (10), it is required that T and F belong to $\mathscr{C}$, (ii) the additional condition is included that $xy = x$ for all x, y in $\mathscr{C}$, and (iii) the condition (7^*) is present there instead of (7). A bit later, in Skordev [1977], the condition from (ii) has been shown to be redundant (see Proposition 2 in this section). Therefore the combinatory spaces, considered in the above-mentioned papers, are exactly those combinatory spaces in the present sense, which are symmetric and satisfy the condition that

[28] For some corrections concerning two examples of combinatory spaces in that papers, cf. Skordev [1980, Chapter II, Section 1.3, Example 12, and Section 5.4, Remark 2] (these examples can be found, in essence, in Exercise 4.22 of this chapter and in Subsection (III) of Section 4 of the Appendix).

T and F belong to $\mathcal{C}$.[29] The same class of combinatory spaces has been studied in the book Skordev [1980], with the unessential differences that quasi-ordered semigroups are used there instead of partially ordered ones,[30] and degenerate spaces are admitted, where all elements of $\mathcal{F}$ are equal each other. When the work on the manuscript of that book was near its end, the author observed that condition (7) is sufficient for some proofs, where (7^*) has been used before. In the thesis Lukanova [1978],[31] many proofs from the manuscript have been examined from this point of view, and it turned out that the essential results of the theory remain valid after such an weakening of the requirements of the definition. Roughly speaking, the present notion of combinatory space coincides with the notion of semicombinatory space from the papers Skordev [1980a, 1984] (the only difference is that the definition from that papers again admits degenerate spaces). The change in the terminology (to replace the adjective "semicombinatory" by "combinatory") was proposed by L. Ivanov, who used the new terminology in the book Ivanov [1986] and in subsequent publications.

Remark 3. From the condition (1), it follows immediately that

$$\forall x(\varphi x = \psi x) \Longrightarrow \varphi = \psi$$

for all φ, ψ in $\mathcal{F}$. Together with (8)-(10), this implies the impossibility of a situation where all elements of $\mathcal{C}$ are equal each other (in particular - the impossibility of the equality $\mathcal{C} = \varnothing$).

Remark 4. From conditions (6), (7), (15) and (16), one easily deduces the following more general properties:

$$(\varphi, \psi x)\theta = (\varphi\theta, \psi x),$$
$$(x, \psi)\theta = (x, \psi\theta),$$
$$(\chi \rightarrow \varphi x, \psi x)\theta = (\chi\theta \rightarrow \varphi x, \psi x),$$
$$\sigma \geq \rho \;\&\; \varphi \geq \psi \;\&\; \theta \geq \chi \Longrightarrow (\sigma \rightarrow \varphi, \theta) \geq (\rho \rightarrow \psi, \chi).$$

The proof of the first three of these more general proper-

---

[29] The last restriction is not very essential, as it can be seen from Corollary 1 below in combination with Proposition 2.

[30] The fact that this difference is really unessential can be seen from Exercise 3 after this section.

[31] An account of the main results from this thesis is given in Lukanova [1986].

ties is straight-forward. For example, the third one can be proved as follows:

$$(\chi \to \varphi x, \psi x)\theta = (I \to \varphi x, \psi x)\chi\theta = (\chi\theta \to \varphi x, \psi x).$$

For proving the fourth one (the monotonicity of Σ), suppose that $\sigma \geq \rho$ & $\varphi \geq \psi$ & $\theta \geq \chi$. Then, for all x in $\mathcal{C}$, we have

$$(\sigma \to \varphi, \theta)x = (\sigma x \to \varphi x, \theta x) = (I \to \varphi x, \theta x)\sigma x \geq$$
$$(I \to \psi x, \chi x)\rho x = (\rho x \to \psi x, \chi x) = (\rho \to \psi, \chi)x,$$

and we can use the condition (1).

Remark 5. If the considered combinatory space is a symmetric one then also the equality $(\varphi x, \psi)\theta = (\varphi x, \psi\theta)$ holds.

The considerations from the previous chapter provide us with some examples of combinatory spaces. Suppose $\mathfrak{U} = \langle M, J, L, R, T, F, H\rangle$ is a computational structure. Then the following examples of combinatory spaces correspond to situations studied in the previous chapter.

Example 1. Let $\mathfrak{S}_{\mathrm{m}}(\mathfrak{U}) = \langle \mathcal{F}_{\mathrm{m}}(M), I_M, \mathcal{C}, \Pi, L, R, \Sigma, T, F\rangle$, where the denotations from Section I.5 (including Exercise I.5.5) are used, and $\mathcal{F}_{\mathrm{m}}(M)$ is supplied with the composition and the partial ordering by inclusion. Then $\mathfrak{S}_{\mathrm{m}}(\mathfrak{U})$ is a symmetric combinatory space (by Exercises 2–7 after that section).

Example 2. Let $\mathfrak{S}_{\mathrm{p}}(\mathfrak{U}) = \langle \mathcal{F}_{\mathrm{p}}(M), I_M, \mathcal{C}, \Pi, L, R, \Sigma, T, F\rangle$, where the denotations from Section I.2 are used, $\mathcal{F}_{\mathrm{p}}(M)$ is supplied with the composition and the partial ordering by inclusion, and $\mathcal{C}$ is the same as in the previous example. Then $\mathfrak{S}_{\mathrm{p}}(\mathfrak{U})$ is also a symmetric combinatory space (by the previous example and by the fact that composition, Π, Σ and the partial ordering predicate on $\mathcal{F}_{\mathrm{p}}(M)$ are restrictions of composition, Π, Σ and the partial ordering predicate on $\mathcal{F}_{\mathrm{m}}(M)$, respectively).

Example 3. Let $\mathfrak{S}_{\mathrm{m}}(\mathfrak{U}, E) = \langle \mathcal{F}_{\mathrm{m}}(M, E), I_M, \mathcal{C}, \Pi, L, R, \Sigma, T, F\rangle$, where the denotations from Section I.8 are used, $\mathcal{F}_{\mathrm{m}}(M, E)$ is supplied with the composition and the partial ordering by inclusion, and $\mathcal{C}$ is the same as in the previous examples. Then $\mathfrak{S}_{\mathrm{m}}(\mathfrak{U}, E)$ is a combinatory space, and this combinatory space is not symmetric, barring the case when $E = \emptyset$ (cf. Exercise 2 after the mentioned section). Note that an equivalent version of this example can be obtained by using the constructions from Exercise I.8.3.

Example 4. Let $\mathfrak{S}_p(\mathfrak{U}, E) = \langle \mathcal{F}_p(M, E), I_M, \mathcal{C}, \Pi, L, R, \Sigma, T, F \rangle$, where again the denotations from Section I.8 are used, $\mathcal{F}_p(M, E)$ is supplied with the composition and the partial ordering by inclusion, and $\mathcal{C}$ is the same as in the previous examples. Then $\mathfrak{S}_p(\mathfrak{U}, E)$ is also a combinatory space, and this combinatory space is not symmetric, barring the case when $E = \emptyset$.

Remark 6. Some modifications of Examples 1 and 3 and of the further examples in this book can be obtained using the following fact: if $\langle \mathcal{F}, I, \mathcal{C}, \Pi, L, R, \Sigma, T, F \rangle$ is a combinatory space, D is an element of $\mathcal{F}$ satisfying the conditions $D\,T = T$, $DF = F$, and the ternary operation Σ_1 in $\mathcal{F}$ is defined by

$$\Sigma_1(\chi, \varphi, \psi) = \Sigma(D\chi, \varphi, \psi),$$

then $\langle \mathcal{F}, I, \mathcal{C}, \Pi, L, R, \Sigma_1, T, F \rangle$ is also a combinatory space. The proof of this statement is immediate, after the statement in Remark 4 is proved.

In the case of Example 2, the application of the above remark does not give anything new, since we could obtain Σ_1 by simply replacing H by $\lambda u. H(D(u))$, which will be also a partial predicate on M. In the case of Examples 1 and 3, however, if $\mathrm{rng}\,T \cup \mathrm{rng}\,F$ is a proper subset of M (for example, if T and F are constant functions), then it is possible to choose D in such a way that both true and false belong to $H(D(u))$ for some u in M. In such a case the corresponding Σ_1 cannot be obtained by a new choice of the partial predicate H (in the case of Example 3, this will be the situation about Σ_1 also every time when $D(u) \cap E \neq \emptyset$ for some u in M).

In the case when $\mathrm{rng}\,J$ is a proper subset of M, some other modifications of Examples 1 and 3 can be given. Namely, we could replace L and R by some elements L_1 and R_1 of the corresponding $\mathcal{F}$ such that

$$L_1(J(s, t)) = \{s\}, \quad R_1(J(s, t)) = \{t\}$$

for all s, t in M, but $\{L_1, R_1\} \not\subseteq \mathcal{F}_p(M)$.

A number of essentially different examples of combinatory spaces will be given further in the book.

In order to prove some elementary general properties of combinatory spaces, let us assume from now on (until the end of the present section) that a combinatory space $\langle \mathcal{F}, I, \mathcal{C}, \Pi, L, R, \Sigma, T, F \rangle$ is given.

Proposition 1. The operation Π is monotonically increasing, i.e.

$$\varphi \geq \psi \ \& \ \theta \geq \chi \Longrightarrow (\varphi, \theta) \geq (\psi, \chi)$$

for all $\varphi, \psi, \theta, \chi$ in $\mathcal{F}$.

Proof. Suppose $\varphi \geq \psi$ & $\theta \geq \chi$. Let x be an arbitrary element of $\mathcal{C}$. In order to apply (1), we shall prove that $(\varphi, \theta)x \geq (\psi, \chi)x$. Since (5) and (6) imply the equalities

$$(\varphi, \theta)x = (I, \theta x)\varphi x, \quad (\psi, \chi)x = (I, \chi x)\psi x,$$

it is sufficient to prove that $(I, \theta x) \geq (I, \chi x)$. The proof of this inequality is carried out by noting that, for all y in $\mathcal{C}$, the conditions (6) and (7) imply

$$(I, \theta x)y = (y, I)\theta x \geq (y, I)\chi x = (I, \chi x)y,$$

and then applying (1). ∎

Proposition 2. For all x and y in $\mathcal{C}$, the equality $xy = x$ holds.

Proof. By (3), the first equality in Remark 4, (5), (7) and again (3), we have

$$xy = L(x, y)y = L(x, Iy)y = L(xy, Iy) = L(x, I)y = L(x, y) = x.$$ ∎

Proposition 3. For all x in $\mathcal{C}$ and all φ, ψ in $\mathcal{F}$, the equalities $(Tx \rightarrow \varphi, \psi) = \varphi$, $(Fx \rightarrow \varphi, \psi) = \psi$ hold.

Proof. The first equality follows from the fact that, for all y in $\mathcal{C}$, condition (14), Proposition 2, again (14), then (11) and again Proposition 2 imply

$$(Tx \rightarrow \varphi, \psi)y = (Txy \rightarrow \varphi y, \psi y) = (Tx \rightarrow \varphi yx, \psi yx) = (T \rightarrow \varphi y, \psi y)x = \varphi yx = \varphi y.$$

The validity of the second equality is seen in a similar way, using (12) instead of (11). ∎

Corollary 1. For all a, b in $\mathcal{C}$, the 9-tuple $\langle \mathcal{F}, I, \mathcal{C}, \Pi, L, R, \Sigma, Ta, Fb \rangle$ is a combinatory space.

In order to formulate easier some more properties of combinatory space, the term "normal element" will be introduced.

Definition 2. An element ζ of $\mathcal{F}$ will be called *normal* iff $\zeta x \in \mathcal{C}$ for all x in $\mathcal{C}$.[92]

---

[92] In our previous publications, we used the term "perfect" for the same notion. We consider this term too pretentious now, and therefore we turn back to the term used many

Proposition 2 and conditions (9), (10) show that all elements of $\mathcal{C}$ and the elements T and F are normal. As an example of a normal element which surely does not belong to $\mathcal{C}$, the element I can be indicated (if we suppose that $I\in\mathcal{C}$, then, using Proposition 2, we get $x=Ix=I$ for arbitrary x in $\mathcal{C}$, and this contradicts Remark 3). It is seen immediately that composition of any two normal elements of $\mathcal{F}$ is a normal element again. In the examples of combinatory spaces considered in this section, the normal elements of the corresponding semigroups $\mathcal{F}$ are just the total mappings of M into itself.

Proposition 4. Let φ and ψ be normal elements of $\mathcal{F}$. Then (φ, ψ) is also a normal element, and the equalities $L(\varphi, \psi)=\varphi$, $R(\varphi, \psi)=\psi$ hold.

Proof. For all x in $\mathcal{C}$, we have the equalities $(\varphi, \psi)x=(\varphi x, \psi x)$, $L(\varphi, \psi)x=\varphi x$, $R(\varphi, \psi)x=\psi x$.∎

Corollary 2. Let θ be an arbitrary element of $\mathcal{F}$, and let ζ be a normal element. Then $L(\theta, \zeta)=R(\zeta, \theta)=\theta$.

Proof. For all x in $\mathcal{C}$, we have

$$L(\theta, \zeta)x=L(\theta x, \zeta x)=L(I, \zeta x)\theta x=I\theta x=\theta x,$$
$$R(\zeta, \theta)x=R(\zeta x, \theta x)=R(\zeta x, I)\theta x=I\theta x=\theta x.$$ ∎

Proposition 5. For all φ, ψ, χ in $\mathcal{F}$ and all normal elements ζ of $\mathcal{F}$, the equalities $(\varphi, \psi)\zeta=(\varphi\zeta, \psi\zeta)$, $(\chi\rightarrow\varphi, \psi)\zeta=(\chi\zeta\rightarrow\varphi\zeta, \psi\zeta)$ hold.

Proof. A straight-forward application of the definition of the notion of normal element, conditions (5), (14) and Remark 3.∎

Remark 7. If θ is a normal element of $\mathcal{F}$ then the equality $(\varphi x, \psi)\theta=(\varphi x, \psi\theta)$ holds for all φ, ψ in $\mathcal{F}$ and all x in $\mathcal{C}$ (without the assumption made in Remark 5 about symmetry of $\mathfrak{S}$). This is easily seen by application of Definition 2, condition (5), Proposition 2 and Remark 3.

The following two propositions generalize Propositions 2 and 3.

Proposition 6. For all x in $\mathcal{C}$ and all normal elements ζ of $\mathcal{F}$, the equality $x\zeta=x$ holds.

Proof. By the definition of the notion of normal element and Proposition 2, $x\zeta y=x=xy$ for all y in $\mathcal{C}$.∎

Proposition 7. For all φ, ψ in $\mathcal{F}$ and all normal elements ζ of $\mathcal{F}$, the equalities $(T\zeta\rightarrow\varphi, \psi)=\varphi$,

---

years ago in our lectures on combinatory spaces.

$(F\zeta \to \varphi, \psi) = \psi$ hold.

Proof. Application of condition (14), the definition of the notion of normal element, Proposition 3 and Remark 3. ∎

Proposition 8. For all $\theta, \chi, \varphi, \psi$ in $\mathcal{F}$ and each normal element ζ of $\mathcal{F}$, the equality

$$(\theta L \to \varphi R, \psi R)(\chi, \zeta) = (\theta\chi \to \varphi\zeta, \psi\zeta)$$

holds.

Proof. For all x in $\mathcal{C}$, we have

$$(\theta L \to \varphi R, \psi R)(\chi, \zeta)x = (\theta L \to \varphi R, \psi R)(\chi x, \zeta x) =$$
$$(\theta L \to \varphi R, \psi R)(I, \zeta x)\chi x = (\theta \to \varphi\zeta x, \psi\zeta x)\chi x =$$
$$(\theta\chi x \to \varphi\zeta x, \psi\zeta x) = (\theta\chi \to \varphi\zeta, \psi\zeta)x. \blacksquare$$

For each subset $\mathcal{A}$ of $\mathcal{C}$, a binary relation $\underset{\mathcal{A}}{\geq}$ in $\mathcal{F}$ will be introduced. Also a notion of invariance of $\mathcal{A}$ with respect to a given element of $\mathcal{F}$ will be defined.

Definition 3. Let $\mathcal{A}$ be some subset of $\mathcal{C}$. If φ, ψ are elements of $\mathcal{F}$ then the inequality $\varphi \underset{\mathcal{A}}{\geq} \psi$ means that $\varphi z \geq \psi z$ for all z in $\mathcal{A}$. For a given σ in $\mathcal{F}$, the set $\mathcal{A}$ is called *invariant with respect to* σ iff for all φ, ψ in $\mathcal{F}$ the implication $\varphi \underset{\mathcal{A}}{\geq} \psi \Longrightarrow \varphi\sigma \underset{\mathcal{A}}{\geq} \psi\sigma$ holds.

In the combinatory spaces from Examples 1, 2, 3 and 4 above, the meaning of $\underset{\mathcal{A}}{\geq}$ is quite clear. For the meaning of invariance in those combinatory spaces, cf. Exercise 37 after this section.

Proposition 9. The inequality $\varphi \underset{\mathcal{C}}{\geq} \psi$ is equivalent to the inequality $\varphi \geq \psi$. The set $\mathcal{C}$ is invariant with respect to each element of $\mathcal{F}$.

Proof. The first statement is an obvious consequence of condition (1) and the monotonicity of the multiplication in $\mathcal{F}$. The second statement follows from the first one and the mentioned monotonicity. ∎

Proposition 10. Let $\mathcal{A} \subseteq \mathcal{C}$, and σ be an element of $\mathcal{F}$ such that $\sigma x \in \mathcal{A}$ for all x in $\mathcal{A}$. Then $\mathcal{A}$ is invariant with respect to σ.

Proof. Immediate. ∎

Corollary 3. Each subset of $\mathcal{C}$ is invariant with respect to its elements and with respect to the element I.

In the examples of combinatory spaces mentioned until now, it was always so that the partially ordered set $\mathcal{F}$ had a least element. However, it is quite easy to give an example, where such a least element does not exist. To obtain such an example, it is sufficient to take an arbitrary combinatory space $\langle\mathcal{F}, I, \mathcal{C}, \Pi, L, R, \Sigma, T, F\rangle$ and to replace the original partial ordering in $\mathcal{F}$ by the equality relation. Since this partial ordering looks not natural in the general case, we give also an example, where the equality relation can be regarded as the natural partial ordering.

Example 5. Let $\mathfrak{U}=\langle M, J, L, R, T, F, H\rangle$ be a computational structure with total L, R, H (i.e. $\mathrm{dom}L=\mathrm{dom}R=\mathrm{dom}H=M$). Let $\mathfrak{S}_t(\mathfrak{U})=\langle\mathcal{F}_t(M), I_M, \mathcal{C}, \Pi, L, R, \Sigma, T, F\rangle$, where $\mathcal{F}_t(M)$ is the set of all total mappings of M into M with the usual composition and with the equality relation taken as partial ordering. Let $\mathcal{C}$ be the same as in the combinatory space $\mathfrak{S}_p(M)$, and Π and Σ be the restrictions to $\mathcal{F}_t(M)$ of the corresponding operations in $\mathcal{F}_p(M)$. Then $\mathfrak{S}_t(\mathfrak{U})$ is a symmetric combinatory space, and obviously there is no least element in $\mathcal{F}_t(M)$.

In case there is a least element in $\mathcal{F}$, the properties of this element are of interest, and we shall prove one such property now.

Proposition 11. Let o be the least element of $\mathcal{F}$. Then the equality $o\zeta=o$ holds for all normal elements ζ of $\mathcal{F}$.

Proof. Let ζ be a normal element of $\mathcal{F}$. To prove the equality $o\zeta=o$, it is sufficient to establish the inequality $o\geq o\zeta$. Its validity follows from the fact that, by Proposition 2, for all x in $\mathcal{C}$ we have $x=x\zeta x$ and hence $ox=ox\zeta x\geq o\zeta x$. ∎

Remark 8. In the general case, it can happen that there is a least element o in $\mathcal{F}$ and the equality $o\rho=o$ is violated for some ρ in $\mathcal{F}$ (see Remark 1 in Section 3). We do not know whether it is possible the equality $\rho o=o$ to be violated if o is the least element of $\mathcal{F}$. A least element o of $\mathcal{F}$ surely exists, and the last equality turns out to be always true in the case which will be of main interest in the further exposition, namely the case of iterative combinatory spaces (see Proposition 3.2).

A lot of additional information about combinatory spaces can be found in the exercises which follow this section. In particular, many specific properties of symmetric combinatory spaces are listed there.

Exercises

1. Let $\langle\mathcal{F}, I, \mathcal{C}, \Pi, L, R, \Sigma, T, F\rangle$ be a combinatory

space. Let Σ_0, Σ_1 and Σ_2 be the ternary operations in $\mathcal{F}$ defined by

$$\Sigma_0(\chi, \varphi, \psi) = \Sigma(\chi, \psi, \varphi)$$

$$\Sigma_1(\chi, \varphi, \psi) = \Sigma(L\chi, \varphi, \psi), \quad \Sigma_2(\chi, \varphi, \psi) = \Sigma(R\chi, \varphi, \psi).$$

Then $\langle \mathcal{F}, I, \mathcal{C}, \Pi, L, R, \Sigma_0, F, T \rangle$, $\langle \mathcal{F}, I, \mathcal{C}, \Pi, L, R, \Sigma_1, (T, I), (F, I) \rangle$ and $\langle \mathcal{F}, I, \mathcal{C}, \Pi, L, R, \Sigma_2, (I, T), (I, F) \rangle$ are also combinatory spaces.

2. If $\langle \mathcal{F}, I, \mathcal{C}, \Pi, L, R, \Sigma, T, F \rangle$ is a symmetric combinatory space, and Π_1 is the ternary operation in $\mathcal{F}$ defined by

$$\Pi_1(\varphi, \psi) = \Pi(\psi. \varphi),$$

then $\langle \mathcal{F}, I, \mathcal{C}, \Pi_1, R, L, \Sigma, T, F \rangle$ is also a symmetric combinatory space.

3. Let $\mathcal{F}$ be a quasi-ordered semigroup in the following sense: $\mathcal{F}$ is a set supplied with a reflexive and transitive relation (denoted as inequality) and a binary operation (denoted as multiplication) such that $(\varphi\psi)\theta \approx \varphi(\psi\theta)$ and $\varphi \geq \psi \Longrightarrow \varphi\theta \geq \psi\theta \;\&\; \theta\varphi \geq \theta\psi$ for all φ, ψ, θ in $\mathcal{F}$, where $\approx$ is the equivalence relation in $\mathcal{F}$ defined as follows: $\varphi \approx \psi$ iff $\varphi \geq \psi \;\&\; \psi \geq \varphi$. Let I be an element of $\mathcal{F}$ such that $I\theta \approx \theta$ for all θ in $\mathcal{F}$. Let $\mathcal{C}, \Pi, L, R, \Sigma, T, F$ satisfy the same conditions as in the definition of the notion of combinatory space, but with $\approx$ instead of $=$ (hence with the negation of $\approx$ instead of $\neq$). Let $\mathcal{F}_1$ be the set of all equivalence classes in $\mathcal{F}$ with respect to $\approx$, supplied with the multiplication operation corresponding to multiplication in $\mathcal{F}$. Let I_1, L_1, R_1, T_1, F_1 be the elements of $\mathcal{F}_1$ containing I, L, R, T, F, respectively, and $\mathcal{C}_1$ be the set of all elements of $\mathcal{F}_1$ which meet $\mathcal{C}$. Prove that $\approx$ is a congruence relation with respect to Π and Σ, and if Π_1, Σ_1 are the operations in $\mathcal{F}_1$ which correspond them, then

$$\langle \mathcal{F}_1, I_1, \mathcal{C}_1, \Pi_1, L_1, R_1, \Sigma_1, T_1, F_1 \rangle$$

is a combinatory space.

Hint. Use $\forall x(\varphi x \approx \psi x) \Longrightarrow \varphi \approx \psi$ to prove that $\theta I \approx \theta$ for all θ in $\mathcal{F}$. Prove the monotonicity of Σ and Π in the same way as in Remark 4 and in the proof of Proposition 1, respectively (with $\approx$ instead of $=$).

4. (Cf. Ivanov [1986, Propositions 27.11 and 27.12]) Let $\langle \mathcal{F}, I, \mathcal{C}, \Pi, L, R, \Sigma, T, F \rangle$ be an arbitrary combinatory space, and let x, y be arbitrary elements of $\mathcal{C}$. Prove the equalities

$((x, y), I) = K_1(x, I)(y, I)$, $(x, I)(y, I) = K_2((x, y), I)$, where $K_1 = ((L, LR), R^2)$, $K_2 = (L^2, (RL, R))$.

5. Let $\langle \mathcal{F}, I, \mathcal{C}, \Pi, L, R, \Sigma, T, F \rangle$ be an arbitrary combinatory space, let θ be a normal element of $\mathcal{F}$, and let φ, ψ be arbitrary elements of $\mathcal{F}$. Prove the equalities

$$(L \rightarrow \varphi R, \psi R)(T\theta, I) = \varphi, \quad (L \rightarrow \varphi R, \psi R)(F\theta, I) = \psi.$$

6. Let $\langle \mathcal{F}, I, \mathcal{C}, \Pi, L, R, \Sigma, T, F \rangle$ be an arbitrary combinatory space, let $\varphi, \psi, \chi, \gamma$ be arbitrary elements of $\mathcal{F}$, and let ζ be a normal element of $\mathcal{F}$. Prove the equalities

$$(\varphi L, \psi R)(\gamma, \zeta) = (\varphi R, \psi L)(\zeta, \gamma) = (\varphi\gamma, \psi\zeta),$$
$$(L, \psi R)(\zeta, \gamma) = (R, \psi L)(\gamma, \zeta) = (\zeta, \psi\gamma).$$

7. (Ivanov [1986, Chapters 27 and 10]) Let $\langle \mathcal{F}, I, \mathcal{C}, \Pi, L, R, \Sigma, T, F \rangle$ be an arbitrary combinatory space, and let the unary operation St in $\mathcal{F}$ (called *storing operation*) be defined by means of the equality

$$St(\theta) = (L, \theta R).$$

For arbitrary θ, α, β in $\mathcal{F}$ and all x in $\mathcal{C}$, prove the equalities

$$\theta = R\, St(\theta)(I, I), \quad St(\theta)(x, I) = (x, I)\theta,$$
$$St(\alpha\beta) = St(\alpha)\, St(\beta),$$
$$(\alpha, \beta) = St(\beta)(R, L)\, St(\alpha)(I, I).$$

8. Let $\langle \mathcal{F}, I, \mathcal{C}, \Pi, L, R, \Sigma, T, F \rangle$ be an arbitrary combinatory space, and let ψ, α, β be arbitrary elements of $\mathcal{F}$. Prove the equality

$$St(\psi)(\alpha, \beta) = (\alpha, \psi\beta),$$

where the operation St is defined as in the previous exercise.

9. (Ivanov [1986, Proposition 27.8]) Let $\langle \mathcal{F}, I, \mathcal{C}, \Pi, L, R, \Sigma, T, F \rangle$ be an arbitrary combinatory space, and let $\varphi, \psi, \rho, \sigma$ be elements of $\mathcal{F}$ satisfying the inequality

$$\sigma(x, I)\rho \geq \psi(x, I)\varphi$$

for all x in $\mathcal{C}$. Prove the inequality

$$\sigma\, St(\rho) \geq \psi\, St(\varphi),$$

where the operation St is defined as in Exercise 7.

10. (Cf. Ivanov [1986, Proposition 10.16]) Let $\langle \mathcal{F}, I, \mathcal{C}, \Pi, L, R, \Sigma, T, F \rangle$ be an arbitrary combinatory space. Let the operation St be defined as in Exercise 7, and let K_1,

K_2 be such as in Exercise 4. Prove that

$$St(St(\theta)) = K_2\, St(\theta)\, K_1\, St(St(I))$$

for all θ in $\mathcal{F}$.

Hint. Using the previous exercise, reduce the problem to proving the equality

$$(x, I)\, St(\theta) = K_2\, St(\theta)\, K_1\, (x, I)\, St(I)$$

for arbitrary x in $\mathcal{C}$, and reduce this new problem to proving the equality

$$(x, I)(y, I)\,\theta = K_2\, St(\theta)\, K_1\, (x, I)(y, I)\, I$$

for arbitrary x, y in $\mathcal{C}$.

11. (Compare with Proposition 10.13 of Ivanov [1986]) Let $\langle \mathcal{F}, I, \mathcal{C}, \Pi, L, R, \Sigma, T, F \rangle$ be an arbitrary combinatory space, and let χ, φ, ψ be arbitrary elements of $\mathcal{F}$. Prove the equality

$$St((\chi \to \varphi, \psi)) = (\chi R \to St(\varphi), St(\psi))\, St(I),$$

where the operation St is defined as in Exercise 7.

12. (Compare with Proposition 27.14 of Ivanov [1986]) Let $\langle \mathcal{F}, I, \mathcal{C}, \Pi, L, R, \Sigma, T, F \rangle$ be an arbitrary combinatory space. Let the operation St be defined as in Exercise 7, and let K_1, K_2 be such as in Exercise 4. Let $\mathcal{B}$ be a fixed subset of $\mathcal{F}$. For an arbitrary element θ of $\mathcal{F}$, prove the equivalence of the following two conditions, where $St(\mathcal{B})$ denotes the set of all elements of the form $St(\beta)$ with β belonging to $\mathcal{B}$:

(i) θ can be generated from elements of the set $\{I, L, R, T, F\} \cup \mathcal{B}$ by means of multiplication and the operations Π, Σ.

(ii) θ can be generated from elements of the set $\{(I, I), R, K_1, K_2, St((I, I)), St((R, L)), St(St(I)), St(L), St(R), St(T), St(F)\} \cup St(\mathcal{B})$ by means of multiplication and the operation Σ.

Hint. To establish the implication (i) $\Longrightarrow$ (ii), prove that $St(\theta)$ can be generated in the way described in condition (ii), whenever θ satisfies condition (i).

13. Let $\langle \mathcal{F}, I, \mathcal{C}, \Pi, L, R, \Sigma, T, F \rangle$ be a symmetric combinatory space, and let $\chi, \varphi, \psi, \alpha, \beta$ be arbitrary elements of $\mathcal{F}$. Prove the equalities

$$(\varphi L, \psi R)(\alpha, \beta) = (\varphi\alpha, \psi\beta),$$

$$(\chi L \to \varphi R, \psi R)(\alpha, \beta) = (\chi\alpha \to \varphi\beta, \psi\beta),$$

$$(\varphi R, \psi L)(\alpha, \beta) = (\varphi\beta, \psi\alpha),$$

$$(\chi R \to \varphi L,\ \psi L)(\alpha,\ \beta) = (\chi\beta \to \varphi\alpha,\ \psi\alpha).$$

Hint. Use Exercise 6 and Proposition 8.

14. Let $\langle \mathcal{F}, I, \mathcal{C}, \Pi, L, R, \Sigma, T, F \rangle$ be an arbitrary combinatory space, and let $\chi, \varphi, \psi, \theta$ be arbitrary elements of $\mathcal{F}$. Let ζ be a normal element of $\mathcal{F}$. Prove the equalities

$$(\varphi,\ \psi R)(\theta,\ \zeta) = (\varphi(\theta,\ \zeta),\ \psi\zeta),$$
$$(\chi \to \varphi R,\ \psi R)(\theta,\ \zeta) = (\chi(\theta,\ \zeta) \to \varphi\zeta,\ \psi\zeta),$$
$$(R,\ \psi)(\theta,\ \zeta) = (\zeta,\ \psi(\theta,\ \zeta)),$$
$$(\varphi,\ \psi L)(\zeta,\ \theta) = (\varphi(\zeta,\ \theta),\ \psi\zeta),$$
$$(\chi \to \varphi L,\ \psi L)(\zeta,\ \theta) = (\chi(\zeta,\ \theta) \to \varphi\zeta,\ \psi\zeta),$$
$$(L,\ \psi)(\zeta,\ \theta) = (\zeta,\ \psi(\zeta,\ \theta)).$$

15. Let $\langle \mathcal{F}, I, \mathcal{C}, \Pi, L, R, \Sigma, T, F \rangle$ be an arbitrary combinatory space, and θ be an element of $\mathcal{F}$ such that the equality (7^*) holds for all φ in $\mathcal{F}$ and all x in $\mathcal{C}$. Let ζ be a normal element of $\mathcal{F}$. Prove the equalities

$$(\varphi R,\ \psi)(\theta,\ \zeta) = (\varphi\zeta,\ \psi(\theta,\ \zeta)),$$
$$(\varphi L,\ \psi)(\zeta,\ \theta) = (\varphi\zeta,\ \psi(\zeta,\ \theta))$$

for all φ, ψ in $\mathcal{F}$.

16. Let $\langle \mathcal{F}, I, \mathcal{C}, \Pi, L, R, \Sigma, T, F \rangle$ be an arbitrary combinatory space, and let $\chi, \varphi, \psi, \theta$ be arbitrary elements of $\mathcal{F}$. Prove the equality

$$((\chi \to \varphi,\ \psi),\ \theta) = (\chi \to (\varphi,\ \theta),\ (\psi,\ \theta))$$

Hint. Use the equality

$$((\chi \to \varphi,\ \psi),\ \theta)x = (I,\ \theta x)(\chi x \to \varphi x,\ \psi x)$$

and condition (13).

17. Let $\langle \mathcal{F}, I, \mathcal{C}, \Pi, L, R, \Sigma, T, F \rangle$ be a symmetric combinatory space, and let $\chi, \varphi, \psi, \theta$ be arbitrary elements of $\mathcal{F}$. Prove the equality

$$(\theta,\ (\chi \to \varphi,\ \psi)) = (\chi \to (\theta,\ \varphi),\ (\theta,\ \psi)).$$

Hint. Use the previous exercise and Exercise 2.

18. Let $\langle \mathcal{F}, I, \mathcal{C}, \Pi, L, R, \Sigma, T, F \rangle$ be an arbitrary combinatory space, let x be an element of $\mathcal{C}$, and let $\varphi, \psi, \chi, \theta$ be arbitrary elements of $\mathcal{F}$. Let τ be an element of $\mathcal{F}$ satisfying the conditions $\tau(Tx,\ I) = \varphi$, $\tau(Fx,\ I) = \psi$.[33] Prove the equality

[33] To see the existence of such a τ, cf. Exercise 5.

$$\tau((\chi \to T, F)x, \theta) = (\chi x \to \varphi\theta, \psi\theta).$$

Hint. Use Exercise 16.

19. Let $\langle \mathcal{F}, I, \mathcal{C}, \Pi, L, R, \Sigma, T, F \rangle$ be an arbitrary combinatory space, and θ be an element of $\mathcal{F}$ such that the equality (7^*) holds for all φ in $\mathcal{F}$ and all x in $\mathcal{C}$. Prove the equalities

$$(\varphi x, \psi)\theta = (\varphi x, \psi\theta), \quad (\chi x \to \varphi, \psi)\theta = (\chi x \to \varphi\theta, \psi\theta)$$

for arbitrary χ, φ, ψ in $\mathcal{F}$ and all x in $\mathcal{C}$.

Hint. For the proof of the first equality, use Exercise 8. For the proof of the second one, take τ as in the previous exercise and represent $(\chi x \to \varphi\theta, \psi\theta)$ in the form $\tau((\chi \to T, F)x, I)\theta$. Then note that, again by the previous exercise, the equality

$$\tau((\chi \to T, F)x, I) = (\chi x \to \varphi, \psi)$$

holds.

20. Let $\langle \mathcal{F}, I, \mathcal{C}, \Pi, L, R, \Sigma, T, F \rangle$ and θ be as in the previous exercise. Let ζ be a normal element of $\mathcal{F}$. Prove the equalities

$$(\chi R \to \varphi, \psi)(\theta, \zeta) = (\chi\zeta \to \varphi(\theta, \zeta), \psi(\theta, \zeta)),$$
$$(\chi L \to \varphi, \psi)(\zeta, \theta) = (\chi\zeta \to \varphi(\zeta, \theta), \psi(\zeta, \theta))$$

for arbitrary χ, φ, ψ in $\mathcal{F}$.

21. Let $\langle \mathcal{F}, I, \mathcal{C}, \Pi, L, R, \Sigma, T, F \rangle$ be an arbitrary combinatory space, and let $\varphi, \psi, \chi, \sigma, \rho$ be arbitrary elements of $\mathcal{F}$. Prove the equality

$$((\chi \to \sigma, \rho) \to \varphi, \psi) = (\chi \to (\sigma \to \varphi, \psi), (\rho \to \varphi, \psi))$$

(compare with McCarthy [1963]).

Hint. For an arbitrary x in $\mathcal{C}$, transform the expression $((\chi \to \sigma, \rho) \to \varphi, \psi)x$ into the expression $(\chi \to (\sigma \to \varphi, \psi), (\rho \to \varphi, \psi))x$ by means of consecutive application of conditions (14), (14), (15), (13), (15), (15), (14), (14).

22. Let $\langle \mathcal{F}, I, \mathcal{C}, \Pi, L, R, \Sigma, T, F \rangle$ be an arbitrary combinatory space, and let φ, ψ be arbitrary elements of $\mathcal{F}$. Prove the equalities

$$((\chi \to T, T) \to \varphi, \psi) = (\chi \to \varphi, \varphi),$$
$$((\chi \to T, F) \to \varphi, \psi) = (\chi \to \varphi, \psi),$$
$$((\chi \to F, T) \to \varphi, \psi) = (\chi \to \psi, \varphi),$$
$$((\chi \to F, F) \to \varphi, \psi) = (\chi \to \psi, \psi).$$

23. Let $\langle \mathcal{F}, I, \mathcal{C}, \Pi, L, R, \Sigma, T, F \rangle$ be a symmetric

combinatory space, and let $\varphi, \psi, \theta, \tau, \chi, \rho$ be arbitrary elements of $\mathcal{F}$. Prove the equality

$$(\chi \to (\rho \to \varphi, \psi), (\rho \to \theta, \tau)) = (\rho \to (\chi \to \varphi, \theta), (\chi \to \psi, \tau))$$

(compare with McCarthy [1963]).

Hint. Take an arbitrary x in $\mathcal{C}$. Making use of the second equality from the previous exercise and of Exercise 19, transform the product $(\chi \to (\rho \to \varphi, \psi), (\rho \to \theta, \tau))x$ into $\delta(\rho x \to T, F)$, where

$$\delta = (\chi x \to (I \to \varphi x, \psi x), (I \to \theta x, \tau x)).$$

Then use the equalities $\delta(\rho x \to T, F) = (\rho x \to \delta T, \delta F)$, $\delta T = (\chi \to \varphi, \theta)x$, $\delta F = (\chi \to \psi, \tau)x$.

24. Let $\langle\mathcal{F}, I, \mathcal{C}, \Pi, L, R, \Sigma, T, F\rangle$ be an arbitrary combinatory space, and let φ, ψ be such elements of $\mathcal{F}$ that $\varphi(x, y) \geq \psi(x, y)$ for all x, y in $\mathcal{C}$. Prove that $\varphi(\alpha, \beta) \geq \psi(\alpha, \beta)$ for all α, β in $\mathcal{F}$.

25. (Generalization of Exercise 24). Let $\langle\mathcal{F}, I, \mathcal{C}, \Pi, L, R, \Sigma, T, F\rangle$ be an arbitrary combinatory space. For each positive integer n, define a set $\mathfrak{M}_n$ of mappings of $\mathcal{F}^n$ into $\mathcal{F}$ by means of the following inductive definition: $\mathfrak{M}_1$ consists of the identity mapping $\lambda\theta.\theta$ and of all constant mappings $\lambda\theta.\varkappa$, where $\varkappa \in \mathcal{F}$, and if $n > 1$ then $\mathfrak{M}_n$ consists of all mappings of the form

(17) $\lambda\alpha_1 \ldots \alpha_k \beta_1 \ldots \beta_l . \rho\Pi(\Gamma_1(\alpha_1, \ldots, \alpha_k), \Gamma_2(\beta_1, \ldots, \beta_l))$,

where k and l are positive integers satisfying the condition $k+l=n$, ρ is some element of $\mathcal{F}$, Γ_1 is a mapping belonging to $\mathfrak{M}_k$, and Γ_2 is a mapping belonging to $\mathfrak{M}_l$. Let Γ be a mapping from $\mathfrak{M}_n$. Prove that whenever φ, ψ are such elements of $\mathcal{F}$ that

$$\varphi\Gamma(z_1, \ldots, z_n) \geq \psi\Gamma(z_1, \ldots, z_n)$$

for all $z_1, \ldots, z_n$ in $\mathcal{C}$, then

$$\varphi\Gamma(\theta_1, \ldots, \theta_n) \geq \psi\Gamma(\theta_1, \ldots, \theta_n)$$

for all $\theta_1, \ldots, \theta_n$ in $\mathcal{F}$.

Hint. Suppose Γ is the mapping (17), where Γ_1 and Γ_2 have the above property. Suppose

$$\varphi\Gamma(x_1, \ldots, x_k, y_1, \ldots, y_l) \geq \psi\Gamma(x_1, \ldots, x_k, y_1, \ldots, y_l)$$

for all $x_1, \ldots, x_k, y_1, \ldots, y_l$ in $\mathcal{C}$. Write the above inequality in the form

$$\bar{\varphi}\Gamma_2(y_1, \ldots, y_l) \geq \bar{\psi}\Gamma_2(y_1, \ldots, y_l),$$

where

$$\bar{\varphi}=\varphi\rho\Pi(\Gamma_1(x_1, \ldots, x_k), I), \quad \bar{\psi}=\psi\rho\Pi(\Gamma_1(x_1, \ldots, x_k), I),$$

and conclude that

$$\bar{\varphi}\Gamma_2(\beta_1, \ldots, \beta_l)\geq\bar{\psi}\Gamma_2(\beta_1, \ldots, \beta_l),$$

for arbitrary $\beta_1, \ldots, \beta_l$ in $\mathcal{F}$. Multiply both sides of this inequality from the right by an arbitrary element z of $\mathcal{C}$, and write the obtained inequality in the form

$$\varphi'\Gamma_1(x_1, \ldots, x_k)\geq\psi'\Gamma_1(x_1, \ldots, x_k),$$

where

$$\varphi'=\varphi\rho\Pi(I, \Gamma_2(\beta_1, \ldots, \beta_l)z), \quad \psi'=\psi\rho\Pi(I, \Gamma_2(\beta_1, \ldots, \beta_l)z).$$

From here conclude that

$$\varphi'\Gamma_1(\alpha_1, \ldots, \alpha_k)\geq\psi'\Gamma_1(\alpha_1, \ldots, \alpha_k),$$

for all $\alpha_1, \ldots, \alpha_k$ in $\mathcal{F}$.

26. Let $\mathfrak{S}=\langle\mathcal{F}, I, \mathcal{C}, \Pi, L, R, \Sigma, T, F\rangle$ be an arbitrary combinatory space. An element φ of $\mathcal{F}$ will be called *constant* iff $\varphi x=\varphi y$ for all x, y in $\mathcal{C}$.

(a) Show that all element of $\mathcal{C}$ are constant, and the element I is not constant.

(b) Find all constant elements of $\mathcal{F}$ in any of the cases when $\mathfrak{S}$ is some of the spaces from Examples 1–4.

(c) Show that, whenever φ is a constant element of $\mathcal{F}$, then $\theta\varphi$ is also a constant element for all θ in $\mathcal{F}$. Prove that the set of all constant elements of $\mathcal{F}$ is closed under the operations Π and Σ.

(d) Prove the mutual equivalence of the following nine conditions, where φ is an arbitrary element of $\mathcal{F}$:

(i) φ is a constant element of $\mathcal{F}$;
(ii) $\varphi z=\varphi$ for all z in $\mathcal{C}$;
(iii) $\varphi z\geq\varphi$ for all z in $\mathcal{C}$;
(iv) $\varphi\geq\varphi z$ for all z in $\mathcal{C}$;
(v) $\varphi x\geq\varphi y$ for all x, y in $\mathcal{C}$;
(vi) $\varphi z=\varphi$ for some z in $\mathcal{C}$;
(vii) $\varphi=\psi z$ for some ψ in $\mathcal{F}$ and some z in $\mathcal{C}$;
(viii) $(I, \varphi)\theta=(\theta, \varphi)$ for all θ in $\mathcal{F}$;
(ix) $(I, \varphi)z=(z, \varphi)$ for all z in $\mathcal{C}$.

27. Let $\langle\mathcal{F}, I, \mathcal{C}, \Pi, L, R, \Sigma, T, F\rangle$ be an arbitrary combinatory space. Prove that for all z in $\mathcal{C}$ and all θ in $\mathcal{F}$ the equalities

$$L(I, \theta)z=R(\theta, I)z=z\theta z,$$

$$L(I, \theta)=R(\theta, I),$$

$$zL(I,\theta)=zR(\theta,I)=z\theta$$

hold.

28. Let $\langle\mathcal{F}, I, \mathcal{C}, \Pi, L, R, \Sigma, T, F\rangle$ be a symmetric combinatory space, and let φ, ψ be arbitrary elements of $\mathcal{F}$. Prove that for all z in $\mathcal{C}$ the equality

$$z\varphi z\psi z=z\psi z\varphi z$$

holds. Prove also the equalities

$$L(I,\varphi)L(I,\psi)=L(I,\psi)L(I,\varphi),$$

$$L(\varphi,\psi)=R(\psi,\varphi).$$

Hint. Use the previous exercise.

29. Let $\langle\mathcal{F}, I, \mathcal{C}, \Pi, L, R, \Sigma, T, F\rangle$ be a symmetric combinatory space. Define a set $\mathfrak{R}$ of mappings of $\mathcal{F}$ into $\mathcal{F}$ by means of the following inductive definition:

(i) $\lambda\theta.\,\theta\in\mathfrak{R}$;

(ii) if $\Gamma\in\mathfrak{R}$ and $\varkappa\in\mathcal{F}$ then $\lambda\theta.\,\varkappa\Gamma(\theta)\in\mathfrak{R}$; $\lambda\theta.\,(\Gamma(\theta),\varkappa)$ and $\lambda\theta.\,(\varkappa,\Gamma(\theta))$ also belong to $\mathfrak{R}$;

(iii) if $\Gamma\in\mathfrak{R}$ and $\sigma,\rho\in\mathcal{F}$ then $\lambda\theta.\,(\Gamma(\theta)\rightarrow\sigma,\rho)$ also belongs to $\mathfrak{R}$;

(iv) if $\Gamma,\Delta\in\mathfrak{R}$ and $\varkappa\in\mathcal{F}$ then $\lambda\theta.\,(\varkappa\rightarrow\Gamma(\theta),\Delta(\theta))$ also belongs to $\mathfrak{R}$.

Assuming $\Gamma,\Delta\in\mathfrak{R}$ and $\Gamma(z)\geq\Delta(z)$ for all z in $\mathcal{C}$, prove that $\Gamma(\theta)\geq\Delta(\theta)$ for all θ in $\mathcal{F}$.

Hint. Use the corollary from Proposition 4, as well as Exercises 14, 15, 20, to prove that each mapping from $\mathfrak{R}$ is representable in the form $\lambda\theta.\,\tau(\theta,I)$ with some fixed τ from $\mathcal{F}$.

30. Let $\mathfrak{S}=\langle\mathcal{F}, I, \mathcal{C}, \Pi, L, R, \Sigma, T, F\rangle$ be a symmetric combinatory space. Apply the previous exercise to give another proof of the equality

$$L(\varphi,\psi)=R(\psi,\varphi)$$

from Exercise 28 and to prove the equalities

$$(\alpha z\varphi\rightarrow\beta z\psi,\gamma z\psi)=(\alpha z\psi\rightarrow\beta z\varphi,\gamma z\varphi),$$

$$(I,I)(\varphi,\psi)=\varkappa((I,I)\varphi,(I,I)\psi), \tag{18}$$

$$((\varphi,\psi),(\theta,\chi))=\varkappa((\varphi,\theta),(\psi,\chi)), \tag{19}$$

where $z\in\mathcal{C}$, $\alpha,\beta,\gamma,\varphi,\psi,\theta,\chi\in\mathcal{F}$ and

$$\varkappa=((L^2,LR),(RL,R^2)).$$

By an appropriate direct proof, show that (18) remains valid without the assumption about symmetry of $\mathfrak{S}$.

31. Let $\langle\mathcal{F}, I, \mathcal{C}, \Pi, L, R, \Sigma, T, F\rangle$ be a symmetric

combinatory space. Define a set $\mathfrak{R}$ of mappings of $\mathcal{F}$ into $\mathcal{F}$ by means of the definition from Exercise 29 with the clause (ii) modified in the following way:

(ii) if $\Gamma \in \mathfrak{R}$ and $\varkappa \in \mathcal{F}$ then $\lambda\theta.\, \varkappa\Gamma(\theta)$ and $\lambda\theta.\, \Gamma(\theta)\varkappa$ also belong to $\mathfrak{R}$.

Assuming $\Gamma, \Delta \in \mathfrak{R}$ and $\Gamma(z) \geq \Delta(z)$ for all z in $\mathcal{C}$, prove that $\Gamma(\theta z) \geq \Delta(\theta z)$ for all θ in $\mathcal{F}$ and all z in $\mathcal{C}$.

Hint. Prove that for each mapping Γ from $\mathfrak{R}$ the equality $\Gamma(\theta z) = \tau(\theta z, I)$ holds with some fixed τ from $\mathcal{F}$. To do this, use the statements mentioned in the hint to Exercise 29, as well as the equality

$$(\theta z, I)\varkappa = (L, \varkappa R)(\theta z, I).$$

32. Let $\langle \mathcal{F}, I, \mathcal{C}, \Pi, L, R, \Sigma, T, F \rangle$ be a symmetric combinatory space. Apply the previous exercise to give another proof of the equality

$$z\varphi z\psi z = z\psi z\varphi z$$

from Exercise 28.

33. Let $\langle \mathcal{F}, I, \mathcal{C}, \Pi, L, R, \Sigma, T, F \rangle$ be an arbitrary combinatory space. Define a set $\mathfrak{T}$ of mappings of $\mathcal{C}$ into $\mathcal{F}$ by means of the following inductive definition:

(i) $\lambda z.\, z$ and all constant mappings $\lambda z.\, \varkappa$, where $\varkappa \in \mathcal{F}$, are elements of $\mathfrak{T}$;

(ii) whenever $\Gamma, \Delta, \mathrm{E}$ are elements of $\mathfrak{T}$, then

$$\lambda z.\, \Gamma(z)\Delta(z), \quad \lambda z.\, (\Gamma(z), \Delta(z)), \quad \lambda z.\, (\mathrm{E}(z) \to \Gamma(z), \Delta(z))$$

also belong to $\mathfrak{T}$.

Prove that each mapping from $\mathfrak{T}$ is representable in the form $\lambda z.\, \tau(z, I)$ with some fixed τ from $\mathcal{F}$.

34. Let $\langle \mathcal{F}, I, \mathcal{C}, \Pi, L, R, \Sigma, T, F \rangle$ be a symmetric combinatory space. Define a set $\mathfrak{R}$ of mappings of $\mathcal{F}$ into $\mathcal{F}$ by means of the definition from Exercise 29 with the clause (i) modified in the following way:

(i) for all $\sigma, \rho \in \mathcal{F}$, the mapping $\lambda\theta.\, (\theta \to \sigma, \rho)$ belongs to $\mathfrak{R}$.

Assuming $\Gamma, \Delta \in \mathfrak{R}$, $\Gamma(T) \geq \Delta(T)$ and $\Gamma(F) \geq \Delta(F)$, prove that $\Gamma(\theta) \geq \Delta(\theta)$ for all θ in $\mathcal{F}$.

Hint. Making use of Condition (13) and Exercises 16, 17, 21, 23, prove that each mapping from $\mathfrak{R}$ is representable in the form $\lambda\theta.\, (\theta \to \sigma, \rho)$ with some fixed σ, ρ from $\mathcal{F}$.

35. Let $\langle \mathcal{F}, I, \mathcal{C}, \Pi, L, R, \Sigma, T, F \rangle$ be a symmetric combinatory space. For all φ, ψ, θ in $\mathcal{F}$, define elements

$\varphi\wedge\psi$, $\varphi\vee\psi$ and $-\theta$ of $\mathcal{F}$ in the following way:

$$\varphi\wedge\psi=(\varphi\rightarrow(\psi\rightarrow T,\ F),\ (\psi\rightarrow F,\ F)),$$
$$\varphi\vee\psi=(\varphi\rightarrow(\psi\rightarrow T,\ T),\ (\psi\rightarrow T,\ F)),$$
$$-\theta=(\theta\rightarrow F,\ T).$$

Prove that

$$\varphi\wedge\psi=\psi\wedge\varphi,\quad \varphi\vee\psi=\psi\vee\varphi,$$
$$(\varphi\wedge\psi)\wedge\theta=\varphi\wedge(\psi\wedge\theta),\quad (\varphi\vee\psi)\vee\theta=\varphi\vee(\psi\vee\theta),$$
$$-(\varphi\wedge\psi)=(-\varphi)\vee(-\psi),\quad -(\varphi\vee\psi)=(-\varphi)\wedge(-\psi),$$
$$--(\varphi\wedge\psi)=\varphi\wedge\psi,\quad --(\varphi\vee\psi)=\varphi\vee\psi,\quad ---\theta=-\theta$$

for all φ, ψ, θ in $\mathcal{F}$.

36. Let $\langle\mathcal{F},\ I,\ \mathcal{C},\ \Pi,\ L,\ R,\ \Sigma,\ T,\ F\rangle$ be an arbitrary combinatory space, and let $\mathcal{A}$ be an arbitrary subset of $\mathcal{C}$. Prove the reflexivity and the transitivity of the relation $\underset{\mathcal{A}}{\geq}$. For arbitrary φ, ψ, θ, χ, σ, ρ in $\mathcal{F}$, prove:

$$\varphi\geq\psi \Longrightarrow \varphi\underset{\mathcal{A}}{\geq}\psi,\quad \varphi\underset{\mathcal{A}}{\geq}\psi\ \&\ \theta\underset{\mathcal{A}}{\geq}\chi \Longrightarrow \varphi\theta\underset{\mathcal{A}}{\geq}\psi\chi,$$
$$\varphi\underset{\mathcal{A}}{\geq}\psi\ \&\ \theta\underset{\mathcal{A}}{\geq}\chi \Longrightarrow (\varphi,\ \theta)\underset{\mathcal{A}}{\geq}(\psi,\ \chi),$$
$$\sigma\underset{\mathcal{A}}{\geq}\rho\ \&\ \varphi\underset{\mathcal{A}}{\geq}\psi\ \&\ \theta\underset{\mathcal{A}}{\geq}\chi \Longrightarrow (\sigma\rightarrow\varphi,\ \theta)\underset{\mathcal{A}}{\geq}(\rho\rightarrow\psi,\ \chi).$$

37. Let $\mathfrak{S}=\langle\mathcal{F},\ I,\ \mathcal{C},\ \Pi,\ L,\ R,\ \Sigma,\ T,\ F\rangle$ be the combinatory space from some of the examples 1 – 4. Let $\mathcal{A}$ be a subset of $\mathcal{C}$, and let σ be an element of $\mathcal{F}$. Let K be the set of the values of the elements of $\mathcal{A}$. In the case when $\mathfrak{S}$ is the combinatory space from Example 1 or Example 2, prove that $\mathcal{A}$ is invariant with respect to σ iff the following implication holds for all u, v in M:

$$u\in K\ \&\ \langle u,\ v\rangle\in\sigma \Longrightarrow v\in K.$$

Otherwise prove the same, but with the implication

$$u\in K\ \&\ \langle u,\ v\rangle\in\sigma \Longrightarrow v\in K\cup E.$$

38. Let $\langle\mathcal{F},\ I,\ \mathcal{C},\ \Pi,\ L,\ R,\ \Sigma,\ T,\ F\rangle$ be an arbitrary combinatory space. Let $\mathcal{A}$ be a subset of $\mathcal{C}$ invariant with respect to each one of the elements σ and ρ of $\mathcal{F}$. Prove that $\mathcal{A}$ is invariant also with respect to the element $\sigma\rho$ and with respect to all elements of the form $(\chi\rightarrow\sigma,\ \rho)$, where $\chi\in\mathcal{F}$.

39. Let $\langle\mathcal{F},\ I,\ \mathcal{C},\ \Pi,\ L,\ R,\ \Sigma,\ T,\ F\rangle$ be an arbitrary combinatory space. Let $\mathcal{A}$ be the set of all elements of $\mathcal{C}$ having the form $(x,\ y)$, where $x,\ y\in\mathcal{C}$. Prove that $\mathcal{A}$ is invariant with respect to each element belonging to the range of Π.

40. Suppose K is an arbitrary non-empty set, and

$$\mathfrak{S} = \langle \mathcal{F}, I, \mathcal{C}, \Pi, L, R, \Sigma, T, F \rangle$$

is an arbitrary combinatory space. Let $\mathcal{F}'$ be the set of all mappings of K into $\mathcal{F}$, considered with the partial ordering and the multiplication induced by the partial ordering and the multiplication in $\mathcal{F}$ in the natural pointwise way, i.e. $\varphi \geq \psi$ in $\mathcal{F}'$ iff $\varphi(k) \geq \psi(k)$ in $\mathcal{F}$ for all k in K, and the equality $\varphi\psi = \lambda k.\, \varphi(k)\psi(k)$ holds. Let I', L', R', T', F' be the constant mappings of K into $\mathcal{F}$ having the values I, L, R, T, F, respectively. Let $\mathcal{C}'$ be the set of all constant mappings of K into $\mathcal{C}$. Set

$$\mathfrak{S}^K = \langle \mathcal{F}', I', \mathcal{C}', \Pi', L', R', \Sigma', T', F' \rangle,$$

where Π' and Σ' are the operations in $\mathcal{F}'$ defined by the equalities

$$\Pi'(\varphi, \psi) = \lambda k.\, \Pi(\varphi(k), \psi(k)),$$

$$\Sigma'(\chi, \varphi, \psi) = \lambda k.\, \Sigma(\chi(k), \varphi(k), \psi(k)).$$

Prove that $\mathfrak{S}^K$ is also a combinatory space.

41. Let $\langle \mathcal{F}, I, \mathcal{C}, \Pi, L, R, \Sigma, T, F \rangle$ be a symmetric combinatory space. An element θ of $\mathcal{F}$ will be called *distributive* iff the equality $(I, I)\theta = (\theta, \theta)$ holds. Prove that

$$(\varphi, \psi)\theta = (\varphi\theta, \psi\theta), \quad (\chi \rightarrow \varphi, \psi)\theta = (\chi\theta \rightarrow \varphi\theta, \psi\theta)$$

for all φ, ψ, χ in $\mathcal{F}$ and all distributive elements θ of $\mathcal{F}$.

Hint. Use Exercise 13.

42. Let $\langle \mathcal{F}, I, \mathcal{C}, \Pi, L, R, \Sigma, T, F \rangle$ be a symmetric combinatory space, and let φ, ψ be distributive elements of $\mathcal{F}$ (in the sense of the previous exercise). Prove that $\varphi\psi$ and (φ, ψ) are also distributive.

Hint. To prove that (φ, ψ) is distributive, use the equalities (18) and (19) from Exercise 30.

43. Let $\langle \mathcal{F}, I, \mathcal{C}, \Pi, L, R, \Sigma, T, F \rangle$ be an arbitrary combinatory space. An element θ of $\mathcal{F}$ will be called *regular* iff the equality $x\theta = x$ holds for all x in $\mathcal{C}$. Prove the following statements:

(a) if θ is an element of $\mathcal{F}$ such that $x\theta = x$ holds for some x in $\mathcal{C}$, then θ is regular;

(b) if φ and ψ are regular elements of $\mathcal{F}$, then so are $\varphi\psi$ and (φ, ψ);

(c) if φ and ψ are regular elements of $\mathcal{F}$, then the element $(\chi \rightarrow \varphi, \psi)$ is regular exactly for those elements χ of $\mathcal{F}$ which satisfy the equality $(\chi \rightarrow I, I) = I$;

(d) if θ is a regular element of $\mathcal{F}$, then the equality $L(\varphi, \theta) = \varphi$ holds for all φ in $\mathcal{F}$;

(e) an element θ of $\mathcal{F}$ is regular iff there is an element λ of $\mathcal{F}$ such that the equalities

$$\lambda(\varphi, I) = \lambda(\varphi, \theta) = \varphi$$

hold for all φ in $\mathcal{F}$.

44. Let $<\mathcal{F}, I, \mathcal{C}, \Pi, L, R, \Sigma, T, F>$ be a symmetric combinatory space satisfying the condition that T and F belong to $\mathcal{C}$. An element θ of $\mathcal{F}$ is called *Boolean* iff the equality $(\theta \to T, F) = \theta$ holds.[34] Prove the following statements:

(a) T and F are Boolean elements;

(b) if θ is a Boolean element, then so is $\theta\varphi$ for all φ in $\mathcal{F}$;

(c) if θ is an element of $\mathcal{F}$ such that θx is a Boolean element for any x in $\mathcal{C}$, then θ is also Boolean;

(d) if σ and ρ are Boolean elements then so is the element $(\chi \to \sigma, \rho)$ for all χ in $\mathcal{F}$;

(e) the operations introduced in Exercise 35 transform arbitrary elements of $\mathcal{F}$ into Boolean ones;

(f) for any Boolean element θ, the equality $--\theta = \theta$ holds;

(g) an element θ of $\mathcal{F}$ is Boolean iff the equality $(I \to T, F)\theta = \theta$ holds;

(h) for any θ in $\mathcal{F}$, the condition that θ is a Boolean element is equivalent to each of the conditions

$$\forall\varphi\,\forall\psi\,(\varphi T \geq \psi T \;\&\; \varphi F \geq \psi F \implies \varphi\theta \geq \psi\theta),$$

$$\forall\varphi\,\forall\psi\,(\varphi T = \psi T \;\&\; \varphi F = \psi F \implies \varphi\theta = \psi\theta).$$

45. Give counter-examples using symmetric combinatory spaces to each of the following equalities (where the vari-

---

[34] This equality corresponds to an equivalence from the paper McCarthy [1963]. In any of the combinatory spaces indicated in Examples 1–5 there are many elements θ violating the equality in question. This divergence between our system and McCarthy's one is caused by an obvious reason, namely our elements θ are not necessarily representations of predicates (even if partial and ambiguous predicates are admitted). The Boolean elements of a combinatory space are those among its elements which can be regarded in some sense as predicate-like. It is proved in Georgieva [1983] that those Boolean elements θ, which satisfy the additional conditions $(\theta \to I, I) = I$, $(\theta \to F, \theta) = F$, form a Boolean algebra with respect to the three operations introduced in Exercise 35.

ables range over the semigroup of the space):[35]

$$(\chi \to \varphi, \varphi) = \varphi,$$

$$(\chi \to (\chi \to \varphi, \psi), \theta) = (\chi \to \varphi, \theta),$$

$$(\chi \to \varphi, (\chi \to \psi, \theta)) = (\chi \to \varphi, \theta),$$

$$(\chi \to (\rho \to \varphi, \psi), \theta) = (\chi \to (\rho \to (\chi \to \varphi, \varphi), (\chi \to \psi, \psi)), \theta),$$

$$(\chi \to \varphi, (\rho \to \psi, \theta)) = (\chi \to \varphi, (\rho \to (\chi \to \psi, \psi), (\chi \to \theta, \theta))).$$

Hint. For the construction of counter-examples to the last two equalities, you may use an appropriate combinatory space of the type considered in Section 6 of the Appendix.

2. The companion operative space of a combinatory space

Combinatory spaces are not the only class of abstract algebraic structures offering a promising uniform way for capturing situations like that ones considered in Chapter I. Another such class of structures, called operative spaces, is introduced in Ivanov [1980, 1980a]; these structures are studied in a number of subsequent publications by the same author, culminating in the monograph Ivanov [1986].[36]

---

[35] All these equalities are counter-parts of equivalences from McCarthy [1963]. It is proved in Georgieva [1979] that these equalities are simultaneously identically satisfied exactly in those of the combinatory spaces fulfilling the assumptions of the previous exercise, which satisfy also the following condition: for all elements χ of $\mathcal{F}$, the equality $(\chi \to \chi, \chi) = \chi$ holds, and the element $(\chi \to T, F)$ is distributive and regular in the sense of Exercises 41 and 43 above (as examples of such combinatory spaces, we indicate the combinatory spaces from Example 5). By the statement (c) in Exercise 43 above, the regularity of $(\chi \to T, F)$ is equivalent to the equality $(\chi \to I, I) = I$, and hence, by condition (13), the equality $(\chi \to \chi, \chi) = \chi$ is a consequence of this regularity.

[36] To be more precise, we must note that not arbitrary combinatory spaces and not arbitrary operative spaces, but so-called iterative ones, are the convenient classes for the mentioned abstract algebraic study. Iterative combinatory spaces will be the main subject of this book, as iterative operative spaces are the main subject of the mentioned Ivanov's publications. A larger class of operative spaces than the iterative ones was independently introduced and studied in Georgieva [1980], but it turned out that only a small part of the theory of iterative operative spaces could be

Presented in slightly modified notations, Ivanov's definition of the notion of operative space reads as follows. An *operative space* is a 5-tuple $\langle \mathcal{F}, I, \Pi_*, L_*, R_* \rangle$, where $\mathcal{F}$ is a partially ordered semigroup, I is an identity of $\mathcal{F}$, Π_* is a monotonically increasing binary operation in $\mathcal{F}$, L_*, R_* are distinct elements of $\mathcal{F}$, and, for all φ, ψ, θ in $\mathcal{F}$, the equalities

$$\theta\Pi_*(\varphi, \psi) = \Pi_*(\theta\varphi, \theta\psi), \quad \Pi_*(\varphi, \psi)L_* = \varphi, \quad \Pi_*(\varphi, \psi)R_* = \psi$$

hold. According to Proposition 27.5 of Ivanov [1986][97], to each combinatory space

$$\mathfrak{S} = \langle \mathcal{F}, I, \mathcal{C}, \Pi, L, R, \Sigma, T, F \rangle,$$

an operative space $\mathfrak{S}_*$ is correlated, namely

$$\mathfrak{S}_* = \langle \mathcal{F}, I, \lambda\varphi\psi. (L \rightarrow \varphi R, \psi R), (T, I), (F, I) \rangle$$

(the straight-forward proof that $\mathfrak{S}_*$ is really an operative space can be based on condition (13), Remark 4 and Propositions 4, 5 of the previous section). The operative space $\mathfrak{S}_*$ is called *the companion operative space of* $\mathcal{F}$. As shown in Proposition 27.19 of Ivanov [1986], not every operative space can be obtained as the companion operative space of some combinatory space. A characterization of those operative spaces, which are companion operative spaces of combinatory spaces, is given in Ivanov [19??] (cf. also Ivanov [1990]). Namely, such operative spaces are characterized by the existence of so-called storing operation in them (for an example of storing operation, cf. Exercise 1.7).

It is convenient to make the following remark: if $\langle \mathcal{F}, I, \mathcal{C}, \Pi, L, R, \Sigma, T, F \rangle$ is a combinatory space, and $\langle \mathcal{F}, I, \Pi_*, L_*, R_* \rangle$ is its companion operative space, then the operation Π_* is obviously injective, whereas the operation Π, although injective on $\mathcal{C}^2$, is, in general, not necessarily injective on the whole $\mathcal{F}$ (e.g., in all examples of combinatory spaces considered in Section 1, we have the equality $\Pi(\emptyset, \theta) = \emptyset$ for all θ in $\mathcal{F}$).

---

developed in that larger class. Nevertheless, Georgieva's notion will be more convenient for our further exposition than the notion of iterative operative space, since we shall not make use of the additional properties possessed by that kind of spaces. The structures studied by Georgieva are closely related to the *programming spaces* introduced by Skordev in 1978, but have an obvious advantage over them, since the definition of the latter notion is more complicated (that definition can be found in Skordev [1982]).

[97] Cf. also p. 71 of Skordev [1980] (at least for the case of a symmetric combinatory space).

In arbitrary operative spaces, the natural numbers are represented in the following way proposed by L. Ivanov. If $\mathfrak{S}_* = \langle \mathcal{F}, I, \Pi_*, L_*, R_* \rangle$ is an operative space, and n is a natural number, then the element $\bar{n} = R_*^n L_*$ of $\mathcal{F}$ is regarded as representing n in $\mathfrak{S}_*$.[38] Of course, if $\mathfrak{S}_*$ is the companion operative space of the combinatory space $\langle \mathcal{F}, I, \mathcal{C}, \Pi, L, R, \Sigma, T, F \rangle$ then the equality $\bar{n} = (F, I)^n (T, I)$ holds, and all elements $\bar{n}$ are normal in the sense of Section 1. We note that in Skordev [1980], where the assumptions $T \in \mathcal{C}$, $F \in \mathcal{C}$ are made, the natural numbers are represented in a different way, namely by certain elements of $\mathcal{C}$ which are in fact the products $\bar{n}T$ (the representation of the natural numbers in Exercise I.2.4 can be regarded as a special case of this, but with exchanged T, F and with the constant mappings replaced by their values).

In the further exposition, when some combinatory space is given and denotations from the theory of operative spaces are used, then we shall always have in mind the companion operative space of the given combinatory space. In particular, this will apply to the denotations $\bar{n}$.

In the sequel, we shall use Π_* with arbitrary number of arguments. By definition, $\Pi_*(\varphi_0) = \varphi_0$, and, for $n > 0$,

$$\Pi_*(\varphi_0, \varphi_1, \ldots, \varphi_n) = \Pi_*(\varphi_0, \Pi_*(\varphi_1, \ldots, \varphi_n)).$$

Obviously,

$$\theta\Pi_*(\varphi_0, \ldots, \varphi_n) = \Pi_*(\theta\varphi_0, \ldots, \theta\varphi_n)$$

for all $\theta, \varphi_1, \ldots, \varphi_n$ in $\mathcal{F}$.

An useful property of the introduced representation of natural numbers is given by the following proposition.

Proposition 1 (Proposition 4.11 of Ivanov [1986]). Let $\langle \mathcal{F}, I, \Pi_*, L_*, R_* \rangle$ be an arbitrary operative space, let n be a natural number, and let $\varphi_0, \varphi_1, \ldots, \varphi_n$ be arbitrary elements of $\mathcal{F}$. Then $\Pi_*(\varphi_0, \ldots, \varphi_n)\bar{r} = \varphi_r$ for all $r < n$, and $\Pi_*(\varphi_0, \ldots, \varphi_n) R_*^n = \varphi_n$.

Proof. Induction on n. ∎

Corollary 1 (Proposition 4.10 of Ivanov [1986]). Let $\langle \mathcal{F}, I, \Pi_*, L_*, R_* \rangle$ be an operative space, and let k, m be such natural numbers that the inequality $\bar{k} \geq \bar{m}$ holds in

---

[38] There is an obvious disagreement between the denotation $\bar{n}$ introduced here and the denotation $\bar{s}$ from Section I.4 and Exercise I.5.5. We hope that the context will prevent the reader from a misunderstanding.

$\mathcal{F}$. Then $k=m$. (Hence the mapping $\lambda n.\bar{n}$ is injective.)

Proof. Suppose $k \neq m$, and take arbitrary elements φ, ψ of $\mathcal{F}$. Then there is an element δ of $\mathcal{F}$ such that $\delta\bar{k}=\varphi$, $\delta\bar{m}=\psi$. From here we conclude that $\varphi \geq \psi$. Since φ and ψ are arbitrary, this implies the conclusion that all elements of $\mathcal{F}$ are equal each other, contrary to the definition of the notion of operative space. ∎

Remark 1. If $<\mathcal{F}, I, \Pi_*, L_*, R_*>$ is an operative space then there are elements σ and π of $\mathcal{F}$ such that

$$\sigma\bar{k}=\overline{k+1}, \quad \pi\overline{k+1}=\bar{k}$$

for all natural numbers k. Namely, we could take $\sigma=R_*$ and, for example, $\pi=\Pi_*(I, I)$. In the special case, when $<\mathcal{F}, I, \Pi_*, L_*, R_*>$ is the companion operative space of the combinatory space $<\mathcal{F}, I, \mathcal{C}, \Pi, L, R, \Sigma, T, F>$, then

$$\Pi_*(I, I)=(L \rightarrow R, R),$$

and a simpler π with the above property can be found, namely $\pi=R$. For this case, we note also the equalities

$$L\overline{k+1}=F\bar{k}, \quad L\bar{0}=T.$$

In the special case mentioned above we shall define one more binary operation in $\mathcal{F}$ in addition to the operation Π_*. This new operation will be denoted by Δ, and it will be introduced by means of the equality

$$\Delta(\varphi, \psi)=(LR \rightarrow \varphi(L, R^2), \psi(L, R^2)).$$

It is easily verified that

$$\Delta(\varphi, \psi)(x, I)=\Pi_*(\varphi(x, I), \psi(x, I))$$

for all φ, ψ in $\mathcal{F}$ and all x in $\mathcal{C}$.[99] More generally, we set

$$\Delta(\varphi_0, \varphi_1, \ldots, \varphi_n)=\Delta(\varphi_0, \Delta(\varphi_1, \ldots, \Delta(\varphi_{n-1}, \varphi_n)\ldots))$$

for all $\varphi_0, \varphi_1, \ldots, \varphi_n$ in $\mathcal{F}$. Then, as an easy induction shows, we have the equality

$$(17) \quad \Delta(\varphi_0, \ldots, \varphi_n)(x, I)=\Pi_*(\varphi_0(x, I), \ldots, \varphi_n(x, I))$$

for all $\varphi_0, \varphi_1, \ldots, \varphi_n$ in $\mathcal{F}$ and all x in $\mathcal{C}$.

Proposition 2. Let $<\mathcal{F}, I, \mathcal{C}, \Pi, L, R, \Sigma, T, F>$ be a combinatory space, let n be a positive integer, and let

---

[99] Another way to prove the above equality is to consider the mapping $\lambda x.\Pi_*(\varphi(x, I), \psi(x, I))$ and to apply the method from the solution of Exercise 1.33. Note that this is a natural way to obtain the defining expression for $\Delta(\varphi, \psi)$.

$\varphi_0, \varphi_1, \ldots, \varphi_n$ be arbitrary elements of $\mathcal{F}$. Then

$$\Delta(\varphi_0, \ldots, \varphi_n)(x, \bar{r}) = \varphi_r(x, I)$$

for all $r < n$, and

$$\Delta(\varphi_0, \ldots, \varphi_n)(x, (F, I)^n) = \varphi_n(x, I).$$

Proof. Right multiplication of both sides of (17) by $\bar{r}$ and by $(F, I)^n$, followed by application of condition (7) from the definition of the notion of combinatory space and of Proposition 1.∎

Proposition 3. Let $\langle\mathcal{F}, I, \mathcal{C}, \Pi, L, R, \Sigma, T, F\rangle$ be a combinatory space. Then for all $\chi, \varphi, \psi, \theta, \alpha, \beta$ in $\mathcal{F}$ and each normal element ζ of $\mathcal{F}$, the following equalities hold:

$$\Pi_*(\varphi, \psi)(\chi, \zeta) = (\chi \to \varphi\zeta, \psi\zeta),$$

$$\Pi_*(\varphi, \psi)(\chi \to L_*\alpha, R_*\beta) = (\chi \to \varphi\alpha, \psi\beta),$$

$$\Pi_*(\varphi, \psi, \theta)(\chi \to \bar{0}\alpha, \bar{1}\beta) = (\chi \to \varphi\alpha, \psi\beta).$$

Proof. The first equality is a special case of the equality in Proposition 1.8. The other ones follow immediately from the properties of Σ and Π_*.∎

Corollary 2. Under the same assumption, for all φ, ψ in $\mathcal{F}$, we have the equality

$$\Delta(\varphi, \psi) = \Pi_*(\varphi(L, R^2), \psi(L, R^2))(LR, I).$$

Exercises

1. (Ivanov [1986, Exercise 4.2]) Let M be an infinite set, and let L_*, R_* be injective mappings of M into M satisfying the condition that $\operatorname{rng} L_* \cap \operatorname{rng} R_* = \emptyset$. For any φ, ψ from $\mathcal{F}_m(M)$, set $\Pi_*(\varphi, \psi) = (\varphi L_*{}^{-1}) \cup (\psi R_*{}^{-1})$. Prove that $\langle\mathcal{F}_m(M), I_M, \Pi_*, L_*, R_*\rangle$ is an operative space.

2. (Ivanov [1986, Example 4.4]) Let $\langle M, J, L, R, T, F, H\rangle$ be a computational structure in the sense of Chapter I, Section 1, such that $\operatorname{dom} L = \operatorname{dom} R = M$. Denote by $\mathcal{F}$ the set of all total mappings of M into itself, and introduce a multiplication and a partial order in $\mathcal{F}$ by adopting that $\varphi\psi$ denotes $\lambda u.\, \psi(\varphi(u))$ and $\varphi \geq \psi$ means $\varphi = \psi$. Let Π be the binary operation in $\mathcal{F}$ defined by means of the equality

$$\Pi(\varphi, \psi) = \lambda u.\, J(\varphi(u), \psi(u)).$$

Prove that $\langle\mathcal{F}, I_M, \Pi, L, R\rangle$ is an operative space.

3. (Ivanov [1986, Proposition 12.1]) Let $\langle\mathcal{F}, I, \Pi_*, L_*,$

$R_*\rangle$ be an arbitrary operative space. Denote by $\mathcal{F}'$ the partially ordered semigroup of all monotonically increasing mappings of $\mathcal{F}$ into itself, where $\varphi'\psi' = \lambda\theta.\,\psi'(\varphi'(\theta))$ for all φ', ψ' in $\mathcal{F}'$, and $\varphi' \geq \psi'$ means that $\varphi'(\theta) \geq \psi'(\theta)$ for all θ in $\mathcal{F}$. Let $I' = \lambda\theta.\,\theta$, $L_*' = \lambda\theta.\,\theta L_*$, $R_*' = \lambda\theta.\,\theta R_*$, and let the binary operation Π_*' in $\mathcal{F}'$ be defined by the equality $\Pi_*'(\varphi', \psi') = \lambda\theta.\,\Pi_*(\varphi'(\theta), \psi'(\theta))$. Prove that $\langle\mathcal{F}', I', \Pi_*', L_*', R_*'\rangle$ is also an operative space.

4. (Ivanov [1986, Exercise 4.4]) Show that the requirement about monotonic increasing of the binary operation Π_* in the definition of the notion of operative space can be replaced by the condition that $\lambda\theta.\,\Pi_*(I, \theta)$ is monotonically increasing or by the condition that $\lambda\theta.\,\Pi_*(\theta, I)$ is monotonically increasing.

Hint. Make use of the equalities

$$\Pi_*(\varphi, \psi) = \Pi_*(I, \varphi)\Pi_*(I, L_*\psi)\Pi_*(L_*R_*, R_*)$$

$$\Pi_*(\psi, \varphi) = \Pi_*(\varphi, \psi)\Pi_*(R_*, L_*).$$

5. Let $\langle\mathcal{F}, I, \Pi_*, L_*, R_*\rangle$ be an arbitrary operative space. Prove that for all natural numbers n and arbitrary $\varphi_0, \varphi_1, \ldots, \varphi_n$ in $\mathcal{F}$ the following equalities hold:

$$\Pi_*(\varphi_0, \varphi_1 \ldots, \varphi_{n-1}, \varphi_n, I) =$$
$$\Pi_*(\varphi_n, I)\Pi_*(R_*\varphi_{n-1}, I) \ldots \Pi_*(R_*^{\;n-1}\varphi_1, I)\Pi_*(R_*^{\;n}\varphi_0, I),$$
$$\Pi_*(\varphi_0, \varphi_1 \ldots, \varphi_n) = \Pi_*(\varphi_0, \varphi_1 \ldots, \varphi_n, I)\Pi_*(\bar{0}, \bar{1}, \ldots, \bar{n}).$$

6. Let $\langle\mathcal{F}, I, \mathcal{C}, \Pi, L, R, \Sigma, T, F\rangle$ be an arbitrary combinatory space, and let $\langle\mathcal{F}, I, \Pi_*, L_*, R_*\rangle$ be its companion operative space. For all θ, φ, ψ in $\mathcal{F}$, prove the equalities

$$\theta\Delta(\varphi, \psi) = \Delta(\theta\varphi, \theta\psi), \quad \Delta(\varphi, \psi) = \Pi_*(\varphi, \psi)\Delta(L_*, R_*).$$

7. Under the same assumptions as in the previous exercise, prove that $\langle\mathcal{F}, I, \lambda\varphi\psi.\,\Delta(\varphi R, \psi R), (\xi, L_*), (\eta, R_*)\rangle$ is an operative space, whenever ξ and η are normal elements of the given combinatory space.

8. Let $\langle\mathcal{F}, I, \mathcal{C}, \Pi, L, R, \Sigma, T, F\rangle$ be an arbitrary combinatory space, and let $\langle\mathcal{F}, I, \Pi_*, L_*, R_*\rangle$ be its companion operative space. Let St be the operation in $\mathcal{F}$ defined in Exercise 7 of the previous section. For all φ, ψ in $\mathcal{F}$, prove the equality

$$St(\Pi_*(\varphi, \psi)) = \Delta(St(\varphi), St(\psi))St(I).$$

Hint. Use Exercise 9 of the previous section.

9. (Compare with Proposition 27.14 of Ivanov [1986]). Assume the premises of the previous exercise together with the premises of Exercise 1.12. Prove that conditions (i) and

(ii) from Exercise 1.12 remain equivalent after replacement the operation Σ in condition (ii) by the operation Π_*.

3. Iteration in combinatory spaces

We introduced the notion of a combinatory space for capturing at least some concrete situations considered in Chapter I. However, an essential feature of that situations remains out of the scope of our general considerations until now. Namely, there is a natural operation of iteration in each one of the structures of functions or function-like objects considered in Chapter I, and this operation plays an essential role in the description of the corresponding notion of computability. So it is desirable to have an abstract algebraic treatment also of iteration. We shall give now such a treatment using least fixed points of some monotonic mappings. In the case of ordinary and multiple-valued functions, the least-fixed-point characterization of iteration is well-known, and also some ways for more general considerations have been noted by several authors (let us note, for example, the papers Blikle [1971], Mazurkiewicz [1971], Scott [1971]). We shall proceed in the way of generalizing the least-fixed-point characterizations given in Chapter I of this book, Sections 2, 5 and 8.

From now on, we suppose that a combinatory space

$$\mathfrak{S} = \langle \mathcal{F}, I, \mathcal{C}, \Pi, L, R, \Sigma, T, F \rangle$$

is given. Let σ and χ be some elements of $\mathcal{F}$. Having in mind the above mentioned characterizations from Chapter I, it is natural to name *iteration of* σ *controlled by* χ and to denote by $[\sigma, \chi]$ the least solution τ of the equation

$$\tau = (\chi \to \tau\sigma, I) \tag{1}$$

if, of course, such a least solution exists. If we adopt this definition, we may, for example, state that an iteration controlled by F is always equal to I, and the iteration of I controlled by T is equal to the least element of $\mathcal{F}$ if such a least element exists (since (1) is equivalent to $\tau = I$ in the case of $\chi = F$, and it is equivalent to $\tau = \tau$ in the case of $\chi = T$, $\sigma = I$). Note also that, whenever $[\sigma, \chi]$ is a solution of (1), then, for each ρ in $\mathcal{F}$, the element $\rho[\sigma, \chi]$ satisfies the equation

$$\tau = (\chi \to \tau\sigma, \rho). \tag{2}$$

It is not difficult to construct $\mathfrak{S}$ with $\mathcal{F}$ having no least element. For example, if $\langle M, J, L, R, T, F, H \rangle$ is a computational structure with total L, R, H (i.e. $\mathrm{dom} L = \mathrm{dom} R = \mathrm{dom} H = M$) then we could take $\mathfrak{S} = \langle \mathcal{F}, I_M, \mathcal{C}, \Pi', L, R, \Sigma', T, F \rangle$, where $\mathcal{F}$ is the subsemigroup of the total elements of $\mathcal{F}_p(M)$, the set $\mathcal{C}$ is the same as in Examples 1.1 and 1.2, and Π', Σ' are the restrictions to $\mathcal{F}$ of the operations Π, Σ from that examples. In such a combinatory space $\mathfrak{S}$, the equation (1) would have no least solution in the case of $\chi = T$, $\sigma = I$. Of course, when we are interested in the algebraic study of iteration in combinatory spaces, it is natural to consider only such ones in which this equation has a least solution for every choice of χ and σ (an obvious corollary of such an assumption would be the existence of a least element in the corresponding $\mathcal{F}$). However, we cannot develop a fruitful theory of iteration on the basis only of this assumption, and we shall formulate stronger assumptions.

First of all, it is useful to remember that the least-fixed-point characterizations of iteration, which were given in Chapter I, draw attention to a stronger property of $[\sigma, \chi]$ than of simply being the least τ satisfying the equation (1). Namely, we observed that, in the cases considered there, the solution $[\sigma, \chi]$ of this equation is also the least τ satisfying the inequality $\tau \geq (\chi \rightarrow \tau\sigma, I)$. Unfortunately, this will be again not enough[40] - a further strengthening of the above minimality condition will be needed. Exercise I.2.3 and the propositions from Sections I.5 and I.8 show that the following condition is also satisfied in the considered cases: for all χ, σ, ρ in $\mathcal{F}$, each solution τ of the inequality

$$\tau \geq (\chi \rightarrow \tau\sigma, \rho) \tag{3}$$

satisfies also the inequality $\tau \geq \rho[\sigma, \chi]$. It turns out that, for some purposes, a convenient decision is to define the notion of iteration in such a style. Namely, we could define the iteration of σ controlled by χ as an element ι of $\mathcal{F}$ satisfying the equality $\iota = (\chi \rightarrow \iota\sigma, I)$ and fulfilling the condition that, for all τ, σ, ρ in $\mathcal{F}$, the inequality (3) implies the inequality $\tau \geq \rho\iota$. Of course, if such an ι exists, it must be unique, since the conditions imposed on ι entail that ι must be the least τ satisfying (1).

The above conditions concerning iteration are not the strongest ones needed for the exposition in this book. The proof of some essential results will make use of somewhat more stronger conditions to be satisfied by iteration. Exercises I.2.8, I.5.9 and I.8.8 can be used for illustrating the spirit of this strengthening before giving the precise formulation. Exercise 1.37 can serve as a bridge from the

---

[40] For example, it seems to be not sufficient for proving that $\rho[\sigma, \chi]$ is the least solution τ of the equation (2).

special cases considered in Chapter I to the general one. And now, let us give finally the precise formulation.

Definition 1. Let σ and χ be some given elements of $\mathcal{F}$. An element ι of $\mathcal{F}$ is said to be *the iteration of* σ *controlled by* χ iff the equality $\iota = (\chi \to \iota\sigma, I)$ holds and, for all τ, ρ in $\mathcal{F}$ and each subset $\mathcal{A}$ of $\mathcal{C}$, whenever $\mathcal{A}$ is invariant with respect to σ, the inequality $\tau \underset{\mathcal{A}}{\geq} (\chi \to \tau\sigma, \rho)$ implies the inequality $\tau \underset{\mathcal{A}}{\geq} \rho\iota$. In the case when such an element ι exists, it will be denoted by $[\sigma, \chi]$.

The above definition really imposes not weaker conditions on ι than in the case when $\geq$ is used instead of $\underset{\mathcal{A}}{\geq}$. This can be easily seen by means of Proposition 1.9. Therefore the iteration of σ controlled by χ is unique if it exists at all, and so the clause concerning its denotation is justified. The terminology and the denotation introduced by the above definition are in concordance with the terminology and the denotations in Chapter I. This follows from the above mentioned Exercises I.2.8, I.5.9 and I.8.8.

Example 1. Let $\mathfrak{S}$ be an arbitrary combinatory space, and let σ be an arbitrary element of $\mathcal{F}$. Then the iteration of σ controlled by F exists, and the equality $[\sigma, F] = I$ holds. Indeed, we have $I = (F \to I\sigma, I)$, and, for all τ, ρ in $\mathcal{F}$ and each subset $\mathcal{A}$ of $\mathcal{C}$, the inequality $\tau \underset{\mathcal{A}}{\geq} (F \to \tau\sigma, \rho)$ is equivalent to the inequality $\tau \underset{\mathcal{A}}{\geq} \rho I$.

The combinatory spaces in Examples 1.1, 1.2, 1.3 and 1.4 (corresponding to the situations considered in Chapter I) are such that $[\sigma, \chi]$ exists for all σ and χ in the corresponding $\mathcal{F}$. In the sequel, we will be interested mainly in combinatory spaces having this property, and such spaces will be called iterative.

Definition 2. The combinatory space $\mathfrak{S}$ is called *iterative* iff the iteration of σ controlled by χ exists for all σ and χ in $\mathcal{F}$.

From now on, until the end of this section, let us suppose the given combinatory space $\mathfrak{S}$ is iterative. Some statements mentioned above before the ultimate definition of iteration (i.e. Definition 1) will be formulated now (possibly enlarged) as explicit propositions.

Proposition 1. For all σ, χ, ρ in $\mathcal{F}$, the element $\rho[\sigma, \chi]$ of $\mathcal{F}$ is the least solution τ of the equation (2) and the least τ in $\mathcal{F}$ satisfying the inequality (3).

Proof. Let $\iota = [\sigma, \chi]$. Then $\iota = (\chi \to \iota\sigma, I)$, and, consequently, $\rho\iota = \rho(\chi \to \iota\sigma, I) = (\chi \to \rho\iota\sigma, \rho)$. So $\tau = \rho\iota$

satisfies (2), and hence (3) is also satisfied if we choose τ in this way. On the other hand, if τ is an arbitrary element of $\mathcal{F}$ satisfying (3) (in particular, if τ is an arbitrary solution of (2))then, taking $\mathcal{A}=\mathcal{C}$ in the condition from Definition 1, we conclude that $\tau \geq \rho\iota$. ∎

Definition 3. The element $[I, T]$ of $\mathcal{F}$ will be denoted by o and will be called *the zero of* $\mathfrak{S}$.

Proposition 2. The zero of $\mathfrak{S}$ is the least element of $\mathcal{F}$, and, for all ρ in $\mathcal{F}$, the equality $\rho o = o$ holds. For all normal elements ζ of $\mathcal{F}$, also the equality $o\zeta = o$ holds[41].

Proof. Let ρ be an arbitrary element of $\mathcal{F}$. By Proposition 1, the element ρo is the least τ in $\mathcal{F}$ satisfying the equation $\tau = (T \rightarrow \tau I, \rho)$. But this equation is equivalent to $\tau = \tau$, hence ρo is the least element of $\mathcal{F}$. Since, in particular, we could take $\rho = I$, the first part of the proposition is thus proven. Let now x be some fixed element of $\mathcal{C}$. The second part follows immediately from Proposition 1.11. ∎

Remark 1. In the general case, it is not possible to prove that $o\rho = o$ for all ρ in $\mathcal{F}$. To have a counterexample, let us consider the special case of Example 1.3 corresponding to Theorem I.8.1, i.e. the case of $M = \mathbb{N}$, $E = \{\bullet\}$. Then, using the denotations from the mentioned theorem, we see that $oP^{\bullet}(0) = \bullet$, and hence $oP^{\bullet} \neq o$.

Remark 2. If the given iterative combinatory space is symmetric, then $o\rho = o$ and $o = [\rho, T]$ for all ρ in $\mathcal{F}$. The first equality can be proven by noting that, for all x in $\mathcal{C}$, we have (using Proposition 2 twice)

$$o\rho x = ox\rho x = L(ox, I)\rho x = L(ox, \rho x) = L(I, \rho x)ox = ox.$$

For proving the second equality, we apply the first one to get the equality $o = (T \rightarrow o\rho, I)$, and then we use the minimal property of iteration to conclude that $o \geq [\rho, T]$.

Two more propositions about iteration will be given.

Proposition 3. The operation of iteration is monotonically increasing, i.e. $[\sigma_1, \chi_1] \geq [\sigma_2, \chi_2]$, whenever $\sigma_1 \geq \sigma_2$, $\chi_1 \geq \chi_2$.

Proof. Let $\sigma_1 \geq \sigma_2$, $\chi_1 \geq \chi_2$. For $k = 1, 2$, set $\iota_k = [\sigma_k, \chi_k]$. Then $\iota_1 = (\chi \rightarrow \iota_1\sigma_1, I) \geq (\chi \rightarrow \iota_1\sigma_2, I)$,

---

[41] In particular, $ox = o$ for all x in $\mathcal{C}$, and hence o is a constant element in the sense of Exercise 1.26.

hence ι_1 is an element τ satisfying the inequality $\tau \geq (\chi \rightarrow, \tau\sigma_2, I)$. On the other hand, ι_2 is the least τ which satisfies the same inequality. Therefore $\iota_1 \geq \iota_2$. ∎

Proposition 4. Let ζ and η be normal elements of $\mathcal{F}$, and σ, χ be arbitrary elements of $\mathcal{F}$. Then the following implications hold:

$$\chi\zeta = T\eta \Longrightarrow [\sigma, \chi]\zeta = [\sigma, \chi]\sigma\zeta,$$

$$\chi\zeta = F\eta \Longrightarrow [\sigma, \chi]\zeta = \zeta.$$

Proof. Using the equality $[\sigma, \chi] = (\chi \rightarrow [\sigma, \chi]\sigma, I)$ and application of Propositions 1.5 and 1.7. ∎

Remark 3. In the proofs of the above propositions, the second condition from the definition of iteration, when used at all, was used only for the case of $\mathcal{A} = \mathcal{C}$. Application of this condition at other choices of $\mathcal{A}$ is needed, for example, for the solution of Exercises 4, 5, 6, 7, 10 after this section.

Before going further on, we should like to discuss the interrelation between the notion of iteration introduced in this section and the notion of iteration used in the book Skordev [1980]. There is an obviously unessential difference between the two notions (up to this difference, the present notion is the same as in Skordev [1980a, 1984]). The difference can be expressed by saying that we consider now a "while"-iteration, and a "while not"-iteration has been considered in the previous author's publications. Namely, the iteration denoted by $[\sigma, \chi]$ in them is an element τ satisfying the equation $\tau = (\chi \rightarrow I, \tau\sigma)$, and the same denotation is used now for an element τ which satisfies the equation $\tau = (\chi \rightarrow \tau\sigma, I)$. Both kinds of iteration (if there are no other differences) can be reduced one to the other by replacing the given combinatory space $\mathfrak{S}$ by the space $\langle \mathcal{F}, I, \mathcal{C}, \Pi, L, R, \Sigma_0, F, T \rangle$ from Exercise 1.1 (in connection with this, cf. also Exercise I.2.6 for the case of the combinatory space from Example 1.2 and Exercise 3 after this section for the general case). However, there is a more essential difference between the two compared notions, and the situation is such that, roughly speaking, the present notion of iteration is weaker than the notion from Skordev [1980] (hence combinatory spaces, which are iterative in the sense of that book, will be iterative also in the present sense, but not necessarily conversely).[42] For explaining

---

[42] The meaning of "weaker" here does not exclude a possible equivalence of the two notions.

this, we shall now recall the former definition of iteration, but with exchanged content of the second and the third argument of Σ. To avoid any confusion, we shall name that other iteration strong. After this change of the name and the above-mentioned exchange, the definition from Skordev [1980] can be formulated as follows:

Definition 4. Let σ and χ be some given elements of $\mathcal{F}$. An element ι of $\mathcal{F}$ is said to be *a strong iteration of σ controlled by* χ iff the equality $\iota=(\chi\rightarrow\iota\sigma, I)$ holds and ι belongs to each set which is closed under the mapping $\lambda\tau.(\chi\rightarrow\tau\sigma, I)$ and can be represented as the intersection of sets of the form $\{\tau\colon \psi\geq\varphi\tau z\}$.[43]

We shall prove now the formulated statement about the connection between both iterations.

Proposition 5. Let σ and χ be some given elements of $\mathcal{F}$, and let ι be a strong iteration of σ controlled by χ. Then ι is the iteration of σ controlled by χ.

Proof. The equality is one and the same in Definitions 1 and 4, so we have only to show that the second condition from Definition 1 is fulfilled. Let $\mathcal{A}$ be a subset of $\mathcal{C}$ invariant with respect to σ, and let τ_0 and ρ_0 be elements of $\mathcal{F}$ satisfying the inequality

$$\tau_0 \underset{\mathcal{A}}{\geq} (\chi\rightarrow\tau_0\sigma, \rho_0).$$

We must prove the inequality $\tau_0 \underset{\mathcal{A}}{\geq} \rho_0\iota$, i.e. prove that $\iota\in\mathcal{B}$, where $\mathcal{B}=\{\tau\colon \tau_0 \underset{\mathcal{A}}{\geq} \rho_0\tau\}$. Obviously, $\mathcal{B}$ is the intersection of all sets of the form $\{\tau\colon \tau_0 z\geq\rho_0\tau z\}$, where $z\in\mathcal{A}$. Therefore it is sufficient to prove that $\mathcal{B}$ is closed under $\lambda\tau.(\chi\rightarrow\tau\sigma, I)$. Let τ be an arbitrary element of $\mathcal{B}$. Since $\mathcal{A}$ is invariant with respect to σ, we may conclude that $\tau_0\sigma \underset{\mathcal{A}}{\geq} \rho_0\tau\sigma$. Then, for all x in $\mathcal{A}$, we shall have

$$\tau_0 x\geq(\chi\rightarrow\tau_0\sigma, \rho_0)x=(\chi x\rightarrow\tau_0\sigma x, \rho_0 x)\geq$$
$$(\chi x\rightarrow\rho_0\tau\sigma x, \rho_0 x)=\rho_0(\chi\rightarrow\tau\sigma, I)x.$$

Hence $\tau_0 \underset{\mathcal{A}}{\geq} \rho_0(\chi\rightarrow\tau\sigma, I)$, i.e. $(\chi\rightarrow\tau\sigma, I)\in\mathcal{B}$. Thus $\mathcal{B}$ is closed indeed under the considered mapping, and the proof is

---

[43] This is a definition which enables reasoning about iteration in the spirit of D. Scott's μ-induction rule (cf., for example, de Bakker and Scott [1969], de Bakker [1971] or Hitchcock and Park [1973]).

completed. ∎

It would be natural to compare the present notion of iteration also with the notion used in author's papers before 1980 (for example, in Skordev [1976b]). However, we shall not give now any details in this direction. We shall mention only that the notion from those papers is stronger than the notion in Skordev [1980][44], and hence it is stronger than the notion used now. We note also that in most examples considered in the present book iteration satisfies the stronger requirements from the definitions previously used.

In Chapter III, Section 4.4 of the book Skordev [1980] several equalities concerning iteration are proved in a way which is not usable under the definition adopted in the present book. We shall discuss these equalities now.

The first of them is given in Proposition 4.4.5 there and looks as follows:

$$[\sigma, \chi] = R[\langle\chi, I\rangle\sigma R, L]\langle\chi, I\rangle.$$

This equality will be proved further (see Corollary 5.2).

The next of the equalities in question is asserted under some assumptions in Proposition 4.4.6 of the mentioned chapter,[45] and then six other equalities are obtained as easy corollaries. Unfortunately, we do not know whether this proposition is always true for the iteration considered now, but still we shall prove the same equality under a certain additional assumption (the condition (*) below), and this will be sufficient for obtaining the above-mentioned corollaries.

Proposition 6. Let ζ be a normal element of $\mathcal{F}$ satisfying the following condition: (*) there is an element ρ of $\mathcal{F}$ such that $\rho\zeta = I$ holds. Let σ, π be elements of $\mathcal{F}$ satisfying the condition $\sigma\zeta = \zeta\pi$. Then

$$[\sigma, \chi]\zeta = \zeta[\pi, \chi\zeta]$$

for all χ in $\mathcal{F}$.

Proof. Let $\iota_1 = [\sigma, \chi]$, $\iota_2 = [\pi, \chi\zeta]$. Then we have to prove the equality $\iota_1\zeta = \zeta\iota_2$. This equality will be estab-

[44] Cf. pp. 252-253 of that book. The meaning of "stronger" here does not exclude equivalence of the compared notions.

[45] A strengthened version of the statement of that proposition is indicated in Exercise 10 after this section.

lished by proving inequalities in both directions. We note first that

$$\iota_1\zeta = (\chi \to \iota_1\sigma, I)\zeta = (\chi\zeta \to \iota_1\sigma\zeta, \zeta) = (\chi\zeta \to \iota_1\zeta\pi, \zeta).$$

Hence (by Proposition 1) the inequality $\iota_1\zeta \geq \zeta\iota_2$ holds. For the proof of the converse inequality, we set $\theta = \zeta\iota_2\rho$. Then $\zeta\iota_2 = \theta\zeta$, and the problem is reduced to proving the inequality $\theta\zeta \geq \iota_1\zeta$. The last inequality is equivalent to the inequality $\theta \underset{\mathcal{A}}{\geq} \iota_1$, where $\mathcal{A}$ is the set of all elements of $\mathcal{C}$ having the form ζz with $z \in \mathcal{C}$. The set $\mathcal{A}$ is invariant with respect to σ, due to the equality $\sigma\zeta = \zeta\pi$. On the other hand,

$$\theta\zeta = \zeta\iota_2 = (\chi\zeta \to \zeta\iota_2\pi, \zeta) = (\chi\zeta \to \theta\zeta\pi, \zeta) =$$
$$(\chi\zeta \to \theta\sigma\zeta, \zeta) = (\chi \to \theta\sigma, I)\zeta,$$

and clearly this implies the inequality $\theta \underset{\mathcal{A}}{\geq} (\chi \to \theta\sigma, I)$.

Now an application of the definition of iteration leads to the needed conclusion. ∎

Corollary 1. For all φ, χ in $\mathcal{F}$ and all x in $\mathcal{C}$, the following equalities hold:

$$[(L, \varphi), \chi](x, I) = (x, [\varphi(x, I), \chi(x, I)]),$$
$$[(\varphi, R), \chi](I, x) = ([\varphi(I, x), \chi(I, x)], x),$$
$$R[(L, \varphi), \chi](x, I) = [\varphi(x, I), \chi(x, I)],$$
$$L[(\varphi, R), \chi](I, x) = [\varphi(I, x), \chi(I, x)].$$

Proof. To obtain the first equality, we set $\zeta = (x, I)$, $\sigma = (L, \varphi)$, $\pi = \varphi(x, I)$, and then we use the equalities

$$R\zeta = I, \quad \sigma\zeta = (L\zeta, \varphi\zeta) = (x, \pi) = \zeta\pi,$$
$$\zeta[\pi, \chi\zeta] = (x, [\pi, \chi\zeta]).$$

The second equality can be obtained in a similar way by setting $\zeta = (I, x)$, $\sigma = (\varphi, R)$, $\pi = \varphi(I, x)$. The third and the fourth equality follow from the first and the second one, respectively, by Corollary 1.2. ∎

Corollary 2. For all σ, χ in $\mathcal{F}$ and all x in $\mathcal{C}$, the following equalities hold:

$$[(L, \sigma R), \chi](x, I) = (x, [\sigma, \chi(x, I)]),$$
$$[(\sigma L, R), \chi](I, x) = ([\sigma, \chi(I, x)], x).$$

Proof. Substitution of σR for φ in the first equality of Corollary 1, and substitution of σL for φ in the second one. ∎

Remark 4. We do not insist on using always the iteration introduced by Definition 1 and avoiding the use of strong iteration. We do not know concrete iterative combinatory spaces in which some iteration is not a strong one. On the other hand, it happens sometimes that a statement is provable for the strong iteration, but no proof of it is known for the other one (cf., for example, Exercise 10). In our opinion, one must have no prejudices against using the strong iteration (and even stronger ones) in the cases when this is appropriate.

Exercises

1. If the equality $\iota = (\chi \to \iota\sigma, I)$ in the definition of iteration is replaced by the inequality $\iota \geq (\chi \to \iota\sigma, I)$, prove that the new definition is equivalent to the original one.

Hint. If ι satisfies the conditions of the new definition then set $\iota_1 = (\chi \to \iota\sigma, I)$. Using the inequality $\iota \geq \iota_1$, conclude that $\iota_1 \geq (\chi \to \iota_1\sigma, I)$.

2. Let $\mathfrak{S} = \langle \mathcal{F}, I, \mathcal{C}, \Pi, L, R, \Sigma, T, F \rangle$ be an iterative combinatory space. Prove that, for all θ, φ, ψ in $\mathcal{F}$, the equalities

$$(o, \theta) = (o \to \varphi, \psi) = o, \quad (\varphi, \psi)o = (\varphi o, \psi o), \quad [\theta, o] = o$$

hold. In the case when $\mathfrak{S}$ is symmetric, prove also that

$$(\theta, o) = (\chi \to o, o) = o$$

for all θ, χ in $\mathcal{F}$ (the equalities concerning Π are from Ivanov [1977]).

3. Let $\mathfrak{S} = \langle \mathcal{F}, I, \mathcal{C}, \Pi, L, R, \Sigma, T, F \rangle$ be an iterative combinatory space, and let $\mathfrak{S}_0 = \langle \mathcal{F}, I, \mathcal{C}, \Pi, L, R, \Sigma_0, F, T \rangle$ be the corresponding space from Exercise 1.1 (i.e.

$$\Sigma_0(\chi, \varphi, \psi) = \Sigma(\chi, \psi, \varphi)$$

for all χ, φ, ψ in $\mathcal{F}$). Prove that $\mathfrak{S}_0$ is also iterative.

Hint. Use Exercise 1.22.

4. Let $\mathfrak{S} = \langle \mathcal{F}, I, \mathcal{C}, \Pi, L, R, \Sigma, T, F \rangle$ be an iterative combinatory space, and let $\{\alpha, \beta\} \subseteq \{L, R\}$. Prove that

$$[(\alpha, \beta), T](x, y) = o$$

for all x, y in $\mathcal{C}$ (of course, the interesting case is that one when $\mathfrak{S}$ is not symmetric, since otherwise we could make an immediate application of Remark 2).

Hint. Use Proposition 1.10 and the fact that

$$o(x, y) = (T \to o(\alpha, \beta), I)(x, y)$$

for all x, y in $\mathcal{C}$.

5. Let $\langle \mathcal{F}, I, \mathcal{C}, \Pi, L, R, \Sigma, T, F \rangle$ be an iterative combinatory space, and let θ, χ be some elements of $\mathcal{F}$. Let the operation St in $\mathcal{F}$ be defined in the same way as in Exercise 1.7. Prove that

$$St([\theta, \chi]) = [St(\theta), \chi R] St(I).$$

Hint. Prove the equalities

$$(\chi \to [St(\theta), \chi R](x, I)\theta, (x, I)) = [St(\theta), \chi R](x, I),$$

$$St([\theta, \chi])(x, y) = (\chi R \to St([\theta, \chi]) St(\theta), I)(x, y).$$

From them (using also Exercise 1.39) conclude that

$$(x, I)[\theta, \chi] = [St(\theta), \chi R](x, I)$$

for all x in $\mathcal{C}$. Then apply Exercise 1.9.

6. Let $\langle \mathcal{F}, I, \mathcal{C}, \Pi, L, R, \Sigma, T, F \rangle$ be an iterative combinatory space, σ and χ be elements of $\mathcal{F}$, $\mathcal{A}$ be a subset of $\mathcal{C}$ invariant with respect to σ. Prove that $\mathcal{A}$ is invariant also with respect to $[\sigma, \chi]$.

Hint. If $\varphi \underset{\mathcal{A}}{\geq} \psi$ then the inequality $\tau \underset{\mathcal{A}}{\geq} (\chi \to \tau\sigma, \psi)$ is satisfied by $\tau = \varphi[\sigma, \chi]$.

7. Let $\langle \mathcal{F}, I, \mathcal{C}, \Pi, L, R, \Sigma, T, F \rangle$ be an iterative combinatory space, $\sigma, \sigma', \chi, \chi'$ be elements of $\mathcal{F}$, $\mathcal{A}$ be a subset of $\mathcal{C}$ invariant with respect to σ, and let the inequalities $\sigma' \underset{\mathcal{A}}{\geq} \sigma$, $\chi' \underset{\mathcal{A}}{\geq} \chi$ hold. Prove the inequality

$$[\sigma', \chi'] \underset{\mathcal{A}}{\geq} [\sigma, \chi].$$

8. Let $\langle \mathcal{F}, I, \mathcal{C}, \Pi, L, R, \Sigma, T, F \rangle$ be an iterative combinatory space, and let

$$\sigma = (RL, R_* R), \quad \varphi = R[\sigma, (L^2 \to F, T)](I, I).$$

Prove that $\varphi\bar{n} = \overline{2n}$ for all natural numbers n. (Compare with Exercises I.2.4 and I.2.5)

9. To the assumptions of Exercise 1.40, add the assumption that $\mathfrak{S}$ is iterative. Prove that $\mathfrak{S}^K$ is also iterative, and the equality $[\sigma, \chi] = \lambda k. [\sigma(k), \chi(k)]$ holds for arbitrary σ, χ in $\mathcal{F}'$.

10. Prove the statement obtained from Proposition 6 by omitting the condition (*) and replacing "for all χ in $\mathcal{F}$" by "for all χ in $\mathcal{F}$ such that a strong iteration of σ controlled by χ exists".

Hint. To prove the inequality $[\sigma, \chi]\zeta \leq \zeta[\pi, \chi\zeta]$, apply the second part of the condition in Definition 4 to the set $\{\theta\in\mathcal{F}:\ \theta\zeta\le\zeta[\pi,\chi\zeta]\}$.

4. On least fixed points in partially ordered sets

If $\langle\mathcal{F}, I, \mathcal{C}, \Pi, L, R, \Sigma, T, F\rangle$ is an iterative combinatory space, σ, χ, ρ are some elements of $\mathcal{F}$, and Γ is the mapping of $\mathcal{F}$ into $\mathcal{F}$ defined by $\Gamma(\tau)=(\chi\rightarrow\tau\sigma, \rho)$, then (by Proposition 3.1) the element $\rho[\sigma,\chi]$ is the least fixed point of Γ, as well as the least solution τ of the inequality $\tau\ge\Gamma(\tau)$. Note also that the mapping Γ is monotonically increasing.

In our further exposition we shall systematically use least fixed points which are also least solutions of the corresponding inequalities of the above form, with different monotonically increasing mappings Γ. More generally, we shall use least solutions $\langle\tau_1, \ldots, \tau_m\rangle$ of systems of equations of the form

(1) $$\tau_i=\Gamma_i(\tau_1, \ldots, \tau_m),\quad i=1, \ldots, m,$$

where $\Gamma_1, \ldots, \Gamma_m$ are monotonically increasing mappings of $\mathcal{F}^m$ into $\mathcal{F}$, and it always will be the situation that the same $\tau_1, \ldots, \tau_m$ form also the least solution of the corresponding system of inequalities

(2) $$\tau_i\ge\Gamma_i(\tau_1, \ldots, \tau_m),\quad i=1, \ldots, m.$$

For such situations some statements will be used whose validity do not really depend on the fact that $\mathcal{F}$ is the partially ordered semigroup of a combinatory space.

From now on in this section, if nothing else is said about $\mathcal{F}$, we shall suppose that $\mathcal{F}$ is some partially ordered set. Of course, a mapping Γ of $\mathcal{F}^m$ into $\mathcal{F}$ will be called *monotonically increasing* iff for all $\varphi_1, \ldots, \varphi_m$, $\psi_1, \ldots, \psi_m$ in $\mathcal{F}$ satisfying the inequalities $\varphi_1\ge\psi_1$, $\ldots, \varphi_m\ge\psi_m$ also the inequality

$$\Gamma(\varphi_1, \ldots, \varphi_m)\ge\Gamma(\psi_1, \ldots, \psi_m)$$

holds. The least solution of some of the systems (1), (2) is by definition, a solution $\langle\varphi_1, \ldots, \varphi_m\rangle$ of this system such that for each solution $\langle\tau_1, \ldots, \tau_m\rangle$ of the system the inequalities $\tau_1\ge\varphi_1, \ldots, \tau_m\ge\varphi_m$ hold (clearly, such a least solution may not exist, and the system may have no solution at all).

We start with a statement whose special case of $m=1$ is implicitly contained in the paper Tarski [1955] (of course, the general case considered below can be reduced to this special one by application of the corresponding result to the partially ordered set $\mathcal{F}^m$).

Proposition 1. If $\Gamma_1, \Gamma_2, \dots, \Gamma_m$ are monotonically increasing mappings of $\mathcal{F}^m$ into $\mathcal{F}$, and the system of inequalities (2) has a least solution, then this solution is also the least solution of the system (1).

Proof. Let $\Gamma_1, \Gamma_2, \dots, \Gamma_m$ be monotonically increasing mappings of $\mathcal{F}^m$ into $\mathcal{F}$, and $\langle\varphi_1, \dots, \varphi_m\rangle$ be the least solution of (2). We set

$$\psi_i = \Gamma_i(\varphi_1, \dots, \varphi_m), \quad i=1, \dots, m.$$

Then we have the inequalities $\varphi_1 \geq \psi_1, \dots, \varphi_m \geq \psi_m$. These inequalities, together with the monotonicity of the mappings Γ_i and with the definition of the elements ψ_i, imply the inequalities

$$\psi_i \geq \Gamma_i(\psi_1, \dots, \psi_m), \quad i=1, \dots, m.$$

Hence $\langle\psi_1, \dots, \psi_m\rangle$ is also a solution of (2). Since $\langle\varphi_1, \dots, \varphi_m\rangle$ is the least solution of this system, the inequalities $\psi_1 \geq \varphi_1, \dots, \psi_m \geq \varphi_m$ follow. But, as we have already seen, also inequalities in the opposite direction hold. Therefore $\varphi_i = \psi_i$, $i=1, \dots, m$, i.e. $\langle\varphi_1, \dots, \varphi_m\rangle$ is a solution of the system (1). Obviously this is its least solution, since all solutions of (1) are solutions of (2).∎

In view of the above proposition, we shall be mainly interested in least solutions of systems of the form (2). In the special case of $m=1$, the system (2) reduces to an inequality of the form $\tau \geq \Gamma(\tau)$, where Γ is a monotonically increasing mapping of $\mathcal{F}$ into $\mathcal{F}$. If this inequality has a least solution φ then the element φ will be denoted by $\mu\tau.\Gamma(\tau)$ (denotation taken, up to some orthographic details, from the paper de Bakker and Scott [1969]). Of course, the introduced denotation will be used also with other variables instead of τ (for example, the same element φ of $\mathcal{F}$ can be denoted also by $\mu\theta.\Gamma(\theta)$). We allow also other expressions instead of $\Gamma(\tau)$, having in mind the mappings which arise when these expressions are regarded as functions of the variable after the symbol μ. These other expressions may depend also on other variables besides the mentioned one, and these other variables remain free in the considered μ-expression. For example, Proposition 3.1 can be expressed by the equality

$$\rho[\sigma, \chi] = \mu\tau.(\chi \to \tau\sigma, \rho).$$

Proposition 2. Let $\Gamma_1, \ldots, \Gamma_m$ be monotonically increasing mappings of $\mathcal{F}^{m+n}$ into $\mathcal{F}$. Suppose that, for all $\tau_1, \ldots, \tau_n$ in $\mathcal{F}$, the system of inequalities

(3) $\tau_{n+i} \geq \Gamma_i(\tau_1, \ldots, \tau_n, \tau_{n+1}, \ldots, \tau_{n+m}), \quad i = 1, \ldots, m.$

has a least solution $\langle\tau_{n+1}, \ldots, \tau_{n+m}\rangle$ in $\mathcal{F}^m$, and denote the components of this solution by

$$\Delta_i(\tau_1, \ldots, \tau_n), \quad i = 1, \ldots, m.$$

Then the mappings Δ_i of $\mathcal{F}^n$ into $\mathcal{F}$ are also monotonically increasing.

Proof. We shall restrict ourselves to the case when $m = 1$ (the general case can be treated in a similar way); we shall write Γ, Δ instead of Γ_1, Δ_1, respectively. Let $\varphi_1, \ldots, \varphi_n, \psi_1, \ldots, \psi_n$ be elements of $\mathcal{F}$ satisfying the inequalities $\varphi_1 \geq \psi_1, \ldots, \varphi_n \geq \psi_n$. Then

$$\Delta(\varphi_1, \ldots, \varphi_n) \geq \Gamma(\varphi_1, \ldots, \varphi_n, \Delta(\varphi_1, \ldots, \varphi_n)) \geq \Gamma(\psi_1, \ldots, \psi_n, \Delta(\varphi_1, \ldots, \varphi_n)).$$ [46]

Hence $\Delta(\varphi_1, \ldots, \varphi_n) \geq \Delta(\psi_1, \ldots, \psi_n)$. ∎

Now the problem will be considered about elimination in systems of the form (2) (in different settings problems of this kind are studied, for example, in Bekic [1969], Leszczyłowski [1971], Wand [1973], Blikle [1974]; Section 1C of Moschovakis [1974] is also relevant to the subject).

Theorem 1. Let $B_1, \ldots, B_n, \Gamma_1, \ldots, \Gamma_m$ be monotonically increasing mappings of $\mathcal{F}^{m+n}$ into $\mathcal{F}$. Suppose that, for all $\tau_1, \ldots, \tau_n$ in $\mathcal{F}$, the system of inequalities (3) has a least solution $\langle\tau_{n+1}, \ldots, \tau_{n+m}\rangle$ in $\mathcal{F}^m$, and denote the components of this solution by $\Delta_i(\tau_1, \ldots, \tau_n)$, $i = 1, \ldots, m$. Then the system of inequalities

$$(4) \quad \begin{aligned} &\tau_j \geq B_j(\tau_1, \ldots, \tau_n, \tau_{n+1}, \ldots, \tau_{n+m}), \quad j = 1, \ldots, n, \\ &\tau_{n+i} \geq \Gamma_i(\tau_1, \ldots, \tau_n, \tau_{n+1}, \ldots, \tau_{n+m}), \quad i = 1, \ldots, m, \end{aligned}$$

has a least solution iff such a solution exists for the system

---

[46] In fact, Proposition 1 enables us to replace the first of the last two inequality signs by an equality sign, but this is not needed for the proof.

(5) $\tau_j \geq B_j(\tau_1, \ldots, \tau_n, \Delta_1(\tau_1, \ldots, \tau_n), \ldots, \Delta_m(\tau_1, \ldots, \tau_n)),$
$j = 1, \ldots, n.$

Moreover, the following two statements hold:

(i) if $\langle \varphi_1, \ldots, \varphi_n, \varphi_{n+1}, \ldots, \varphi_{n+m} \rangle$ is the least solution of (4) then $\langle \varphi_1, \ldots, \varphi_n \rangle$ is the least solution of (5);

(ii) if $\langle \varphi_1, \ldots, \varphi_n \rangle$ is the least solution of (5) then $\langle \varphi_1, \ldots, \varphi_n, \Delta_1(\varphi_1, \ldots, \varphi_n), \ldots, \Delta_m(\varphi_1, \ldots, \varphi_n) \rangle$ is the least solution of (4).

Proof. Again, we shall restrict ourselves to the case when $m = 1$; we shall write Δ instead of Δ_1. If $\langle \tau_1, \ldots, \tau_n, \tau_{n+1} \rangle$ is a solution of (4) then from the last inequality of (4) we conclude that

(6) $$\tau_{n+1} \geq \Delta(\tau_1, \ldots, \tau_n),$$

and therefore (by the monotonicity of the mappings B_j) the other inequalities of (4) imply the inequalities (5). Thus, whenever $\langle \tau_1, \ldots, \tau_n, \tau_{n+1} \rangle$ is a solution of (4), then $\langle \tau_1, \ldots, \tau_n \rangle$ is a solution of (5), and the inequality (6) holds. Conversely, if $\langle \tau_1, \ldots, \tau_n \rangle$ is a solution of (5) then $\langle \tau_1, \ldots, \tau_n, \Delta(\tau_1, \ldots, \tau_n) \rangle$ is a solution of (4) (the last inequality of (4) is satisfied according to the definition of Δ).

Now suppose $\langle \varphi_1, \ldots, \varphi_n, \varphi_{n+1} \rangle$ is the least solution of (4). Then, by the above reasoning, $\langle \varphi_1, \ldots, \varphi_n \rangle$ is a solution of (5). Let $\langle \tau_1, \ldots, \tau_n \rangle$ be an arbitrary solution of (5). Then, again by the above reasoning, the $n+1$-tuple $\langle \tau_1, \ldots, \tau_n, \Delta(\tau_1, \ldots, \tau_n) \rangle$ is a solution of (4), and hence the inequalities $\tau_1 \geq \varphi_1, \ldots, \tau_n \geq \varphi_n$ hold. Thus $\langle \varphi_1, \ldots, \varphi_n \rangle$ is the least solution of (5), and statement (i) is proved.

For proving (ii), suppose that $\langle \varphi_1, \ldots, \varphi_n \rangle$ is the least solution of (5). Then $\langle \varphi_1, \ldots, \varphi_n, \Delta(\varphi_1, \ldots, \varphi_n) \rangle$ is a solution of (4). Let $\langle \tau_1, \ldots, \tau_n, \tau_{n+1} \rangle$ be an arbitrary solution of (4). Then $\langle \tau_1, \ldots, \tau_n \rangle$ is a solution of

(5), and therefore the inequalities $\tau_1 \geq \varphi_1, \ldots, \tau_n \geq \varphi_n$ hold. By Proposition 2, these inequalities imply that

$$\Delta(\tau_1, \ldots, \tau_n) \geq \Delta(\varphi_1, \ldots, \varphi_n).$$

Since we have also the inequality (6), we conclude that also $\tau_{n+1} \geq \Delta(\varphi_1, \ldots, \varphi_n)$. Thus $\langle\varphi_1, \ldots, \varphi_n, \Delta(\varphi_1, \ldots, \varphi_n)\rangle$ is the least solution of (4). ∎

Corollary 1. Under the premises of the above theorem, if $\langle\varphi_1, \ldots, \varphi_n, \varphi_{n+1}, \ldots, \varphi_{n+m}\rangle$ is the least solution of (4) then

$$\varphi_{n+i} = \Delta_i(\varphi_1, \ldots, \varphi_n), \quad i = 1, \ldots, m.$$

Corollary 2. Let $B_1, \ldots, B_n$ be monotonically increasing mappings of $\mathcal{F}^{m+n}$ into $\mathcal{F}$, and let $\Delta_1, \ldots, \Delta_m$ be monotonically increasing mappings of $\mathcal{F}^n$ into $\mathcal{F}$. Then the system of inequalities

$$\begin{aligned} &\tau_j \geq B_j(\tau_1, \ldots, \tau_n, \tau_{n+1}, \ldots, \tau_{n+m}), \quad j = 1, \ldots, n, \\ &\tau_{n+i} \geq \Delta_i(\tau_1, \ldots, \tau_n), \quad i = 1, \ldots, m, \end{aligned} \tag{7}$$

has a least solution iff such a solution exists for the system (5). Moreover, the following two statements hold:

(i) if $\langle\varphi_1, \ldots, \varphi_n, \varphi_{n+1}, \ldots, \varphi_{n+m}\rangle$ is the least solution of (7) then $\langle\varphi_1, \ldots, \varphi_n\rangle$ is the least solution of (5);

(ii) if $\langle\varphi_1, \ldots, \varphi_n\rangle$ is the least solution of (5) then $\langle\varphi_1, \ldots, \varphi_n, \Delta_1(\varphi_1, \ldots, \varphi_n), \ldots, \Delta_m(\varphi_1, \ldots, \varphi_n)\rangle$ is the least solution of (7).

Corollary 3. Under the premises of Theorem 1, an $m+n$-tuple of elements of $\mathcal{F}$ is the least solution of the system (4) iff this $m+n$-tuple is the least solution of the corresponding system (7).

As an application of Theorem 1, we shall obtain a result giving the interconnection between $\mu\theta.\, B(\Gamma(\theta))$ and $\mu\tau.\, \Gamma(B(\tau))$ for monotonically increasing mappings B, Γ of $\mathcal{F}$ into $\mathcal{F}$.

Theorem 2. Let B, Γ be monotonically increasing mappings of $\mathcal{F}$ into $\mathcal{F}$, and let

$$\mu\theta.\, B(\Gamma(\theta)) = \varphi. \tag{8}$$

Then

(9) $$\mu\tau.\Gamma(B(\tau))=\Gamma(\varphi).$$

Proof. We consider the system of the two inequalities

(10) $$\begin{aligned}\theta&\geq B(\tau),\\ \tau&\geq\Gamma(\theta).\end{aligned}$$

Theorem 1 will be applied to this system in two different ways: the first time eliminating τ, and the second time eliminating θ.

The first application of Theorem 1 shows that (10) has a least solution iff there is such a solution for the inequality

(11) $$\theta\geq B(\Gamma(\theta)).$$

Moreover, we can assert that if φ is the least solution of (11) then the least solution of (10) is

(12) $$\tau=\Gamma(\varphi),\quad \theta=\varphi.$$

Since we have the assumption (8), we conclude that (10) has the least solution (12). Now the second application of Theorem 1 to the system (10) shows that the τ-component of (12) must be the least solution of the inequality

$$\tau\geq\Gamma(B(\tau)),$$

and this statement is exactly the statement (9).∎

Corollary 4. Let $\langle\mathcal{F}, I, \mathcal{C}, \Pi, L, R, \Sigma, T, F\rangle$ be an iterative combinatory space, and let $\chi, \sigma, \rho, \alpha$ be elements of $\mathcal{F}$. Then the equality

$$\mu\tau.(\chi\rightarrow\tau\sigma, \rho)\alpha=\rho[\alpha\sigma, \chi]\alpha$$

holds.

Proof. We apply Theorem 2 to the mappings B and Γ of $\mathcal{F}$ into $\mathcal{F}$, which are defined as follows:

$$B(\tau)=(\chi\rightarrow\tau\sigma, \rho),\quad \Gamma(\theta)=\theta\alpha.$$∎

In an obvious sense (mentioned in the paragraph preceding Proposition 1), a system of the form (2) in an arbitrary partially ordered set $\mathcal{F}$ can always be reduced to a single equation of the form $\tau\geq\Gamma(\tau)$ in the partially ordered set $\mathcal{F}^n$. However, a reduction to a single equation of this form in the initially given partially ordered set $\mathcal{F}$ turns out to be also possible in the special case when $\mathcal{F}$ is the semigroup of an operative space (in particular, when $\mathcal{F}$ is the semigroup of a combinatory space).

Proposition 3. Let $\langle\mathcal{F}, I, \Pi_*, L_*, R_*\rangle$ be an operative space, and let $\Gamma_1, \ldots, \Gamma_m$ be monotonically increasing mappings of $\mathcal{F}^m$ into $\mathcal{F}$. Let Γ be the monotonically increasing mapping of $\mathcal{F}$ into $\mathcal{F}$ defined by the equality

$$\Gamma(\tau)=\Gamma_0(\tau\overline{0},\ \tau\overline{1},\ \dots,\ \tau\overline{m-2},\ \tau R_*^{\ m-1}),$$

where

$$\Gamma_0(\tau_1,\ \dots\ \tau_m)=\Pi_*(\Gamma_1(\tau_1,\ \dots\ \tau_m),\ \dots,\ \Gamma_m(\tau_1,\ \dots\ \tau_m)),$$

and let $\mu\tau.\Gamma(\tau)=\varphi$. Then $\langle\varphi\overline{0},\ \varphi\overline{1},\ \dots,\ \varphi\overline{m-2},\ \varphi R_*^{\ m-1}\rangle$ is the least solution of the system (2).

Proof. Set $\zeta_1=0,\ \zeta_2=1,\ \dots,\ \zeta_{m-1}=m-2,\ \zeta_m=R_*^{\ m-1}$, for short. Then, for $i=1,\ \dots,\ m$, we have (using $\varphi\geq\Gamma(\varphi)$ and Proposition 2.1)

$$\varphi\zeta_i\geq\Gamma_0(\varphi\zeta_1,\ \dots,\ \varphi\zeta_m)\zeta_i=\Gamma_i(\varphi\zeta_1,\ \dots,\ \varphi\zeta_m).$$

Thus $\langle\varphi\zeta_1,\ \dots,\ \varphi\zeta_m\rangle$ is a solution of (2). Consider now an arbitrary solution $\langle\tau_1,\ \dots,\ \tau_m\rangle$ of (2), and set

$$\tau=\Pi_*(\tau_1,\ \dots,\ \tau_m).$$

Then, $\tau\zeta_i=\tau_i,\ i=1,\ \dots,\ m$, and, using the inequalities (2), we get

$$\tau\geq\Gamma_0(\tau_1,\ \dots\ \tau_m)=\Gamma(\tau).$$

From here, by the minimality of φ, the inequality $\tau\geq\varphi$ follows. Hence $\tau_i\geq\varphi\zeta_i,\ i=1,\ \dots,\ m.$ ∎

In the above proposition, the existence of a least solution of the inequality $\tau\geq\Gamma(\tau)$ has been assumed, and the existence of a least solution of the system (2) has been established as a consequence. An implication in the opposite direction can also be proven, namely: if $\langle\varphi_1,\ \dots,\ \varphi_m\rangle$ is the least solution of (2) then $\mu\tau.\Gamma(\tau)=\Pi_*(\varphi_1,\ \dots,\ \varphi_m)$. This statement, as well as the statement of Proposition 3, can be regarded as a special case of a much more general statement (see Exercises 7 and 8 below).

In the preceding considerations in this section, we usually assumed the existence of some least fixed points and carried certain reasonings on the base of this assumption. Sometimes this existence follows easily from certain well-known sufficient conditions of a quite general nature. We shall recall now two such results. Some relevant references concerning these results are Knaster [1928], Birkhoff [1948, pp. 44, 54], Bourbaki [1949-50], Kleene [1952, § 66], Tarski [1955], Abian and Brown [1961], Platek [1966], Markowsky [1976] (the list is surely not complete, and we do not attribute the results only to those people whose names occur below).

Theorem 3 (Knaster – Tarski – Kleene Theorem). Let $\mathcal{F}$ be a partially ordered set having a least element o, let each monotonically increasing infinite sequence of elements of $\mathcal{F}$ has a least upper bound, and let Γ be a monotonically increasing mapping of $\mathcal{F}$ into itself such that Γ is continuous with respect to least upper bounds of monotonically increasing infinite sequences[47]. Then the sequence $\{\Gamma^k(o)\}_{k=0}^{\infty}$ is monotonically increasing and the equality

$$\sup\{\Gamma^k(o)\}_{k=0}^{\infty} = \mu\tau.\,\Gamma(\tau)$$

holds.

Proof. One proves by induction that $\Gamma^k(o) \le \Gamma^{k+1}(o)$ for each natural number k. Thus the sequence $\{\Gamma^k(o)\}_{k=0}^{\infty}$ is monotonically increasing. Let $\varphi = \sup\{\Gamma^k(o)\}_{k=0}^{\infty}$. Then

$$\Gamma(\varphi) = \sup\{\Gamma^{k+1}(o)\}_{k=0}^{\infty} = \varphi.$$

On the other hand, if τ is an arbitrary solution of the inequality $\tau \ge \Gamma(\tau)$ then, again by induction, one proves that $\tau \ge \Gamma^k(o)$ for each natural number k, and from here the inequality $\tau \ge \varphi$ follows. ∎

Theorem 4 (Knaster – Tarski – Platek Theorem). Let $\mathcal{F}$ be a partially ordered set such that each chain in $\mathcal{F}$ (including the empty one) has a least upper bound[48], and let Γ be an arbitrary monotonically increasing mapping of $\mathcal{F}$ into itself. Then an element φ_α of $\mathcal{F}$ can be defined for each ordinal number α so that the equality

$$\varphi_\alpha = \sup\{\Gamma(\varphi_\beta)\}_{\beta<\alpha} \tag{13}$$

holds for all α, the transfinite sequence $\{\varphi_\alpha\}$ is monotonically increasing, and there is some ordinal number γ such that

$$\varphi_\gamma = \mu\tau.\,\Gamma(\tau).$$

Proof. Let $\mathcal{G}$ be the set of all elements θ of $\mathcal{F}$ which satisfy the inequality $\theta \le \Gamma(\theta)$. Making use of the

---

[47] This means the following: whenever a monotonically increasing sequence $\{\theta_k\}_{k=0}^{\infty}$ of elements of $\mathcal{F}$ has a least upper bound, then $\Gamma(\sup\{\theta_k\}_{k=0}^{\infty}) = \sup\{\Gamma(\theta_k)\}_{k=0}^{\infty}$.

[48] Of course, the least upper bound of the empty chain will be the least element of $\mathcal{F}$.

monotonic increasing of Γ, it is easy to see that $\mathcal{G}$ is closed under Γ and under least upper bounds of chains. These properties of $\mathcal{G}$ enable the definition (by transfinite recursion) of a monotonically increasing transfinite sequence $\{\varphi_\alpha\}$ of elements of $\mathcal{G}$ with the property (13). The members of this sequence remain stationary from some place on, and hence there is an ordinal number γ such that $\varphi_\gamma=\varphi_{\gamma+1}$, i.e. $\varphi_\gamma=\Gamma(\varphi_\gamma)$. On the other hand, if τ is an arbitrary solution of the inequality $\tau\geq\Gamma(\tau)$ then, by transfinite induction, one proves that $\tau\geq\varphi_\alpha$ for each ordinal number α; in particular, $\tau\geq\varphi_\gamma$. ∎

We shall apply now Theorems 3 and 4 for obtaining some conditions sufficient for the existence of iteration in a given combinatory space.

Proposition 4 (Level Omega Iteration Lemma). Let

$$\mathfrak{S}=\langle\mathcal{F}, I, \mathcal{C}, \Pi, L, R, \Sigma, T, F\rangle$$

be a combinatory space, and let the following conditions be satisfied:

(i) there is a least element o in $\mathcal{F}$, and $\tau o=o$ for all τ in $\mathcal{F}$;

(ii) each monotonically increasing infinite sequence of elements of $\mathcal{F}$ has a least upper bound;

(iii) for every fixed $\varkappa$ in $\mathcal{F}$, the mappings $\lambda\tau.\varkappa\tau$, $\lambda\tau.\tau\varkappa$ and $\lambda\tau.(\varkappa\rightarrow\tau, I)$ are continuous with respect to least upper bounds of monotonically increasing infinite sequences.

Then:

(a) the combinatory space $\mathfrak{S}$ is iterative, and, for all σ, χ from $\mathcal{F}$, the equality

$$[\sigma, \chi]=\sup\{\Gamma^k(o)\}_{k=0}^{\infty} \tag{14}$$

holds, where $\Gamma=\lambda\tau.(\chi\rightarrow\tau\sigma, I)$;

(b) $\mathfrak{S}$ turns into an iterative combinatory space with the same iteration after any replacement of the original partial ordering $\geq$ in $\mathcal{F}$ by some partial ordering $\geq'$ not violating the requirements of the definition of the notion of combinatory space and such that whenever an infinite sequence of elements of $\mathcal{F}$ is monotonically increasing with respect to $\geq$, then the least upper bound of this sequence with respect to $\geq$ is also its least upper bound with respect to $\geq'$.

Proof. Let χ and σ be arbitrary elements of $\mathcal{F}$. Then the monotonically increasing mapping $\Gamma=\lambda\tau.(\varkappa\rightarrow\tau\sigma, I)$ is continuous with respect to least upper bounds of monotonically increasing infinite sequences (as a composition of two monotonically increasing mappings with this sort of continuity). By the Knaster-Tarski-Kleene Theorem, the sequence $\{\Gamma^k(o)\}_{k=0}^{\infty}$ is monotonically increasing and for the element $\iota=\sup\{\Gamma^k(o)\}_{k=0}^{\infty}$ the equality $\iota=(\varkappa \rightarrow \iota\sigma, I)$ holds. Suppose now that $\mathfrak{S}'$ is an arbitrary combinatory space obtained from $\mathfrak{S}$ by replacing the original partial ordering $\geq$ in $\mathcal{F}$ by some partial ordering $\geq'$ with the property formulated in (b) (in particular, $\mathfrak{S}'$ may be $\mathfrak{S}$ itself). We shall show that ι is the iteration of σ controlled by χ in the combinatory space $\mathfrak{S}'$. By Proposition 3.5 and Definition 3.4, it is sufficient to show that ι belongs to each set closed under Γ and representable as the intersection of sets of the form $\{\tau: \psi\geq'\varphi\tau z\}$. Let $\mathcal{B}$ be a set with these properties. From Proposition 1.11, it follows that $oz=o$ for each z in $\mathcal{C}$. Hence $\varphi oz=o$ for all φ in $\mathcal{F}$ and all z in $\mathcal{C}$. On the other hand, $\psi\geq'o$ for all ψ in $\mathcal{F}$ (since ψ is the least upper bound of the sequence $o, \psi, \psi, \psi, \ldots$ with respect to $\geq$). Therefore $o\in\mathcal{B}$, and hence $\Gamma^k(o)\in\mathcal{B}$ for each natural number k. For any φ in $\mathcal{F}$ and any z in $\mathcal{C}$, the element $\varphi\iota z$ is the least upper bound of the sequence $\{\varphi\Gamma^k(o)z\}_{k=0}^{\infty}$ with respect to $\geq$, and consequently $\varphi\iota z$ is the least upper bound of this sequence also with respect to $\geq'$. Therefore, if $\psi\geq'\varphi\tau z$ for all τ in $\mathcal{B}$, then $\psi\geq'\varphi\iota z$. Making use of this, we conclude that $\iota\in\mathcal{B}$. ∎

Proposition 5 (Unrestricted Iteration Lemma). Let

$$\mathfrak{S}=\langle\mathcal{F}, I, \mathcal{C}, \Pi, L, R, \Sigma, T, F\rangle$$

be a combinatory space, and let the following conditions be satisfied:

(i) each chain in $\mathcal{F}$ (including the empty one) has a least upper bound;

(ii) the mappings $\lambda\tau.\varphi\tau$, with fixed φ in $\mathcal{F}$, and the mappings $\lambda\tau.\tau z$, with fixed z in $\mathcal{C}$, are continuous with respect to least upper bounds of arbitrary chains (including the empty one).[49]

Then:

(a) the combinatory space $\mathfrak{S}$ is iterative;

---

[49] I.e., whenever $\mathcal{D}$ is a chain in $\mathcal{F}$, and $\delta=\sup\mathcal{D}$, then $\varphi\delta=\sup\{\varphi\tau: \tau\in\mathcal{D}\}$, $\delta z=\sup\{\tau z: \tau\in\mathcal{D}\}$ for all φ in $\mathcal{F}$ and all z in $\mathcal{C}$. Taking $\mathcal{D}=\emptyset$, we conclude that $\varphi o=oz=o$, where o is the least element of $\mathcal{F}$.

(b) $\mathfrak{S}$ turns into an iterative combinatory space with the same iteration after any replacement of the original partial ordering $\geq$ in $\mathcal{F}$ by some partial ordering $\geq'$ not violating the requirements of the definition of the notion of combinatory space and such that whenever a subset of $\mathcal{F}$ is a chain with respect to $\geq$, then the least upper bound of this subset with respect to $\geq$ is also its least upper bound with respect to $\geq'$.

Proof. Let χ and σ be arbitrary elements of $\mathcal{F}$. The mapping $\Gamma=\lambda\tau.(\varkappa\rightarrow\tau\sigma, I)$ is again monotonically increasing (although it is possibly not continuous). Making use of the Knaster-Tarski-Platek Theorem, we take a transfinite sequence $\{\varphi_\alpha\}$ and an ordinal number γ with the properties listed there (hence the equality $\varphi_\gamma=\Gamma(\varphi_\gamma)$ holds). Let $\mathfrak{S}'$ be a combinatory space obtained from $\mathfrak{S}$ by changing the partial ordering in $\mathcal{F}$ in such a way, as described in (b) (in particular, $\mathfrak{S}'$ may be $\mathfrak{S}$ itself). We shall show that ι is the iteration of σ controlled by χ in the combinatory space $\mathfrak{S}'$. For that purpose, it is sufficient to show that φ_γ belongs to each set closed under Γ and representable as the intersection of sets of the form $\{\tau\colon \psi\geq'\varphi\tau z\}$. Let $\mathcal{B}$ be a set with these properties. Making use of the assumption (ii) and of the assumption about $\geq'$, we see that, whenever a subset of $\mathcal{B}$ is a chain with respect to $\geq$, then the least upper bound of this subset with respect to $\geq$ belongs to $\mathcal{B}$. This enables a transfinite recursion showing that all φ_α belong to $\mathcal{B}$. ∎

Remark 1. In almost all applications of the above two propositions, only part (a) of their conclusions will be used. Part (b) is needed for Remark 8.9 in the Appendix.

Remark 2. As seen from the proofs of these propositions, an iteration, whose existence is established on the basis of some of them, is surely a strong one.

Until now, we gave only such examples of iterative combinatory spaces (namely, Examples 1.1-1.4) which satisfy the assumptions of both the Level Omega and the Unrestricted Iteration Lemma. Exercise 10 after this section gives an example of iterative combinatory space which satisfies the assumptions of none of these propositions. Exercise 12 gives an example of combinatory space satisfying the assumptions of the Level Omega Iteration Lemma, but not satisfying the assumptions of the Unrestricted Iteration Lemma, and Exercise 14 shows that the latter assumptions imply neither the assumptions of the Level Omega Iteration Lemma nor the equality (14) in its conclusion.

The Level Omega Iteration Lemma enables not only proving

that certain combinatory spaces are iterative, but also, thanks to the equality (14) in it, making some conclusions about the explicit form of the iteration (the proof of the Unrestricted Iteration Lemma also enables making conclusions of this sort). Some additional considerations can be added to this, which facilitate in many cases the making of the mentioned conclusions. We note that most examples of combinatory spaces exposed until now in this book have the following features. The elements of the semigroup $\mathcal{F}$ in them are sets, the empty set belongs to $\mathcal{F}$, and, for any choice of σ, χ in $\mathcal{F}$, the mapping $\Gamma = \lambda\tau.(\chi \rightarrow \tau\sigma, I)$ occurring in (14) is representable in the form

$$\Gamma(\tau) = \Gamma(\emptyset) \cup A(\tau),$$

where A is a mapping of $\mathcal{F}$ into itself such that $A(\emptyset) = \emptyset$ and $A(\tau_1 \cup \tau_2) = A(\tau_1) \cup A(\tau_2)$, whenever τ_1, τ_2 and $\tau_1 \cup \tau_2$ belong to $\mathcal{F}$. In the simpler cases, we have

$$A(\tau) = \Sigma(\chi, \tau\sigma, \emptyset) = \tau\Sigma(\chi, \sigma, o),$$
$$o\varkappa = o, \quad (\tau_1 \cup \tau_2)\varkappa = (\tau_1\varkappa \cup \tau_2\varkappa),$$

and in the more complicated Examples 1.3 and 1.4 we have

$$A(\tau) = \{\langle u, w\rangle : \langle u, \mathrm{true}\rangle \in H\chi \ \& \ \exists v(\langle u, v\rangle \in \sigma \ \& \ \langle v, w\rangle \in \tau)\},$$

where $H\chi$ does not depend on τ. Now we shall show that a representation of Γ in the above form leads to a useful representation of the elements $\Gamma^k(o)$. Generalizing the role of the set-theoretical operation of union in the above situation, we shall consider the situation when a certain partial binary operation playing this role is given.

Proposition 6. Suppose o is an element of $\mathcal{F}$, and a partial (possibly total) binary operation of addition is defined in $\mathcal{F}$ such that $\theta + o = \theta$ for all θ in $\mathcal{F}$. Let A be a mapping of $\mathcal{F}$ into itself such that $A(o) = o$ and $A(\tau_1 + \tau_2) = A(\tau_1) + A(\tau_2)$, whenever τ_1, τ_2 belong to $\mathcal{F}$ and $\tau_1 + \tau_2$ is defined. Let Γ be a mapping of $\mathcal{F}$ into itself having the form

$$\Gamma(\tau) = \alpha + A(\tau),$$

where α is a fixed element of $\mathcal{F}$. Then for each natural number k the following equality holds

$$(15) \qquad \Gamma^k(o) = \alpha + A(\alpha) + A^2(\alpha) + \ldots + A^{k-1}(\alpha),$$

where associativity to the right is adopted, i.e.

$$\theta_0 + \theta_1 + \ldots + \theta_n \underset{\mathrm{def}}{=} \theta_0 + (\theta_1 + \ldots + \theta_n),$$

as well the natural conventions that a sum having only one term is equal to it, and a sum without terms is equal to o.

Proof. An easy induction shows that

$$A(\theta_1 + \ldots + \theta_m) = A(\theta_1) + \ldots A(\theta_m),$$

whenever $\theta_1 + \ldots + \theta_m$ is defined. Now the validity of (15) can be shown by induction on k. The case $k=0$ is trivial. Suppose now k is a natural number such that (15) is true. Then

$$\Gamma^{k+1}(o) = \Gamma(\Gamma^k(o)) = \alpha + A(\Gamma^k(o)) =$$
$$\alpha + (A(\alpha) + A^2(\alpha) + A^3(\alpha) + \ldots + A^k(\alpha)) =$$
$$\alpha + A(\alpha) + A^2(\alpha) + A^3(\alpha) + \ldots + A^k(\alpha). \blacksquare$$

Remark 3. The partial ordering in $\mathcal{F}$ is obviously not used in the above proof.

As an illustration, we shall apply Proposition 6 to the case of the combinatory space from Example 1.2, and thus we shall indicate another way to see that the iteration introduced in Section I.2 coincides with the iteration in this combinatory space. In this case o is the empty function, $+$ is the partial operation of union of functions, α is the restriction of I_M to the set $\{u \in M : H(\chi(u)) = \mathrm{false}\}$, and $A(\tau)$ is the restriction of the function $\tau\sigma$ to the set $\{u \in M : H(\chi(u)) = \mathrm{true}\}$. Then an induction shows that, for each natural number m and each u and w in M, the equality $A^m(\alpha)(u) = w$ is equivalent to the existence of elements $v_0, v_1, \ldots, v_m$ of M such that

$$v_0 = u \;\&\; v_m = w \;\&\; \forall_{j<m} j\,(H(\chi(v_j)) = \mathrm{true} \;\&\; v_{j+1} = \sigma(v_j)) \;\&\; H(\chi(v_m)) = \mathrm{false}.$$

The equality (15) shows that $\Gamma^k(o)$ is the union of all functions $A^m(\alpha)$ with $m < k$, and therefore $\sup\{\Gamma^k(o)\}_{k=0}^{\infty}$ is the union of all these functions. Hence $[\sigma, \chi](u) = w$ (where $[\sigma, \chi]$ denotes the iteration in the combinatory space) is equivalent to the existence of a natural number m for which elements $v_0, v_1, \ldots, v_m$ of M with the above property can be found.

Example 1.1 can be treated in essentially the same way, and the application of Proposition 6 to the combinatory spaces from Examples 1.3 and 1.4 is left as an exercise for the reader (Exercise 15 after this section). Of course, Proposition 6 can be useful in this respect only in connection with the Level Omega Iteration Lemma, and therefore one has to use other ways of reasoning for the explicit characterization of iteration in the cases when this lemma is not applicable (see, for example, Exercise 17 after this section).

Exercises

1. For each natural number t, let $\pi(t)$ denote the number of the primes which are less than or equal to t. Let a and b be arbitrary natural numbers. Prove that there is a natural number t satisfying the equation

$$t = a\pi(t) + b.$$

Hint. Use the known fact that $\lim\limits_{t\to\infty} \frac{\pi(t)}{t} = 0$.

2. Let a, b, c, d, e, f be arbitrary natural numbers. Prove the existence of natural numbers t, u satisfying the system of equations

$$t = au^2 + b[\sqrt{t}] + c,$$
$$u = d[\sqrt[3]{t}] + e[\sqrt{u}] + f,$$

where $[A]$ denotes the greatest integer which is less than or equal to A.

3. Give a direct proof of Corollary 4.

Hint. To show that $\rho[\alpha\sigma, \chi]\alpha$ satisfies the equation $\tau = (\chi \to \tau\sigma, \rho)\alpha$, simply substitute in the equation. To prove that the inequality $\tau \geq (\chi \to \tau\sigma, \rho)\alpha$ implies $\tau \geq \rho[\alpha\sigma, \chi]\alpha$, set $\theta = (\chi \to \tau\sigma, \rho)$ and prove that $\theta \geq \rho[\alpha\sigma, \chi]$ is implied by the first inequality.

4. Let $\langle \mathcal{F}, I, \mathcal{C}, \Pi, L, R, \Sigma, T, F \rangle$ be the combinatory space from Example 1.1. Find elements χ, σ and α of $\mathcal{F}$ such that

$$\mu\tau.(\chi \to \tau\sigma, \rho)\alpha \neq \mu\tau.(\chi\alpha \to \tau\sigma\alpha, \rho\alpha).$$

5. Let $\langle \mathcal{F}, I, \mathcal{C}, \Pi, L, R, \Sigma, T, F \rangle$ be an iterative combinatory space, and let σ, χ be arbitrary elements of $\mathcal{F}$. Prove that

$$\mu\tau.((\chi \to L\tau, I), I)\sigma = ([\sigma, \chi], I)\sigma.$$

6. Let $\langle \mathcal{F}, I, \mathcal{C}, \Pi, L, R, \Sigma, T, F \rangle$ be an iterative combinatory space, let $\chi_0, \chi_1, \ldots, \chi_l, \sigma_0, \sigma_1, \ldots, \sigma_l, \alpha_0, \alpha_1, \ldots, \alpha_l$ be elements of $\mathcal{F}$, and let the mappings $\Gamma_0, \Gamma_1, \ldots, \Gamma_l$ of $\mathcal{F}^2$ into $\mathcal{F}$ and the elements $\delta_0, \delta_1, \ldots, \delta_l$ of $\mathcal{F}$ be defined by the equalities

$$\Gamma_0(\tau, \theta) = \theta, \quad \delta_0 = I,$$
$$\Gamma_{k+1}(\tau, \theta) = \Gamma_k(\tau, (\chi_k \to \tau\sigma_k, \theta)\alpha_k),$$
$$\delta_{k+1} = [\alpha_k\delta_k\sigma_k, \chi_k]\alpha_k\delta_k$$

($k = 0, 1, \ldots, l-1$). Prove that

$$\mu\tau.\,\Gamma_k(\tau,\theta)=\theta\delta_k,\quad k=0,1,\ldots,l.$$

7. Let $\mathcal{F}$ and $\mathcal{F}'$ be some partially ordered sets, Π be a monotonically increasing mapping of $\mathcal{F}'$ into $\mathcal{F}$, and Π' be a monotonically increasing mapping of $\mathcal{F}$ into $\mathcal{F}'$ such that

$$\Pi'(\Pi(\tau'))=\tau'$$

for all τ' into $\mathcal{F}'$. Let Γ' be a monotonically increasing mapping of $\mathcal{F}'$ into itself, and Γ be the monotonically increasing mapping of $\mathcal{F}$ into itself defined by the equality

$$\Gamma(\tau)=\Pi(\Gamma'(\Pi'(\tau))).$$

Prove that $\mu\tau.\,\Gamma(\tau)=\varphi$ implies $\mu\tau'.\,\Gamma'(\tau')=\Pi'(\varphi)$. Show that this is a generalization of Proposition 3.

Hint. To obtain Proposition 3 as a special case, take $\mathcal{F}'$ to be $\mathcal{F}^n$, and Π to be Π_*.

8. Under the same premises as in the previous exercise, prove that $\mu\tau'.\,\Gamma'(\tau')=\varphi'$ implies $\mu\tau.\,\Gamma(\tau)=\Pi(\varphi')$.

Hint. To show that $\tau\geq\Gamma(\tau)$ implies $\tau\geq\Pi(\varphi')$, make use of the equality $\varphi'=\Pi(\varphi')$.

9. Let $\mathcal{F}$ be a partially ordered set, and Γ be a monotonically increasing mapping of $\mathcal{F}$ into $\mathcal{F}$. Suppose

$$\varphi=\mu\tau.\,\Gamma^n(\tau),$$

where n is some positive integer. Prove that

$$\varphi=\mu\tau.\,\Gamma(\tau).$$

Hint. Use the equality $\Gamma(\varphi)=\Gamma^n(\Gamma(\varphi))$ to conclude that $\Gamma(\varphi)\geq\varphi$ and hence $\varphi=\Gamma^n(\varphi)\geq\Gamma(\varphi)$. Use also the fact that $\tau\geq\Gamma(\tau)$ implies $\tau\geq\Gamma^n(\tau)$.

10. Let $\mathfrak{U}=\langle\mathbb{N},J,L,R,T,F,H\rangle$ be a standard computational structure on the national numbers (in the sense of Section I.3). Let $\mathcal{F}$ be the sub-semigroup of $\mathcal{F}_p(\mathbb{N})$ consisting of all one-argument partial recursive functions, and $\mathcal{C}$ be the set of all constant functions from $\mathbb{N}$ into $\mathbb{N}$. Let Π and Σ be the binary operations in $\mathcal{F}_p(\mathbb{N})$ corresponding to $\mathfrak{U}$ in the way described in Section I.2). Let $\mathfrak{S}=\langle\mathcal{F},I_{\mathbb{N}},\mathcal{C},\Pi_0,L,R,\Sigma_0,T,F\rangle$, where Π_0 and Σ_0 are the restrictions of Π and Σ to $\mathcal{F}^2$ and to $\mathcal{F}^3$, respectively. Prove that $\mathfrak{S}$ is an iterative combinatory space, but assumption (ii) of the Level Omega Iteration Lemma is not satisfied for $\mathfrak{S}$ (hence assumption (i) of the Unrestricted Iteration Lemma is also not satisfied).

11. (Cf. Skordev [1980, Section II.5.7 and Example 17 in Section III.3.2) Let $\langle M, J, L, R, T, F, H\rangle$ be an arbitrary computational structure (cf. Section I.1), let $\mathbb{L}$ be a lattice having a greatest element $\mathbb{1}$ and a least element $\mathbf{0}$, where $\mathbb{1} \neq \mathbf{0}$, and let the range of each mapping μ of M into $\mathbb{L}$ have a least upper bound in $\mathbb{L}$ with the property that

$$l \wedge \sup \operatorname{rng} \mu = \sup\{l \wedge \mu(u) :\ u \in M\}$$

for all l in $\mathbb{L}$. Denote by $\mathcal{F}$ the set of all $\mathbb{L}$-fuzzy binary relations in M (cf. Goguen [1967]), i.e. all mappings of M^2 into $\mathbb{L}$.[50] The set $\mathcal{F}$ is considered with the composition operation defined by means of the equality

$$\varphi\psi = \lambda u w.\ \sup\{\psi(u, v) \wedge \varphi(v, w) :\ v \in M\}$$

and with the partial ordering defined by means of the equivalence

$$\varphi \geq \psi \Longleftrightarrow \forall u v\, (\varphi(u, v) \geq \psi(u, v)).$$

For each subset f of M^2, let $f^\sim$ be the element of $\mathcal{F}$ defined by

$$f^\sim(u, v) = \begin{cases} \mathbb{1} & \text{if } \langle u, v\rangle \in f, \\ \mathbf{0} & \text{if } \langle u, v\rangle \notin f. \end{cases}$$

Let $\mathcal{C}$ be the set of all elements of $\mathcal{F}$ having the form $(M \times \{s\})^\sim$, where $s \in M$. Let Π and Σ be the binary and the ternary operation in $\mathcal{F}$ defined in the following way:

$$\Pi(\varphi, \psi)(u, v) = \begin{cases} \varphi(u, L(v)) \wedge \psi(u, R(v)) & \text{if } v \in \operatorname{rng} J, \\ \mathbf{0} & \text{if } v \notin \operatorname{rng} J, \end{cases}$$

$$\Sigma(\chi, \varphi, \psi)(u, v) = ((H\chi)(u, \mathbf{true}) \wedge \varphi(u, v)) \vee ((H\chi)(u, \mathbf{false}) \wedge \psi(u, v)),$$

where

$$(H\chi)(u, p) = \sup\{\chi(u, s) :\ s \in H^{-1}(p)\}.$$

---

[50]Here is an idea about a possible use of $\mathbb{L}$-fuzzy relations with suitable lattices $\mathbb{L}$. Suppose a set $\mathbb{S}$ of formal systems is given for proving statements in a language expressing properties of elements of M. Let $\mathbb{L}$ be the set of all subsets of $\mathbb{S}$ with the partial ordering by inclusion. Suppose a binary relation between elements of M is given, and for any fixed pair $\langle u, v\rangle$ of elements of M some formula $\Phi_{u,v}$ of the mentioned language expresses that $\langle u, v\rangle$ is in this relation. Then it is natural to consider an $\mathbb{L}$-fuzzy relation φ such that, for all u, v in M, $\varphi(u, v)$ is the set of those systems from $\mathbb{S}$ which have $\Phi_{u,v}$ among their theorems.

Prove that $\langle \mathcal{F}, I_M^{\sim}, \mathcal{C}, \Pi, L^{\sim}, R^{\sim}, \Sigma, T^{\sim}, F^{\sim} \rangle$ is a symmetric and iterative combinatory space satisfying the assumptions of the Level Omega Iteration Lemma.

12. Show that the set M and the lattice $\mathbb{L}$ in the previous exercise can be chosen so that the corresponding combinatory space does not satisfy the assumption (i) of the Unrestricted Iteration Lemma.

Hint. Take $M = \mathbb{N}$, and choose $\mathbb{L}$ to be a suitable linearly ordered set.

13. (For some relevant references, cf. Exercise 19 below and the first footnote to Exercise I.8.3). Let $\mathfrak{A} = \langle M, J, L, R, T, F, H \rangle$ be a computational structure, the predicate H being assumed total. Let $\mathcal{F}$ be the set of all pairs $\langle f, A \rangle$, where $f \in \mathcal{F}_m(M)$, $A \subseteq M$ (i.e. the set $\mathcal{F}$ from Exercise I.8.3).[51] The set $\mathcal{F}$ is considered with the same composition operation as in Exercise I.8.3 and with a different partial ordering which is defined by means of the following convention:

$$\langle f, A \rangle \geq \langle g, B \rangle \iff f \geq g \ \& \ A \geq B \ \& \\ \forall u \in B \, \forall v \, (\langle u, v \rangle \in f \implies \langle u, v \rangle \in g).$$

For each f in $\mathcal{F}_m(M)$, let $f^{\sim} = \langle f, \operatorname{dom} f \rangle$, and let $\mathcal{C}$ be the set of all elements of $\mathcal{F}$ having the form $(M \times \{s\})^{\sim}$, where $s \in M$. Let Π and Σ be the binary and the ternary operation in $\mathcal{F}$ defined in the same way as in Exercise I.8.3. Prove that $\langle \mathcal{F}, I_M^{\sim}, \mathcal{C}, \Pi, L^{\sim}, R^{\sim}, \Sigma, T^{\sim}, F^{\sim} \rangle$ is an iterative combinatory space satisfying the assumptions of the Unrestricted Iteration Lemma.

14. For the combinatory space from the previous exercise, show that the requirement from the Level Omega Iteration Lemma is violated about the continuity of the mappings $\lambda\tau.\tau\varkappa$ with respect to least upper bounds of monotonically increasing infinite sequences. Show also that the equality (14) is violated for some elements σ, χ of this combinatory space.

---

[51] The intuitive idea about the pairs $\langle f, A \rangle$ belonging to $\mathcal{F}$ is now the following one. We consider f to be the input-output relation of some non-deterministic computational procedure, and A to be the set of those input data for which all possible variants of execution of the procedure terminate. Using the terminology from Manna [1971], we could say that A consists of those input data for which the given computational procedure is $\forall$*-defined*. For an equipollent mathematical model, cf. Egli [1975], Chen [1984].

Hint. Take elements $s_0, s_1, s_2, \ldots$ and s_ω of M such that $s_i \neq s_j$, whenever $i \neq j$. Let $\sigma = \langle f, A \setminus \{s_0\} \rangle$, $\chi = \langle h, A \rangle$, where A is the set of all elements s_i, f consists of all pairs $\langle s_i, s_j \rangle$ with $i > j$ (the inequality $\omega > j$ is adopted to be true for all j in $\mathbb{N}$), h is a function such that $\mathrm{dom}\, h = A$, $H(h(u)) = \mathrm{true}$ for all u in $A \setminus \{s_0\}$ and $H(h(s_0)) = \mathrm{false}$. Let ι be the least upper bound on the right-hand side of (14) for these σ, χ. Show that s_ω belongs to the second component of $\Gamma(\iota)$ without belonging to the second component of ι.

15. Apply Propositions 4 and 6 to obtain the explicit characterization of iteration for the combinatory spaces from Examples 1.3 and 1.4.

16. Apply Propositions 4 and 6 to obtain the following characterization of iteration in the combinatory space from Exercise 11:

$$[\sigma, \chi](u, w) = \sup\{\rho_m(u, w) : m \in \mathbb{N}\},$$

where

$$\rho_m(u, w) = \sup\{ \bigwedge_{j=0}^{m-1} ((H\chi)(v_j, \mathrm{true}) \wedge \sigma(v_j, v_{j+1})) \wedge (H\chi)(v_m, \mathrm{false}) : v_0, v_1, \ldots, v_m \in M,\ v_0 = u,\ v_m = w\}.$$

17. Let $\mathfrak{S} = \langle \mathcal{F}, I_M^{\sim}, \mathcal{C}, \Pi, L^{\sim}, R^{\sim}, \Sigma, T^{\sim}, F^{\sim} \rangle$ be the combinatory space from Exercise 13, and let $\sigma = \langle f, A \rangle$, $\chi = \langle h, C \rangle$ be elements of $\mathcal{F}$. An element u of M will be called *σ, χ-regular* iff the following condition is satisfied:

$$u \in C \ \&\ (\langle u, \mathrm{true} \rangle \in Hh \Longrightarrow u \in A)$$

(compare with the definition of Σ). An element w of M will be called a *σ, χ-successor of* the element u iff

$$\langle u, \mathrm{true} \rangle \in Hh \ \&\ \langle u, w \rangle \in f.$$

Let D be the intersection of all subsets Q of M having the following property: whenever an element u of M is σ, χ-regular and all σ, χ-successors of u belong to Q, then $u \in Q$. Prove the equality

$$[\sigma, \chi] = \langle [f, h], D \rangle,$$

where $[\sigma, \chi]$ and $[f, h]$ is understood in the sense of the combinatory spaces $\mathfrak{S}$ and $\mathfrak{S}_m(\mathfrak{U})$, respectively.

Hint. Prove that $\langle [f, h], D \rangle = \mu\tau.(\chi \rightarrow \tau\sigma, I_M^{\sim})$.

18. In the conditions of the previous exercise, a se-

quence $\{\omega_j\}$ (finite or infinite) of elements of M will be called a σ,χ*-path* iff, whenever ω_j and ω_{j+1} are two consecutive members of this sequence, then ω_{j+1} is a σ,χ-successor of ω_j (if the sequence has only one term then this sequence is also considered a σ,χ-path). A σ,χ-path is called to *begin at* a given element u of M iff u is the initial member of this σ,χ-path. Prove the following characterization of the set D defined in that exercise: an element u of M belongs to D iff all σ,χ-paths beginning at u consist only of σ,χ-regular elements, and among these σ,χ-paths there is no infinite one.[52]

19. (Skordev [1980, Section II.5.1, Example 11 in Section III.3.2 and Example 9 in Section IV.1.2]). Show that the statements of Exercises 13, 14, 17 and 18 remain valid if the smaller set $\mathcal{F}$ is considered which is obtained by replacing the requirement $A\subseteq M$ in Exercise 13 by the stronger requirement $A\subseteq \mathrm{dom} f$. Show that the combinatory space $\langle\mathcal{F}, I_M^{\sim}, \mathcal{C}, \Pi, L^{\sim}, R^{\sim}, \Sigma, T^{\sim}, F^{\sim}\rangle$ is symmetric in this case.

20. Do the same as in the previous exercise, except for proving symmetry, in the case when, in addition to the requirement $A\subseteq M$, the requirement is imposed that the set $\{v: \langle u, v\rangle\in f\}$ is finite for all u in A. Prove that the combinatory space $\langle\mathcal{F}, I_M^{\sim}, \mathcal{C}, \Pi, L^{\sim}, R^{\sim}, \Sigma, T^{\sim}, F^{\sim}\rangle$ satisfies the assumptions of the Level Omega Iteration Lemma in this case. Show also that the condition from Exercise 18 about non-existence of infinite σ,χ-paths beginning at u can be replaced in this case by the condition that there is a finite upper bound for the lengths of the σ,χ-paths beginning at u.

21. Let $\mathfrak{U}=\langle M, J, L, R, T, F, H\rangle$ be a computational structure, where M is a topological space, the sets $\mathrm{dom}L$, $\mathrm{dom}R$, $H^{-1}(\mathrm{true})$, $H^{-1}(\mathrm{false})$ are open, and the mappings J, L, R, T, F are continuous.[53] Let $\mathcal{F}$ be the set of those el-

---

[52] A comparison of this characterization with the definition of S. Nikolova's iteration considered in Exercise I.8.3 is appropriate at this moment. The difference is that actually only the first of the two conditions about the σ,χ-paths beginning at u is present in the definition of Nikolova's iteration.

[53] Cf., e.g., Kelley [1975] for the necessary information about the topological notions. As to examples satis-

ements of $\mathcal{F}_{p}(M)$ which are continuous and have open domains. Show that $\mathcal{F}$ is closed under composition, $\mathfrak{U}$-combination and $\mathfrak{U}$-branching. If $\mathfrak{S}$ is obtained from $\mathfrak{S}_{p}(\mathfrak{U})$ by replacing $\mathcal{F}_{p}(M)$ by $\mathcal{F}$ (taking the induced partial ordering and multiplication) and replacing Π and Σ by their restrictions to $\mathcal{F}^{2}$ and $\mathcal{F}^{3}$, respectively, show that $\mathfrak{S}$ is an iterative combinatory space, and the $\mathfrak{S}$-iteration is the restriction of the $\mathfrak{S}_{p}(\mathfrak{U})$-iteration to $\mathcal{F}^{2}$.

22. Let M be a topological space, and $\mathcal{F}$ be the set of the elements θ of $\mathcal{F}_{m}(M)$ such that the corresponding set-valued mapping $\lambda u.\{v: \langle u, v\rangle\in\theta\}$ is lower semicontinuous.[54] Let the following assumptions be satisfied: the set M is infinite, J is a continuous injection of M^{2} into M, L and R are such elements of $\mathcal{F}$ that

$$\langle J(s, t), v\rangle\in L \Leftrightarrow v=s, \quad \langle J(s, t), v\rangle\in R \Leftrightarrow v=t$$

for all s, t, v in M, T and F are continuous mappings of M into M, and two open subsets M^{t} and M^{f} of M are given such that $T(u)\in M^{\mathrm{t}}\setminus M^{\mathrm{f}}$, $F(u)\in M^{\mathrm{f}}\setminus M^{\mathrm{t}}$ for all u in M.[55] Let

$$\mathfrak{S}=\langle\mathcal{F}, I_{M}, \mathcal{C}, \Pi, L, R, \Sigma, T, F\rangle,$$

where $\mathcal{F}$ is considered with the composition and the partial ordering inherited from $\mathcal{F}_{m}(M)$, $\mathcal{C}$ is the set of all constant total mappings of M into itself, and the operations Π and Σ with domains $\mathcal{F}^{2}$ and $\mathcal{F}^{3}$, respectively, are defined by means of the equalities

$$\Pi(\varphi, \psi)=\{\langle u, w\rangle: \exists s\exists t(\langle u, s\rangle\in\varphi \ \&\ \langle u, t\rangle\in\psi \ \&\ J(s, t)=w)\},$$

---

fying the assumptions of this exercise, cf. the second footnote to Theorem 5.2 in the Appendix of the present book.

[54] If M is a topological space, and f is a mapping of M into the set of the subsets of M, then f is called *lower semicontinuous* iff the set $\{u\in M: f(u)\cap V\neq\emptyset\}$ is open for any open subset V of M (cf., e.g., Berge [1966, Chapter VI, § 1]).

[55] These assumptions will be satisfied, for instance, if $\langle M, J, L, R, T, F, H\rangle$ is a computational structure satisfying the assumptions of the previous exercise, and we set $M^{\mathrm{t}}=H^{-1}(\mathrm{true})$, $M^{\mathrm{f}}=H^{-1}(\mathrm{false})$.

$$\Sigma(\chi, \varphi, \psi) = \{\langle u, w\rangle :\ \exists s\,(\langle u, s\rangle \in \chi\ \&\ (s \in M^{t}\ \&\ \langle u, w\rangle \in \varphi \vee s \in M^{f}\ \&\ \langle u, w\rangle \in \psi))\}.$$

Prove that $\mathfrak{S}$ is an iterative combinatory space, and, for all elements σ, χ of $\mathcal{F}$ and all u, w in M, the condition $\langle u, w\rangle \in [\sigma, \chi]$ is equivalent to the existence of a finite sequence $v_0, v_1, \ldots, v_m$ of elements of M such that

$$v_0 = u\ \&\ v_m = w\ \&\ \forall j_{j<m}\,(\exists s \in M^{t}\,(\langle v_j, s\rangle \in \chi)\ \&\ \langle v_j, v_{j+1}\rangle \in \sigma)\ \&\ \exists s \in M^{f}\,(\langle v_m, s\rangle \in \chi).$$

5. The companion operative space of an iterative combinatory space

In Section 2, we defined the notion of iteration in a combinatory space. Now we are going to define a similar notion for the case of an operative space.

Definition 1. Let $\langle \mathcal{F}, I, \Pi_*, L_*, R_*\rangle$ be an operative space, and let σ be an element of $\mathcal{F}$. An element ι of $\mathcal{F}$ will be called *the iteration of* σ iff for each ρ in $\mathcal{F}$ the equality

$$\mu\tau.\,\Pi_*(\tau\sigma, \rho) = \rho\iota$$

holds. If ι is the iteration of σ then ι will be denoted by $[\sigma]$.[56]

Remark 1. It is reasonable to compare the introduced notion of iteration with that one used in Ivanov [1986] (cf. condition ($\mathcal{LL}$) in Chapter 5 of that book). Given an arbitrary operative space $\langle \mathcal{F}, I, \Pi_*, L_*, R_*\rangle$ and an element σ of $\mathcal{F}$, then, according to Ivanov's definition, $[\sigma]$ as an element ι of $\mathcal{F}$ satisfying the condition that

$$\mu\tau.\,\Pi_*(\rho, \tau\sigma) = \rho\iota$$

for all ρ in $\mathcal{F}$. So we see an exchange of the contents of the arguments of the operation Π_*, and, of course, the difference between the two notions caused by this exchange must be considered unessential (cf. also Exercise 4 after this section in connection with this).

One more notion concerning iteration in operative spaces will be used in our further exposition. By introducing it, we shall in fact describe the class of the operative spaces studied in Georgieva [1980] (up to the above-mentioned ex-

---

[56] The last clause in the given definition is justified by the fact that if ι is the iteration of σ then the equality $\iota = \mu\tau.\,\Pi_*(\tau\sigma, I)$ holds.

change between the arguments of Π_*).[57]

Definition 2. We shall call a *G-space* any operative space $<\mathcal{F}, I, \Pi_*, L_*, R_*>$ satisfying the condition that $[\sigma]$ exists for each σ in $\mathcal{F}$.

Remark 2. It would be natural to call the operative spaces satisfying the above condition iterative. However, this would be not convenient due to the fact that Ivanov's definition of the notion of an iterative operative space requires not only existence of iteration, but also existence of so-called translation. Leaving aside more subtle conditions which are imposed on translation, we shall mention only that *the translation of* σ, where σ is some given element of $\mathcal{F}$, must be equal to $\mu\tau.\Pi_*(L_*\sigma, R_*\tau)$.

The using of G-spaces in our study of iterative combinatory spaces is based on the following fact.

Proposition 1. (Cf. Proposition 27.15 of Ivanov [1986]). Let $\mathfrak{S}=<\mathcal{F}, I, \mathcal{C}, \Pi, L, R, \Sigma, T, F>$ be an iterative combinatory space. Then its companion operative space $\mathfrak{S}_*=<\mathcal{F}, I, \Pi_*, L_*, R_*>$ is a G-space, and for all σ in $\mathcal{F}$ the equality $[\sigma]=R[\sigma R, L]$ holds.

Proof. For all τ, σ, ρ in $\mathcal{F}$, we have the equality

$$\Pi_*(\tau\sigma, \rho)=(L\rightarrow\tau\sigma R, \rho R),$$

hence

$$\mu\tau.\Pi_*(\tau\sigma, \rho)=\rho R[\sigma R, L]. \blacksquare$$

When an iterative combinatory space $\mathfrak{S}$ is considered, and σ is an element of its semigroup $\mathcal{F}$, then the element $[\sigma]$ of $\mathcal{F}$ will be called *the* $\mathfrak{S}_*$*-iteration of* σ, to make more easy the distinction between the two iteration operations present in $\mathcal{F}$ in this case (for the same reason, the iteration operation in $\mathfrak{S}$ will be called $\mathfrak{S}$*-iteration*).

Remark 3. The existence of $\mu\tau.\Pi_*(L_*\sigma, R_*\tau)$ also can be proved in the case considered in the above proposition, but this is not easy. The mentioned existence will be established by application of the First Recursion Theorem for iterative combinatory spaces (to be proven later in this book).

Now we shall note some general properties of iteration in G-spaces.

---

[57] Instead of such operative spaces, another kind of structures, called spaces of Böhm-Jacopini type, have been used in the book Skordev [1980]. The so-called programming spaces, mentioned in the first footnote to Section 2, also could be used for the same purposes.

Proposition 2. Let $\langle \mathcal{F}, I, \Pi_*, L_*, R_*\rangle$ be a G-space. Then, for all σ in $\mathcal{F}$, the equalities

$$[\sigma]=\Pi_*([\sigma]\sigma, I), \quad [\sigma]L_*=[\sigma]\sigma, \quad [\sigma]R_*=I,$$
$$[R_*\sigma]=\Pi_*(\sigma, I)$$

hold.[58]

Proof. The first equality follows immediately from the definition of $[\sigma]$ (taking $\rho=I$ in this definition). The second and the third equality are consequences of the first one and of the definition of the notion of operative space. To prove the last equality, we make an use of the first and the third one in the following way:

$$[R_*\sigma]=\Pi_*([R_*\sigma]R_*\sigma, I)=\Pi_*(I\sigma, I)=\Pi_*(\sigma, I).\blacksquare$$

Proposition 3. (Cf. Proposition 6.10 of Ivanov [1986]). Let $\langle \mathcal{F}, I, \Pi_*, L_*, R_*\rangle$ be a G-space. Then, for all σ, ρ, α in $\mathcal{F}$, the equality

$$\mu\tau.\Pi_*(\tau\sigma, \rho)\alpha = \rho[\alpha\sigma]\alpha$$

holds.[59]

Proof. Application of Theorem 4.2 to the mappings B and Γ of $\mathcal{F}$ into $\mathcal{F}$ defined as follows:

$$\mathrm{B}(\tau)=\Pi_*(\tau\sigma, \rho), \quad \Gamma(\theta)=\theta\alpha.\blacksquare$$

Now we shall show how the operation Σ and the iteration in an iterative combinatory space $\mathfrak{S}$ can be expressed by means of the $\mathfrak{S}_*$-iteration, composition, Π and some fixed elements of $\mathcal{F}$.

Proposition 4. Let $\langle \mathcal{F}, I, \mathcal{C}, \Pi, L, R, \Sigma, T, F\rangle$ be an iterative combinatory space. Then for all χ, φ, ψ in $\mathcal{F}$, the equality

$$(\chi \rightarrow \varphi, \psi)=[R_*\psi][R_*^{\,2}\varphi R](\chi, L_*)$$

holds.[60]

Proof. By application of Proposition 2 and Proposition 2.3, we get

---

[58] Compare the second and the third equalities with the equalities in Lemma 1 of Georgieva [1980] and in Proposition 5.12 of Ivanov [1986].

[59] Compare with Corollary 4.4.

[60] Compare with the expression for $\Sigma(\chi, \varphi, \psi)$ in Exercise I.2.2. Another representation of Σ, not using Π and not making an explicit use of L, R, but using $\Sigma(\chi, \bar{0}, \bar{1})$, will be given in Exercise 1 after this section.

$$[R_*\psi][R_*^{\,2}\varphi R](\chi, L_*) = [R_*\psi]\Pi_*(R_*\varphi R, I)(\chi, L_*) =$$
$$[R_*\psi](\chi \rightarrow R_*\varphi R L_*, L_*) = [R_*\psi](\chi \rightarrow R_*\varphi, L_*) =$$
$$(\chi \rightarrow [R_*\psi]R_*\varphi, [R_*\psi]L_*) = (\chi \rightarrow I\varphi, \Pi_*(\psi, I)L_*) =$$
$$(\chi \rightarrow \varphi, \psi). \blacksquare$$

Corollary 1. If $\langle \mathcal{F}, I, \mathcal{C}, \Pi, L, R, \Sigma, T, F \rangle$ is an iterative combinatory space, then for all φ, ψ in $\mathcal{F}$ the equality

$$\Pi_*(\varphi, \psi) = [R_*\psi R][R_*^{\,2}\varphi R^2](L, L_*)$$

holds.

Remark 4. Since (L, L_*) is a fixed element of $\mathcal{F}$, the above equality gives a representation of Π_* by means of $\mathfrak{S}_*$-iteration, composition and some fixed elements of $\mathcal{F}$ (in the considered case when $\mathfrak{S}_*$ is the companion operative space of an iterative combinatory space). Another representation of Π_* (due to N. Georgieva) which is valid in all G-spaces (and consequently makes no use of Π, L, R) will be given in Exercise 2 after this section.

Proposition 5. Let $\langle \mathcal{F}, I, \mathcal{C}, \Pi, L, R, \Sigma, T, F \rangle$ be an iterative combinatory space. Then for all σ, χ in $\mathcal{F}$, the equality

$$[\sigma, \chi] = [(\chi, I)\sigma](\chi, I)$$

holds.

Proof. By Proposition 1.8, Corollary 4.4 and Proposition 1, we have

$$[\sigma, \chi] = \mu\tau.(\chi \rightarrow \tau\sigma, I) =$$
$$\mu\tau.(L \rightarrow \tau\sigma R, R)(\chi, I) = R[(\chi, I)\sigma R, L](\chi, I) =$$
$$[(\chi, I)\sigma](\chi, I). \blacksquare$$

Corollary 2. If $\langle \mathcal{F}, I, \mathcal{C}, \Pi, L, R, \Sigma, T, F \rangle$ is an iterative combinatory space then, for all σ, χ in $\mathcal{F}$, the equality

$$[\sigma, \chi] = R[(\chi, I)\sigma R, L](\chi, I)$$

holds. ■

Besides the representation of $[\sigma, \chi]$ from Proposition 5, some other ones will be given which again make use of composition, $\mathfrak{S}_*$-iteration and some fixed elements of $\mathcal{F}$, but the operation Σ is used in them instead of Π. These representations will be obtained in the next section by application of a theorem about least solutions of a certain kind of inequalities in G-spaces.

Exercises

1. Let $\mathfrak{S}=\langle\mathcal{F}, I, \mathcal{C}, \Pi, L, R, \Sigma, T, F\rangle$ be an iterative combinatory space. Prove that, for all χ, φ, ψ in $\mathcal{F}$, the equality

$$(\chi\rightarrow\varphi, \psi)=[R_*\psi][R_*^{\,2}\varphi](\chi\rightarrow\bar{0}, \bar{1})$$

holds.

2. Let $\langle\mathcal{F}, I, \Pi_*, L_*, R_*\rangle$ be an arbitrary G-space. Prove that for all natural numbers n and arbitrary $\varphi_0, \varphi_1, \ldots, \varphi_n$ in $\mathcal{F}$ the following equality holds:[61]

$$\Pi_*(\varphi_0, \varphi_1 \ldots, \varphi_{n-1}, \varphi_n)=$$
$$[R_*\varphi_n][R_*^{\,2}\varphi_{n-1}]\ldots[R_*^{\,n}\varphi_1][R_*^{\,n+1}\varphi_0]\Pi_*(\bar{0}, \bar{1}, \ldots, \bar{n}).$$

Hint. Use Exercise 2.5.

3. Let $\langle\mathcal{F}, I, \Pi_*, L_*, R_*\rangle$ be an arbitrary G-space. Prove that a least element o in $\mathcal{F}$ exists, and the equality $\theta o=o$ holds for all θ in $\mathcal{F}$. Write an explicit expression for the element o.

4. Let $\langle\mathcal{F}, I, \Pi_*, L_*, R_*\rangle$ be an arbitrary G-space, and let Π_*' be the binary operation in $\mathcal{F}$ defined by

$$\Pi_*'(\varphi, \psi)=\Pi_*(\psi, \varphi).$$

Prove that $\langle\mathcal{F}, I, \Pi_*', R_*, L_*\rangle$ is also a G-space.

6. Left-homogeneous mappings and least fixed points connected with them

For the time being, we shall suppose that a semigroup $\mathcal{F}$ is given. Two definitions will be formulated under this assumption.

Definition 1. Let m be a positive integer, and Γ be a mapping of $\mathcal{F}^m$ into $\mathcal{F}$. The mapping Γ is called *left-homogeneous* iff

$$\Gamma(\varkappa\tau_1, \ldots, \varkappa\tau_m)=\varkappa\Gamma(\tau_1, \ldots, \tau_m)$$

for all $\varkappa, \tau_1, \ldots, \tau_m$ in $\mathcal{F}$.

For example, if $\mathcal{F}$ is the semigroup of an operative space $\langle\mathcal{F}, I, \Pi_*, L_*, R_*\rangle$ then the operation Π_* (in its initial form - with only two arguments) is a left-homogene-

---

[61] The case of $n=1$ corresponds to Lemma 3 in the paper Georgieva [1980].

ous mapping of $\mathcal{F}^2$ into $\mathcal{F}$ (and Π_* with m arguments is a left-homogeneous mapping of $\mathcal{F}^m$ into $\mathcal{F}$).

Definition 2. Let m be a positive integer. An *m-ary join mechanism in* $\mathcal{F}$ is a $m+1$-tuple

(1) $$\langle \Xi, \zeta_1, \ldots, \zeta_m \rangle,$$

where Ξ is a mapping of $\mathcal{F}^m$ into $\mathcal{F}$, $\zeta_1, \ldots, \zeta_m$ are elements of $\mathcal{F}$, and

$$\Xi(\tau_1, \ldots, \tau_m)\zeta_i = \tau_i, \quad i = 1, \ldots, m,$$

for all $\tau_1, \ldots, \tau_m$ in $\mathcal{F}$.

For example, if again $\mathcal{F}$ is the semigroup of an operative space $\langle \mathcal{F}, I, \Pi_*, L_*, R_* \rangle$, and Ξ is Π_*, considered as an m-ary operation, then $\langle \Xi, \overline{0}, \overline{1}, \ldots, \overline{m-2}, R_*^{m-1} \rangle$ is an m-ary join mechanism.

Here is a statement which connects the notions introduced by the above definitions.

Proposition 1. Let m be a positive integer, Γ be a left-homogeneous mapping of $\mathcal{F}^m$ into $\mathcal{F}$, and $\langle \Xi, \zeta_1, \ldots, \zeta_m \rangle$ be an m-ary join mechanism in $\mathfrak{S}_*$. Then for all $\tau_1, \ldots, \tau_m$ in $\mathcal{F}$ the equality

$$\Gamma(\tau_1, \ldots, \tau_m) = \Xi(\tau_1, \ldots, \tau_m)\Gamma(\zeta_1, \ldots, \zeta_m)$$

holds.

Proof. Let $\tau_1, \ldots, \tau_m$ be arbitrary elements of $\mathcal{F}$, and let $\varkappa = \Xi(\tau_1, \ldots, \tau_m)$. Then

$$\Gamma(\tau_1, \ldots, \tau_m) = \Gamma(\varkappa\zeta_1, \ldots, \varkappa\zeta_m) = \varkappa\Gamma(\zeta_1, \ldots, \zeta_m). \blacksquare$$

Corollary 1. In an operative space, each left-homogeneous mapping is monotonically increasing.

Proof. We can use the operation Π_* as Ξ, and Π_* is monotonically increasing. $\blacksquare$

From now on in this section, a G-space

$$\mathfrak{S}_* = \langle \mathcal{F}, I, \Pi_*, L_*, R_* \rangle,$$

(in the sense of Definition 5.2) is supposed to be given.

Theorem 1. Let n be a positive integer, let Γ be a left-homogeneous mapping of $\mathcal{F}^{n+1}$ into $\mathcal{F}$, and let

(1) $$\gamma = \Gamma(\overline{1}, \ldots, \overline{n}, \overline{0}).$$

Then, for all $\theta_1, \ldots, \theta_n, \sigma$ in $\mathcal{F}$, $\mu\tau.\Gamma(\theta_1, \ldots, \theta_n, \tau\sigma)$ exists, and the following equality holds

$$\mu\tau.\,\Gamma(\theta_1, \ldots, \theta_n, \tau\sigma) = \Pi_*(\theta_1, \ldots, \theta_n)[\gamma\sigma]\gamma.$$

In particular,

(2) $$\mu\tau.\,\Gamma(\theta_1, \ldots, \theta_n, \tau) = \Pi_*(\theta_1, \ldots, \theta_n)[\gamma]\gamma.$$

Proof. One easily checks the equality

$$\Gamma(\theta_1, \ldots, \theta_n, \tau\sigma) = \Pi_*(\tau\sigma, \Pi_*(\theta_1, \ldots, \theta_n))\gamma.$$

Using this equality and Proposition 5.3, we get the needed conclusion.■

Corollary 2. If n is a positive integer and Γ is a left-homogeneous mapping of $\mathcal{F}^{n+1}$ into $\mathcal{F}$, then the mapping Δ of $\mathcal{F}^n$ into $\mathcal{F}$, defined by

$$\Delta(\theta_1, \ldots, \theta_n) = \mu\tau.\,\Gamma(\theta_1, \ldots, \theta_n, \tau),$$

is also left-homogeneous.

Corollary 3. Let $\langle\mathcal{F}, I, \mathcal{C}, \Pi, L, R, \Sigma, T, F\rangle$ be an iterative combinatory space. Then, for all σ, χ in $\mathcal{F}$, the equalities

$$[\sigma, \chi] = [\gamma_1\sigma]\gamma_1 = [\gamma_2]\gamma_2$$

hold, where $\gamma_1 = (\chi \rightarrow \bar{0}, \bar{1})$, $\gamma_2 = (\chi \rightarrow \bar{0}\sigma, \bar{1})$.

Proof. We apply Theorem 1 to the mappings Γ_1 and Γ_2 defined by $\Gamma_1(\theta, \tau) = (\chi \rightarrow \tau, \theta)$, $\Gamma_2(\theta, \tau) = (\chi \rightarrow \tau\sigma, \theta)$, and we set $\theta = I$ in the obtained equalities.■

In the proof of the above corollary, we applied Theorem 1 to a left-homogeneous mapping Γ such that it was clear how to find the corresponding least solution without application of this theorem (the theorem was used only for obtaining a new expression for the solution). Now we shall give an example, where the situation is different.

Example 1. Let again $\langle\mathcal{F}, I, \mathcal{C}, \Pi, L, R, \Sigma, T, F\rangle$ be an iterative combinatory space, and let Γ be the mapping of $\mathcal{F}^3$ into $\mathcal{F}$ defined by the equality

$$\Gamma(\theta_1, \theta_2, \tau) = (\chi_0 \rightarrow (\chi_1 \rightarrow \tau\sigma_1, \theta_1), (\chi_2 \rightarrow \tau\sigma_2, \theta_2)),$$

where $\chi_0, \chi_1, \chi_2, \sigma_1, \sigma_2$ are some given elements of $\mathcal{F}$. Then the application of Theorem 1 gives the equality

$$\mu\tau.\,\Gamma(\theta_1, \theta_2, \tau) = \Pi_*(\theta_1, \theta_2)[\gamma]\gamma,$$

where

$$\gamma = \Gamma(\bar{1}, \bar{2}, \bar{0}) = (\chi_0 \rightarrow (\chi_1 \rightarrow \bar{0}\sigma_1, \bar{1}), (\chi_2 \rightarrow \bar{0}\sigma_2, \bar{2})).$$

However, it is not seen, say, how $\mu\tau.\,\Gamma(\theta_1, \theta_2, \tau)$ could be expressed by means of composition, branching and iteration using only $\chi_0, \chi_1, \chi_2, \sigma_1, \sigma_2, \theta_1, \theta_2$ and possibly L, R, T, F (the definition of $\bar{0}, \bar{1}, \bar{2}$ makes use of combination).

Theorem 1 and Corollary 2, together with Theorem 4.1, allow to find by successive elimination the least solution $\langle \tau_1, \ldots, \tau_m \rangle$ of an arbitrary system of the form

(3) $\quad \tau_i \geq \Gamma_i(\theta_1, \ldots, \theta_n, \tau_1, \ldots, \tau_m), \quad i=1, \ldots, m,$

where $\Gamma_1, \ldots, \Gamma_m$ are left-homogeneous mappings of $\mathcal{F}^{m+n}$ into $\mathcal{F}$. The application of Theorem 4.1 shows that such a least solution exists, and the obtained expressions for $\tau_1, \ldots, \tau_m$ are left-homogeneous with respect to $\theta_1, \ldots, \theta_m$. However, there is a shorter way to reach a similar result, and with simpler (in some respect) expressions for $\tau_1, \ldots, \tau_m$.

Theorem 2 (Generalization of equality (2)). Let m and n be positive integers, and let $\Gamma_1, \ldots, \Gamma_m$ be left-homogeneous mappings of $\mathcal{F}^{m+n}$ into $\mathcal{F}$. Let

$$\gamma_i = \Gamma\,(\bar{1}, \ldots, \bar{n}, \bar{0}\,\bar{0}, \ldots, \bar{0}\,\overline{m-2}, \bar{0}\,R_*^{\,m-1}), \qquad i=1, \ldots, m,$$

$$\gamma = \Pi_*(\gamma_1, \ldots, \gamma_m).$$

Then, for every choice of $\theta_1, \ldots, \theta_n$ in $\mathcal{F}$, the system of inequalities (3) has a least solution with respect to $\tau_1, \ldots, \tau_m$, and this least solution is given by the expressions

$$\tau_i = \Pi_*(\theta_1, \ldots, \theta_n)[\gamma]\gamma\,\overline{i-1}, \quad i=1, \ldots, m-1,$$
$$\tau_m = \Pi_*(\theta_1, \ldots, \theta_n)[\gamma]\gamma R_*^{\,m-1}.$$

Proof. In order to apply Proposition 4.3 (with $\theta_1, \ldots, \theta_n$ as parameters), we define a mapping Γ of $\mathcal{F}^{n+1}$ into $\mathcal{F}$ by means of the equality

$$\Gamma(\theta_1, \ldots, \theta_n, \tau) = \Gamma_0(\theta_1, \ldots, \theta_n, \tau\bar{0}, \tau\bar{1}, \ldots, \tau\,\overline{m-2}, \tau R_*^{\,m-1}),$$

where

$$\Gamma_0(\theta_1, \ldots, \theta_n, \tau_1, \ldots \tau_m) = \Pi_*(\Gamma_1(\theta_1, \ldots, \theta_n, \tau_1, \ldots \tau_m), \ldots, \Gamma_m(\theta_1, \ldots, \theta_n, \tau_1, \ldots \tau_m)).$$

It is easy to check that equality (1) holds. Since Γ is left-homogeneous, Theorem 1 can be applied, and we get the equality (2). Now an application of Proposition 4.3 (having in mind Corollary 1) immediately yields the needed result. ∎

Corollary 4. Let m and n be positive integers, and let $\Gamma_1, \ldots, \Gamma_m$ be left-homogeneous mappings of $\mathcal{F}^{m+n}$ into $\mathcal{F}$. Then there are left-homogeneous mappings $\Delta_1, \ldots, \Delta_m$ of $\mathcal{F}^n$ into $\mathcal{F}$ such that, for every choice of $\theta_1, \ldots, \theta_n$ in $\mathcal{F}$, $\langle\Delta_1(\theta_1, \ldots, \theta_n), \ldots, \Delta_m(\theta_1, \ldots, \theta_n)\rangle$ is the least solution of the system of inequalities (3).

Example 2. Suppose again that $\langle\mathcal{F}, I, \mathcal{C}, \Pi, L, R, \Sigma, T, F\rangle$ is an iterative combinatory space. Consider the system of the two inequalities

$$\tau_1 \geq (\chi_1 \to \tau_1\sigma_1, \theta),$$
$$\tau_2 \geq (\chi_2 \to \tau_2\sigma_2, \tau_1\alpha),$$

where $\chi_1, \chi_2, \sigma_1, \sigma_2, \alpha$ are some given elements of $\mathcal{F}$. It is not difficult to find the least solution of this system by the elimination method based on Theorem 4.1. Namely, we can eliminate τ_1 making use of the fact that

$$\mu\tau_1 . (\chi_1 \to \tau_1\sigma_1, \theta) = \theta[\sigma_1, \chi_1].$$

Thus we reduce the system to the inequality

$$\tau_2 \geq (\chi_2 \to \tau_2\sigma_2, \theta[\sigma_1, \chi_1]\alpha),$$

and then we can use the fact that

$$\mu\tau_2 . (\chi_2 \to \tau_2\sigma_2, \theta[\sigma_1, \chi_1]\alpha) = \theta[\sigma_1, \chi_1]\alpha[\sigma_2, \chi_2].$$

Hence the least solution of the given system is

$$\tau_1 = \theta[\sigma_1, \chi_1], \quad \tau_2 = \theta[\sigma_1, \chi_1]\alpha[\sigma_2, \chi_2].$$

Note however that the expression for τ_2 contains two applications of iteration, and, on the other hand, if we find the least solution by application of Theorem 2, the corresponding expression will contain only one application of iteration. Indeed, the least solution according to Theorem 2 is

$$\tau_1 = \theta[\gamma]\gamma\bar{0}, \quad \tau_2 = \theta[\gamma]\gamma R_*,$$

where $\gamma = \Pi_*((\chi_1 \to \bar{0}^2\sigma_1, \bar{1}), (\chi_2 \to \bar{0}R_*\sigma_2, \bar{0}^2))$. Of course, a comparison of the two expressions for τ_2 in the case $\theta = I$ gives the equality

$$[\sigma_1, \chi_1]\alpha[\sigma_2, \chi_2] = [\gamma]\gamma R_*,$$

i.e. the equality

$$[\sigma_1, \chi_1]\alpha[\sigma_2, \chi_2] = R[\gamma R, L]\gamma R_*.$$

There is a connection of some of the considered systems with goto-programs. Suppose $\mathcal{F}$ is the partially ordered semigroup $\mathcal{F}_p(M)$ of all partial mappings of a given set M into itself, and, of course, $I=I_M$ holds. Suppose a variable V for elements of M and a finite set L of labels are chosen, L having more than one element, and a label e from L is chosen to be the terminal one (we do not suppose a choice of initial label to be made, since this is unessential for our purpose). We shall consider any program whose instructions are of the forms

(i) l: $V:=\varphi(V)$; goto l'

(ii) l: if $P(V)$ then goto l' else goto l''

(iii) e: end

with $l, l', l'' \in L$, assuming that, for each label, there is exactly one instruction beginning with this label, and all φ and P occurring in the instructions are elements of $\mathcal{F}$ and partial predicates on M, respectively. For transforming the above description of the considered program into description of a mathematical object, we shall denote by $\mathcal{P}$ the set of all partial predicates om M, and we shall represent the program by a function A whose domain is the set $L\setminus\{e\}$ and whose values belong to the union of the sets $\mathcal{F}\times L$ and $\mathcal{P}\times L^2$, assuming that $A(l)=\langle\varphi, l'\rangle$ iff instruction (i) occurs in the program, $A(l)=\langle P, l', l''\rangle$ iff instruction (ii) occurs in the program. For defining the semantics of such a program, we consider the partial mapping S of the set $L\times M$ into itself defined as follows:

$$S(\langle l, s\rangle)=\begin{cases}\langle l', t\rangle \text{ if } A(l)=\langle\varphi, l'\rangle, & \varphi(s) = t,\\ \langle l', s\rangle \text{ if } A(l)=\langle P, l', l''\rangle, & P(s) =\text{true},\\ \langle l'', s\rangle \text{ if } A(l)=\langle P, l', l''\rangle, & P(s) =\text{false}\end{cases}$$

($S(\langle l, s\rangle)$ is considered to be not defined if $l=e$, or $A(l)=\langle\varphi, l'\rangle$ $s\notin\mathrm{dom}\,\varphi$, or $A(l)=\langle P, l', l''\rangle$ and $s\notin\mathrm{dom}\,P$). Suppose u, v are elements of M, and $l\in L\setminus\{e\}$. Then v is called *the result of execution of the program starting with initial state* $\langle l, u\rangle$ iff there is a finite sequence of elements of $L\times M$ beginning with $\langle l, u\rangle$ and ending with $\langle e, v\rangle$ such that each term in this sequence after the initial one is equal to the value of S at the previous one.

Now suppose that $L=\{l_0, l_1, \ldots, l_m\}$, where $l_0, l_1, \ldots, l_m$ are distinct, and $l_0=e$. For each i in the set $\{1, \ldots, m\}$, let ι_i be the partial function in M defined in the following way: $\iota_i(u)=v$ iff v is the result of execution of the program starting with initial state

$\langle l_i, u\rangle$. The functions $\iota_1, \ldots, \iota_m$ are called *the tail functions of the given program* (cf. the papers Mazurkiewicz [1971], Blikle [1972, 1972a, 1973]). They form the least solution of a system of inequalities in $\mathcal{F}$, namely

(4) $\quad \tau_i \geq \Gamma_i(I, \tau_1, \ldots, \tau_m), \quad i=1, \ldots, m,$

where the mappings $\Gamma_1, \ldots, \Gamma_m$ of $\mathcal{F}^{m+1}$ into $\mathcal{F}$ are defined in the following way:

a) if $A(l_i)=\langle\varphi, l_j\rangle$ then

$$\Gamma_i(\tau_0, \tau_1, \ldots, \tau_m)=\tau_j\varphi;$$

b) if $A(l_i)=\langle P, l_j, l_k\rangle$ then

$$\Gamma_i(\tau_0, \tau_1, \ldots, \tau_m)=(P \rightarrow \tau_j, \tau_k),$$

assumed the arrow here has the usual meaning as denotation of branching controlled by a predicate. We shall call (4) *the characteristic system* of the given program.

It is clear that the mappings Γ_i defined in the above way are left-homogeneous. Hence the results proven in this section are applicable to them. In particular, Theorem 2 can be applied to the system (4). It is seen thus that the tail functions can be expressed by means of composition, Π_* and iteration using only some relatively simple elements of $\mathcal{F}$, which are constructed correspondingly to the instructions of the program. Therefore it is justified to regard Theorem 2 as a generalization of a result from the paper Böhm and Jacopini [1966] (cf. also Cooper [1967]) about the equivalence of goto-programs to structured ones.

Example 3. Suppose $\mathfrak{S}_*$ is the companion operative space of a combinatory space $\mathfrak{S}_p(\mathfrak{U})$ of the kind considered in Example 1.2. Let us set $\theta=I$ in the system of two inequalities from Example 2, and let us introduce additional unknowns τ_3, τ_4, τ_5 for the expressions $\tau_1\sigma_1, \tau_2\sigma_2, \tau_1\alpha$, respectively, together with corresponding inequalities (written below). Then we obtain the following system of five inequalities (equivalent to the initial system in the sense of the elimination from Section 4):

$$\tau_1 \geq (\chi_1 \rightarrow \tau_3, I),$$
$$\tau_2 \geq (\chi_2 \rightarrow \tau_4, \tau_5),$$
$$\tau_3 \geq \tau_1\sigma_1,$$
$$\tau_4 \geq \tau_2\sigma_2,$$
$$\tau_5 \geq \tau_1\alpha.$$

This new system is the characteristic system of the following program:

l_1: if $H(\chi_1(V))$ then goto l_3 else goto l_6
l_2: if $H(\chi_2(V))$ then goto l_4 else goto l_5
l_3: $V:=\sigma_1(V)$; goto l_1
l_4: $V:=\sigma_2(V)$; goto l_2
l_5: $V:=\alpha(V)$; goto l_1
e: end

The flow diagram of this program is shown on Figure 1.[62] This diagram makes intuitively visible the fact that

$$\iota_1=[\sigma_1,\chi_1],\quad \iota_2=[\sigma_1,\chi_1]\alpha[\sigma_2,\chi_2]$$

for the considered program.

Remark 1. In our treatment of goto-programs, we restricted ourselves only to instructions of the forms (i), (ii), (iii). If more complicated instructions were allowed, it would be possible to write a program whose characteristic system is the system of inequalities from Example 2 itself (without additional unknowns introduced). Here is such a program (we omit the general exposition of the syntax and semantics of the larger class of programs to which this program belongs):

l_1: if $H(\chi_1(V))$ then $V:=\sigma_1(V)$; goto l_1 else end
l_2: if $H(\chi_2(V))$ then $V:=\sigma_2(V)$; goto l_2 else
$V:=\alpha(V)$; goto l_1

The discussed connection of the considered systems of inequalities with the algebraic study of goto-programs supports a point of view that such systems can be in some sense regarded as goto-programs in the G-spaces in question. We note also that the case of Theorem 2 with $n>1$ can be regarded as corresponding to goto-programs with more than one exit point.

We shall prove one more result which, in the case of iterative combinatory spaces, enables obtaining for the least solutions of systems of the form (3) certain expressions different from the expressions given by Theorem 2. For the sake of simplicity, we shall restrict ourselves to the

[62] No starting point for the execution of the program is indicated on the diagram, since none of the labels of the program is chosen to be the initial one; of course, it would be convenient to start execution from the uppermost if-statement, and thus to choose l_2 to be the initial label.

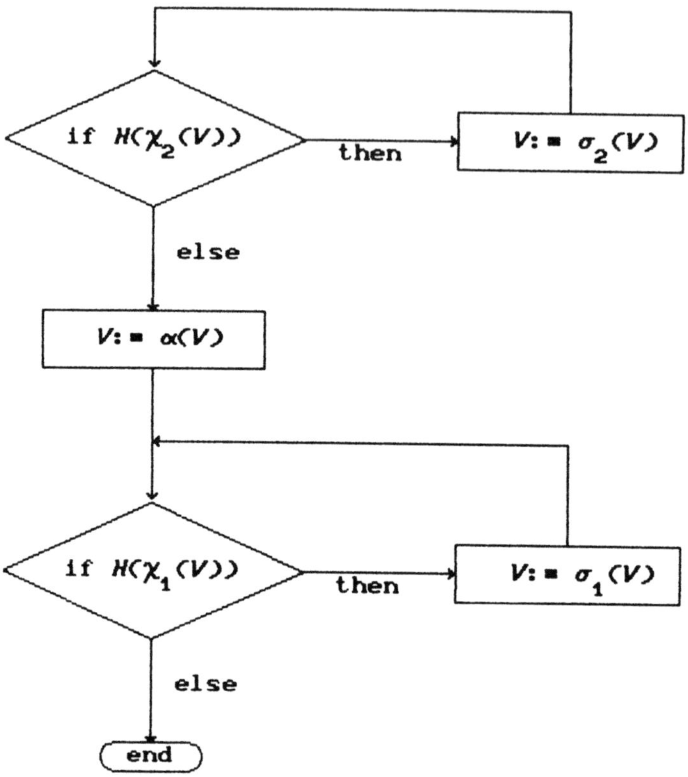

Figure 1. Flow diagram of the program in Example 3

case of $n=1$.

Theorem 3. Let $\mathfrak{S} = \langle\mathcal{F}, I, \mathcal{C}, \Pi, L, R, \Sigma, T, F\rangle$ be an iterative combinatory space, m be a positive integer, $\Gamma_1, \ldots, \Gamma_m$ be left-homogeneous mappings of $\mathcal{F}^{m+1}$ into $\mathcal{F}$, $\langle\Xi, \zeta_0, \zeta_1, \ldots, \zeta_m\rangle$ be an $m+1$-ary join mechanism in $\mathcal{F}$, the elements $\zeta_0, \zeta_1, \ldots, \zeta_m$ being normal. Let ε, χ and ρ be elements of $\mathcal{F}$ which satisfy the following conditions:

(i) $\varepsilon\zeta_0 = \zeta_0$ and

$\varepsilon\zeta_i = \Gamma_i(\zeta_0, \zeta_1, \ldots, \zeta_m)$, $i=1, \ldots, m$;

(ii) there are normal elements $\eta_0, \ldots, \eta_m$ such that $\chi\zeta_0 = F\eta_0$ and $\chi\zeta_i = T\eta_i$, $i=1, \ldots, m$;

(iii) $\rho\xi_0 = I$.[63]

Then the elements

$$\varphi_i = \rho[\varepsilon, \chi]\xi_i, \quad i=1, \ldots, m,$$

form the least solution $\langle\tau_1, \ldots, \tau_m\rangle$ of the system of inequalities

$$(5) \qquad \tau_i \geq \Gamma_i(I, \tau_1, \ldots, \tau_m), \quad i=1, \ldots, m.$$

Proof. Let $\iota = [\varepsilon, \chi]$. Then, by Proposition 3.4, we have the equalities

$$\iota\xi_0 = \xi_0,$$
$$\iota\xi_i = \iota\varepsilon\xi_i, \quad i=1, \ldots, m.$$

From here, we get

$$\varphi_i = \rho\iota\varepsilon\xi_i = \rho\iota\Gamma_i(\xi_0, \xi_1, \ldots, \xi_m) = \Gamma_i(\rho\iota\xi_0, \rho\iota\xi_1, \ldots, \rho\iota\xi_m) = \Gamma_i(I, \varphi_1, \ldots, \varphi_m)$$

for $i=1, \ldots, m$. Thus $\langle\varphi_1, \ldots, \varphi_m\rangle$ is a solution of the system (5). Suppose now $\langle\psi_1, \ldots, \psi_m\rangle$ is an arbitrary solution of this system, i.e.

$$\psi_i \geq \Gamma_i(I, \psi_1, \ldots, \psi_m), \quad i=1, \ldots, m.$$

We have to prove that $\psi_i \geq \varphi_i$, $i=1, \ldots, m$. In order to do this, we set

$$\psi = \Xi(I, \psi_1, \ldots, \psi_m)$$

and denote by $\mathcal{A}$ the set of all elements of $\mathcal{C}$ having the form $\xi_i x$, where $i \in \{0, 1, \ldots, m\}$, $x \in \mathcal{C}$. We shall prove the inequality

$$\psi \underset{\mathcal{A}}{\geq} (\chi \to \psi\varepsilon, \rho),$$

and the fact will be established that $\mathcal{A}$ is invariant with respect to ε. The validity of the above inequality is seen from the following equalities and inequalities:

$$\psi\xi_0 x = x = (\chi \to \psi\varepsilon, \rho)\xi_0 x,$$
$$\psi\xi_i x = \psi_i x \geq \Gamma_i(I, \psi_1, \ldots, \psi_m)x = \psi\Gamma_i(\xi_0, \xi_1, \ldots, \xi_m)x = \psi\varepsilon\xi_i x =$$

---

[63] The existence of such elements ε, χ, ρ follows from the assumption that $\langle\Xi, \xi_0, \xi_1, \ldots, \xi_m\rangle$ is a join mechanism in $\mathcal{F}$.

$$(\chi \to \psi\varepsilon, \rho)\xi_i x, \quad i=1, \ldots, m.$$

To prove the other statement concerning $\mathcal{A}$, let us suppose that λ_1 and λ_2 are elements of $\mathcal{F}$ satisfying the inequality $\lambda_1 \underset{\mathcal{A}}{\geq} \lambda_2$. Then the inequality $\lambda_1 \varepsilon \underset{\mathcal{A}}{\geq} \lambda_2 \varepsilon$ holds, as seen from the inequalities

$$\lambda_1 \xi_i \geq \lambda_2 \xi_i, \quad i=0, 1, \ldots, m,$$

using the monotonic increasing of the mappings $\Gamma_1, \ldots, \Gamma_m$ and the following equalities which are valid for $t=1, 2$:

$$\lambda_t \varepsilon \xi_0 x = \lambda_t \xi_0 x,$$

$$\lambda_t \varepsilon \xi_i x = \lambda_t \Gamma_i(\xi_0, \xi_1, \ldots, \xi_m)x =$$

$$\Gamma_i(\lambda_t \xi_0, \lambda_t \xi_1, \ldots, \lambda_t \xi_m)x, \quad i=1, \ldots, m.$$

Now we are in a position to apply the definition of iteration, and its application shows that

$$\psi \underset{\mathcal{A}}{\geq} \rho\iota.$$

From here, we get

$$\psi_i = \psi\xi_i \geq \rho\iota\xi_i = \varphi_i, \quad i=1, \ldots, m. \blacksquare$$

Corollary 5. Let $\mathfrak{S} = \langle \mathcal{F}, I, \mathcal{C}, \Pi, L, R, \Sigma, T, F \rangle$ be an iterative combinatory space, and $\langle \Xi, \xi_0, \xi_1, \xi_2 \rangle$ be a ternary join mechanism in $\mathcal{F}$, the elements ξ_0, ξ_1, ξ_2 being normal. Let $\alpha_0, \varepsilon_0, \chi_0, \rho_0, \varepsilon, \chi, \rho$ be elements of $\mathcal{F}$ satisfying the following conditions:

(i) $\varepsilon\xi_0 = \xi_0$,

$\varepsilon\xi_1 = \xi_2\alpha_0$,

$\varepsilon\xi_2 = (\chi_0 \to \xi_2\varepsilon_0, \xi_0\rho_0)$;

(ii) there are normal elements η_0, η_1, η_2 such that $\chi\xi_0 = F\eta_0$ and $\chi\xi_i = T\eta_i$, $i=1, 2$;

(iii) $\rho\xi_0 = I$.

Then the equalities

$$\rho_0[\varepsilon_0, \chi_0]\alpha_0 = \rho[\varepsilon, \chi]\xi_1, \tag{6}$$

$$\rho_0[\varepsilon_0, \chi_0] = \rho[\varepsilon, \chi]\xi_2 \tag{7}$$

hold.

Proof. Using the elimination method, it is easy to see that $\langle \rho_0[\varepsilon_0, \chi_0]\alpha_0, \rho_0[\varepsilon_0, \chi_0] \rangle$ is the least solution $\langle \tau_1, \tau_2 \rangle$ of the system

$$\tau_1 \geq \tau_2\alpha_0,$$
$$\tau_2 \geq (\chi_0 \to \tau_2\varepsilon_0, \rho_0).$$

This system can be written in the form

$$\tau_1 \geq \Gamma_1(I, \tau_1, \tau_2),$$
$$\tau_2 \geq \Gamma_2(I, \tau_1, \tau_2),$$

where

$$\Gamma_1 = \lambda\theta\tau_1\tau_2 . \tau_2\alpha_0, \quad \Gamma_2 = \lambda\theta\tau_1\tau_2 . (\chi_0 \to \tau_2\varepsilon_0, \theta\rho_0).$$

But conditions (i)-(iii) in the corollary are exactly the conditions (i)-(iii) of Theorem 3 for the case of $m = 2$ and for the above Γ_1, Γ_2. An application of the theorem for this case yields the equalities (6), (7).■

Example 4. Let $\mathfrak{S} = \langle \mathcal{F}, I, \mathcal{C}, \Pi, L, R, \Sigma, T, F \rangle$ be an iterative combinatory space, and let

$$\xi_0 = (F, I), \quad \xi_1 = (T, (T, I)), \quad \xi_2 = (T, (F, I)).$$

It is easy to construct a mapping Ξ of $\mathcal{F}^3$ into $\mathcal{F}$ such that $\langle \Xi, \xi_0, \xi_1, \xi_2 \rangle$ is a join mechanism, namely the mapping Ξ defined by

$$(8) \qquad \Xi(\tau_0, \tau_1, \tau_2) = (L \to (LR \to \tau_1 R^2, \tau_2 R^2), \tau_0 R).$$

Conditions (ii) and (iii) in the above corollary are obviously satisfied if we set $\chi = L$, $\rho = R$. Suppose now some elements α_0, ε_0, χ_0, ρ_0 of $\mathcal{F}$ are given. Then, by the corollary, the equalities

$$\rho_0[\varepsilon_0, \chi_0]\alpha_0 = R[\varepsilon, L](T, (T, I)),$$
$$\rho_0[\varepsilon_0, \chi_0] = R[\varepsilon, L](T, (F, I))$$

hold with

$$\varepsilon = \Xi(\xi_0, \xi_2\alpha_0, (\chi_0 \to \xi_2\varepsilon_0, \xi_0\rho_0)).$$

We shall not write explicitly the result of the actual substitution in the right-hand expression in (8), but we note that the following slightly different ε also satisfies the conditions (1):

$$\varepsilon = (L \to (LR \to \xi_2\alpha_0 R^2, (\chi_0 R^2 \to \xi_2 \gamma_0 R^2, \xi_0\rho_0 R^2)), \xi_0 R)$$

(this ε is equal to the other one in the cases when R is a normal element).

Of course, infinitely many other examples of a similar nature are possible. For some of them, see Exercise 5.

Exercises

1. Construct a G-space with a binary join mechanism whose first component is not monotonically increasing (hence not left-homogeneous).

Hint. In an appropriate operative space of the kind considered in Exercise 2.1, set

$$\Xi(\varphi, \psi) = \Pi_*(\varphi, \psi) \cup Z(\varphi, \psi),$$

where Z is a suitable mapping of $\mathcal{F}_m(M)$ into itself.

2. Let $\mathfrak{S}_* = \langle \mathcal{F}, I, \Pi_*, L_*, R_* \rangle$ be the companion operative space of an iterative combinatory space $\mathfrak{S} = \langle \mathcal{F}, I, \mathcal{C}, \Pi, L, R, \Sigma, T, F \rangle$. For any φ, ψ in $\mathcal{F}$, set

$$\Xi(\varphi, \psi) = [R_* \psi][R_*^{\,2} \varphi R]$$

(compare with Proposition 5.4). Prove that $\langle \Xi, L_* R_*, R_* L_* \rangle$ is a binary join mechanism in $\mathfrak{S}_*$, the mapping Ξ is monotonically increasing, but it is not left-homogeneous.

Hint. For proving that Ξ is not left-homogeneous, consider $\Xi(\varphi, \psi) R_*^{\,2}$.

3. Let $\mathfrak{S} = \langle \mathcal{F}, I, \mathcal{C}, \Pi, L, R, \Sigma, T, F \rangle$ be an iterative combinatory space, m be a positive integer, and $\langle \Xi, \xi_1, \ldots, \xi_m \rangle$ be an m-ary join mechanism in $\mathcal{F}$, the elements $\xi_1, \ldots, \xi_m$ being normal. Show that elements $\chi_1, \ldots, \chi_{m-1}$ and ρ of $\mathcal{F}$ can be generated from I, T, F by means of Ξ so that $\langle \Xi', \xi, \ldots, \xi \rangle$ is also an m-ary join mechanism in $\mathcal{F}$ if Ξ' is the mapping of $\mathcal{F}^m$ into $\mathcal{F}$ defined by

$$\Xi'(\tau_1, \ldots, \tau_m) = (\chi_1 \to \tau_1 \rho, (\chi_2 \to \tau_2 \rho, \ldots, (\chi_{m-2} \to \tau_{m-2} \rho, (\chi_{m-1} \to \tau_{m-1} \rho, \tau_m \rho)) \ldots)).$$

4. Write a goto-program corresponding to the inequality

$$\tau \geq (\chi_0 \to (\chi_1 \to \tau \sigma_1, \alpha_1), (\chi_2 \to \tau \sigma_2, \alpha_2))$$

in the case when $\mathfrak{S}_*$ is such as in Example 3. Draw also the corresponding flow diagram.

5. To obtain other examples of the sort of Example 4, apply Corollary 4 to the following cases:

(a) $\xi_0 = (F, I)$, $\xi_1 = (T, I)^2$, $\xi_2 = (T, I)(F, I)$;

(b) $\xi_0 = ((F, I), I)$, $\xi_1 = ((T, T), I)$, $\xi_2 = ((T, F), I)$;

(c) $\xi_0 = \bar{0}$, $\xi_1 = \bar{1}$, $\xi_2 = \bar{2}$.

7. Some formal systems for the theory of iterative combinatory spaces

The definition of the notion of a combinatory space given in Section 1 can be formalized in a certain first-order language with variables for the elements of the set $\mathcal{F}$ and variables for the elements of the set $\mathcal{C}$ from the combinatory space $\langle\mathcal{F}, I, \mathcal{C}, \Pi, L, R, \Sigma, T, F\rangle$. Unfortunately, the definition of iteration (Definition 3.1) uses a quantifier on arbitrary subsets of $\mathcal{C}$. In order to obtain a first-order formalization comprising also iteration we shall add to the above-mentioned language also variables for such subsets. The formalization which will be exposed below is essentially one which is used in the papers Skordev [1984 a, 1989].

Let f_0, f_1, f_2, ..., c_0, c_1, c_2, ... and s_0, s_1, s_2, ... be the variables for the elements of $\mathcal{F}$, for the elements of $\mathcal{C}$ and for the subsets of $\mathcal{C}$, respectively. The alphabet of the formal system contains also the letters L, R, T, F, the sign $\supset$, round and square brackets, the comma sign, the equality sign, the inequality sign $\geq$, the sign $\in$, the propositional connectives $\neg$, $\&$, $\vee$, $\Longrightarrow$, $\Longleftrightarrow$ and the quantifiers $\forall$ and $\exists$. The notion of a functional expression is defined by means of the following inductive definition:

(i) the empty string Λ is a *functional expression*;

(ii) whenever $\mathcal{Z}$, $\mathcal{U}$, $\mathcal{V}$, $\mathcal{W}$ are functional expressions, then the strings $\mathcal{Z}f_i$, $\mathcal{Z}c_i$ (for $i=0, 1, 2, \ldots$), $\mathcal{Z}$L, $\mathcal{Z}$R, $\mathcal{Z}$T, $\mathcal{Z}$F, $\mathcal{Z}(\mathcal{U}, \mathcal{V})$, $\mathcal{Z}(\mathcal{U}\supset\mathcal{V}, \mathcal{W})$ and $\mathcal{Z}[\mathcal{U}, \mathcal{V}]$ are also *functional expressions*.

The equalities and the inequalities between functional expressions, as well as the strings $c_i \in s_j$, $i, j=0, 1, 2, \ldots$, are the atomic formulas of the system, and arbitrary formulas are constructed from the atomic ones by using the propositional connectives and the quantifiers (quantification is permitted with respect to each of the three sorts of variables).

Let $\mathfrak{S}=\langle\mathcal{F}, I, \mathcal{C}, \Pi, L, R, \Sigma, T, F\rangle$ be an iterative combinatory space. A *valuation (of the variables) in* $\mathfrak{S}$ is an arbitrary mapping v having the set of all variables as its domain and transforming, for all i in $\mathbb{N}$, the variables f_i, c_i and s_i into elements of $\mathcal{F}$, elements of $\mathcal{C}$ and subsets of $\mathcal{C}$, respectively. The value $|\mathcal{Z}|$ of an arbitrary functional expression $\mathcal{Z}$ at a given valuation v in $\mathfrak{S}$ is an element of $\mathcal{F}$ defined recursively by means of the equalities

$$|\Lambda| = I, \quad |\mathcal{Z} f_i| = |\mathcal{Z}|\, v(f_i), \quad |\mathcal{Z} c_i| = |\mathcal{Z}|\, v(c_i),$$
$$|\mathcal{Z} L| = |\mathcal{Z}|\, L, \quad |\mathcal{Z} R| = |\mathcal{Z}|\, R, \quad |\mathcal{Z} T| = |\mathcal{Z}|\, T, \quad |\mathcal{Z} F| = |\mathcal{Z}|\, F,$$
$$|\mathcal{Z}(\mathcal{U}, \mathcal{V})| = |\mathcal{Z}|\, \Pi(|\mathcal{U}|, |\mathcal{V}|),$$
$$|\mathcal{Z}(\mathcal{U} \supset \mathcal{V}, \mathcal{W})| = |\mathcal{Z}|\, \Sigma(|\mathcal{U}|, |\mathcal{V}|, |\mathcal{W}|),$$
$$|\mathcal{Z}[\mathcal{U}, \mathcal{V}]| = |\mathcal{Z}|\, [|\mathcal{U}|, |\mathcal{V}|].$$

An easy induction shows that, for any two functional expressions $\mathcal{X}$ and $\mathcal{Y}$, the result $\mathcal{X}\mathcal{Y}$ of their concatenation is also a functional expression, and the equality $|\mathcal{X}\mathcal{Y}| = |\mathcal{X}|\,|\mathcal{Y}|$ holds.

The truth definition for atomic formulas is obvious: the formulas $\mathcal{U} = \mathcal{V}$, $\mathcal{U} \geq \mathcal{V}$ and $c_i \in s_j$ are regarded to be *true at the valuation* v iff $|\mathcal{U}| = |\mathcal{V}|$ in $\mathfrak{S}$, $|\mathcal{U}| \geq |\mathcal{V}|$ in $\mathfrak{S}$ and $v(c_i) \in v(s_j)$, respectively. Starting from the truth notion for atomic formulas, we expand it on arbitrary formulas in the usual way.

The axioms of the considered formal system are divided to *logical* and *special* ones. The logical axioms have the traditional forms for a Hilbert-style formalization of the predicate calculus. Namely, we take as logical axioms all formulas of the following kinds, where Φ, Ψ, Θ are arbitrary formulas of the considered system, ζ is a variable, and $\mathcal{Z}$ is a expression of the same type as ζ (i.e. $\mathcal{Z}$ is a functional expression in the case when ζ is a variable of the form f_i, and $\mathcal{Z}$ is a variable of the form c_i or of the form s_i in the case when ζ is a variable of the same form):

$$\Phi \Rightarrow (\Psi \Rightarrow \Phi), \quad (\Phi \Rightarrow \Psi) \Rightarrow ((\Phi \Rightarrow (\Psi \Rightarrow \Theta)) \Rightarrow (\Phi \Rightarrow \Theta)),$$
$$\Phi \,\&\, \Psi \Rightarrow \Phi, \quad \Phi \,\&\, \Psi \Rightarrow \Psi, \quad \Phi \Rightarrow \Phi \vee \Psi, \quad \Psi \Rightarrow \Phi \vee \Psi,$$
$$\Phi \Rightarrow (\Psi \Rightarrow \Phi \,\&\, \Psi), \quad (\Phi \Rightarrow \Theta) \Rightarrow ((\Psi \Rightarrow \Theta) \Rightarrow (\Phi \vee \Psi \Rightarrow \Theta)),$$
$$(\Phi \Rightarrow \Psi) \Rightarrow ((\Phi \Rightarrow \neg\Psi) \Rightarrow \neg\Phi), \quad \neg\neg\Phi \Rightarrow \Phi,$$
$$(\Phi \Leftrightarrow \Psi) \Rightarrow (\Phi \Rightarrow \Psi), \quad (\Phi \Leftrightarrow \Psi) \Rightarrow (\Psi \Rightarrow \Phi),$$
$$(\Phi \Rightarrow \Psi) \Rightarrow ((\Psi \Rightarrow \Phi) \Rightarrow (\Phi \Leftrightarrow \Psi)),$$
$$\forall \zeta\, \Phi \Rightarrow \Phi(\mathcal{Z}/\zeta), \quad \Phi(\mathcal{Z}/\zeta) \Rightarrow \exists \zeta\, \Phi.$$

The special axioms are a finite number of formulas, expressing in some sense the definition of the notion of an iterative combinatory space, and an infinite variety of formulas obtained from a suitable comprehension scheme. The special axioms of the first sort are the following ones, where, for the last of them, the convention is adopted that

$$\mathcal{U} \underset{s_0}{\geq} \mathcal{V}$$

is an abbreviation for the formula

$$\forall c_0 (c_0 \in s_0 \Rightarrow \mathcal{U} c_0 \geq \mathcal{V} c_0),$$

whenever $\mathcal{U}$ and $\mathcal{V}$ are functional expressions not containing c_0 (note that (1) - (16) correspond to 1.(1) - 1.(16)):

(0_0) $$f_0 \geq f_0 ,$$

(0_1) $$f_0 \geq f_1 \;\&\; f_1 \geq f_2 \Rightarrow f_0 \geq f_1 ,$$

(0_2) $$f_1 = f_2 \Leftrightarrow f_1 \geq f_2 \;\&\; f_2 \geq f_1 ,$$

(0_3) $$f_0 \geq f_1 \;\&\; f_2 \geq f_3 \Rightarrow f_0 f_2 \geq f_1 f_3 ,$$

(1) $$\forall c_0 (f_1 c_0 \geq f_2 c_0) \Rightarrow f_1 \geq f_2 ,$$

(2) $$\exists c_0 ((c_1, c_2) = c_0),$$

(3) $$L(c_1, c_2) = c_1 ,$$

(4) $$R(c_1, c_2) = c_2 ,$$

(5) $$(f_1, f_2) c_0 = (f_1 c_0, f_2 c_0),$$

(6) $$(\, , f_2 c_0) f_0 = (f_0, f_2 c_0),$$

(7) $$(c_0, \,) f_0 = (c_0, f_0),$$

(8) $$\neg (T = F),$$

(9) $$\exists c_0 (T c_1 = c_0),$$

(10) $$\exists c_0 (F c_1 = c_0),$$

(11) $$(T \supset f_1, f_2) = f_1 ,$$

(12) $$(F \supset f_1, f_2) = f_2 ,$$

(13) $$f_3 (f_0 \supset f_1, f_2) = (f_0 \supset f_3 f_1, f_3 f_2),$$

(14) $$(f_0 \supset f_1, f_2) c_0 = (f_0 c_0 \supset f_1 c_0, f_2 c_0),$$

(15) $$(\supset f_1 c_0, f_2 c_0) f_3 = (f_3 \supset f_1 c_0, f_2 c_0),$$

(16) $$f_0 \geq f_1 \;\&\; f_2 \geq f_3 \Rightarrow (\supset f_0, f_2) \geq (\supset f_1, f_3),$$

(17) $$[f_1, f_0] = (f_0 \supset [f_1, f_0] f_1, \,),$$

(18) $$\forall f_4 \forall f_5 (f_4 \underset{s_0}{\geq} f_5 \Rightarrow f_4 f_1 \underset{s_0}{\geq} f_5 f_1) \;\&$$
$$f_2 \underset{s_0}{\geq} (f_0 \supset f_2 f_1, f_3) \Rightarrow f_2 \underset{s_0}{\geq} f_3 [f_1, f_0].$$

The special axioms of the second sort are all formulas of the form

(19) $$\exists s_0 \forall c_0 (c_0 \in s_0 \Leftrightarrow \Phi),$$

where Φ is any formula without free occurrences of the

variable s_0.

The rules of inference of the considered formal system are the usual rules for a Hilbert-style formalization of the predicate calculus, namely modus ponens and the rules

$$\frac{\Theta \Longrightarrow \Phi}{\Theta \Longrightarrow \forall\zeta\,\Phi} \qquad \frac{\Phi \Longrightarrow \Theta}{\exists\zeta\,\Phi \Longrightarrow \Theta}$$

where Φ can be an arbitrary formula, ζ can be an arbitrary variable, and Θ can be an arbitrary formula without free occurrences of ζ.

The formal system described above will be denoted by A. Since all axioms of this system are identically true in any iterative combinatory space, the same holds for every formula deducible in A, i.e. the system is correct. We claim this system is sufficient for the formalization of most proofs in this book which concern iterative combinatory spaces.[64] First of all we shall note some properties of the equality which are deducible in A.

Proposition 1. The following formulas (expressing reflexivity, symmetry and transitivity of equality and certain special instances of the the replacement property) are deducible in A:

$$f_0=f_0,\ \ f_0=f_1 \Longrightarrow f_1=f_0,\ \ f_0=f_1\ \&\ f_1=f_2 \Longrightarrow f_0=f_2,$$
$$f_0=f_1\ \&\ f_2=f_3 \Longrightarrow (f_0 \ge f_2 \Longrightarrow f_1 \ge f_3),$$
$$f_0=f_1\ \&\ f_2=f_3 \Longrightarrow f_0f_2=f_1f_3.$$

Proof. The first three of the above formulas can be easily derived by using the special axioms $(O_0)-(O_2)$. The forth one can be derived by using (O_2) and (O_3). ∎

Proposition 2 (formalization of the statements of Remarks 1.3 and 1.4). The following five formulas are deducible in A:

---

[64] A problem arises in connection with the fact that some results about iterative combinatory spaces are obtained by using results about G-spaces (operative spaces with an iteration), since the language of the G-spaces is not a part of the language of the iterative combinatory spaces. The problem can be solved by restriction only to the companion operative spaces of the considered combinatory spaces. Also when results concerning more or less arbitrary mappings in $\mathcal{F}$ are used, one can restrict himself to mappings definable by means of functional expressions.

$$\forall c_0 (f_1 c_0 = f_2 c_0) \Longrightarrow f_1 = f_2,$$
$$(f_1, f_2 c_0) f_0 = (f_1 f_0, f_2 c_0),$$
$$(c_0, f_2) f_0 = (c_0, f_2 f_0),$$
$$(f_0 \supset f_1 c_0, f_2 c_0) f_3 = (f_0 f_3 \supset f_1 c_0, f_2 c_0),$$
$$f_0 \geq f_1 \ \& \ f_2 \geq f_3 \ \& \ f_4 \geq f_5 \Longrightarrow (f_0 \supset f_2, f_4) \geq (f_1 \supset f_3, f_5).$$

Proof. The deducibility of the first of the above formulas follows from the presence of axioms (0_2) and (1). The deducibility of the next three ones follows from Proposition 1 and the presence of the axioms (6), (7) and (15). The deducibility of the last formula can be seen by means of formalization of the proof of the corresponding implication in Remark 1.4 (Proposition 1 and the presence of axioms (0_0), (0_3) and (14) – (16) are used). ∎

Corollary 1. The formula

$$f_0 = f_1 \ \& \ f_2 = f_3 \ \& \ f_4 = f_5 \Longrightarrow (f_0 \supset f_2, f_4) = (f_1 \supset f_3, f_5)$$

is deducible in **A**.

Proposition 3 (monotonicity of combination). The formula

$$f_0 \geq f_1 \ \& \ f_2 \geq f_3 \Longrightarrow (f_0, f_2) \geq (f_1, f_3),$$

is deducible in **A**.

Proof. Formalization of the proof of Proposition 1.1. ∎

Corollary 2. The formula

$$f_0 = f_1 \ \& \ f_2 = f_3 \Longrightarrow (f_0, f_2) = (f_1, f_3)$$

is deducible in **A**.

A formalization of the proofs of Propositions 1.2 and 1.3 becomes now also possible, and we conclude that the formulas

$$c_0 c_1 = c_0, \quad (T c_0 \supset f_1, f_2) = f_1, \quad (F c_0 \supset f_1, f_2) = f_2$$

are also deducible in **A**.

Proposition 4 (minimality of iteration). The formula

$$f_2 \geq (f_0 \supset f_2 f_1, f_3) \Longrightarrow f_2 \geq f_3 [f_1, f_0]$$

is deducible in **A**.

Proof. Using a suitable axiom of the form (19), we see the deducibility of the formula

$$\exists s_0 \forall c_0 (c_0 \in s_0). \tag{20}$$

The following formula (expressing the statement of Proposition 1.9) is also deducible in **A**:

(21) $$\forall c_0 (c_0 \in s_0) \Longrightarrow (f_0 \underset{s_0}{\geq} f_1 \Longleftrightarrow f_0 \geq f_1)$$

(this can be seen using the presence of the axioms (0_0), (0_3) and (1)). From the deducibility of (21), using again the presence of the axioms (0_0) and (0_3), we infer the deducibility of the formula

$$\forall c_0 (c_0 \in s_0) \Longrightarrow \forall f_4 \forall f_5 (f_4 \underset{s_0}{\geq} f_5 \Longrightarrow f_4 f_1 \underset{s_0}{\geq} f_5 f_1).$$

From here, taking into account the axiom (18), we see the deducibility of the formula

$$\forall c_0 (c_0 \in s_0) \Longrightarrow (f_2 \underset{s_0}{\geq} (f_0 \supset f_2 f_1, f_3) \Longrightarrow f_2 \underset{s_0}{\geq} f_3 [f_1, f_0]).$$

Now the proof can be completed by using the deducibility of (20) and (21). ∎

Proposition 5 (monotonicity of iteration). The formula

$$f_0 \geq f_1 \ \& \ f_2 \geq f_3 \Longrightarrow [f_0, f_2] \geq [f_1, f_3],$$

is deducible in A.

Proof. Formalization of the proof of Proposition 3.3. ∎

Corollary 3. The formula

$$f_0 = f_1 \ \& \ f_2 = f_3 \Longrightarrow [f_0, f_2] = [f_1, f_3]$$

is deducible in A.

It is desirable to have also the formula

(22) $$c_1 = c_2 \Longrightarrow (c_1 \in s_0 \Longrightarrow c_2 \in s_0)$$

at our disposal. Unfortunately this formula is not deducible in the system A (cf. Exercise 2). This fact is no serious obstacle for the formalization of the proofs we are interested in, but anyway it makes a certain additional degree of carefulness necessary when treating problems of formalizability of proofs in A. Therefore the non-deducibility of the formula (22) can be regarded as a defect of the system A.

From the point of view of convenience for the formalization, the system A has also another defect. Suppose, for example, we have to express a statement of the form $\varphi x \in \mathcal{A}$, where φ is some element of $\mathcal{F}$, x is some element of $\mathcal{C}$, and $\mathcal{A}$ is some subset of $\mathcal{C}$ (an iterative combinatory space $\langle \mathcal{F}, I, \mathcal{C}, \Pi, L, R, \Sigma, T, F \rangle$ being given). Suppose also that φ, x and $\mathcal{A}$ are the values assigned to the variables f_0, c_0 and s_0, respectively. Then it would be natural to write $f_0c_0\in s_0$ for expressing the above statement. However, such an way of writing is not admissible in the system A, since $f_0c_0\in s_0$ is not a formula of that system. Therefore we would be forced to use a more complicated way of writing in the considered situation (for example, the formula $\exists c_1(f_0c_0=c_1\,\&\,c_1\in s_0)$ could be used).

Now an extension A′ of A will be indicated, such that both mentioned defects will be removed, and the extension will be shown to be conservative with respect to formulas not containing the sign $\in$. For obtaining A′, the syntax of A is extended by adopting atomic formulas of the form $\mathfrak{X}\in s_j$, where $\mathfrak{X}$ is an arbitrary functional expression, instead of the atomic formulas $c_i\in s_j$. The logical axioms and the rule of inference of the system A′ have the same form as the logical axioms and the rules of inference of A, with the difference that arbitrary formulas of A′ can be used in them instead of formulas of A. As to the special axioms of A′, they comprise the formulas (0_0-18), all formulas of the form (19), where Φ is any formula of A′ without free occurrences of s_0, and, in addition, the following two formulas:

$$f_1=f_2 \Longrightarrow (f_1\in s_0 \Longrightarrow f_2\in s_0),$$
$$f_0\in s_0 \Longrightarrow \exists c_0(f_0=c_0).$$

Since the formula (22) is deducible in the system A′, this system is not a conservative extension of A. However, the following weaker conservativeness property (mentioned above) is present:

Theorem 1. Whenever a formula of A′ not containing the sign $\in$ is deducible in A′, this formula is deducible in A.

Proof. Using induction on the construction of the formulas of the system A′, to each such formula the notion of a translation in A is defined. The definition consists of the following clauses:

1) Each atomic formula having the form of an equality or an inequality is its own translation.

2) If c_i does not enter in the functional expression $\mathfrak{X}$ then $\exists c_i(\mathfrak{X}=c_i\,\&\,c_i\in s_j)$ is a translation of the atomic formula $\mathfrak{X}\in s_j$.

3) If Φ and Ψ are translations of the formulas Φ' and Ψ', respectively, then the formulas $\neg\Phi$, $\Phi\,\&\,\Psi$, $\Phi\vee\Psi$, $\Phi\Longrightarrow\Psi$, $\Phi\Longleftrightarrow\Psi$, $\forall\zeta\,\Phi$ and $\exists\zeta\,\Phi$ are translations of the formulas $\neg\Phi'$, $\Phi'\,\&\,\Psi'$, $\Phi'\vee\Psi'$, $\Phi'\Longrightarrow\Psi'$, $\Phi'\Longleftrightarrow\Psi'$, $\forall\zeta\,\Phi'$

and $\exists\zeta\Phi'$, respectively.

Obviously, each formula of the system A′ has at least one translation in A, all translations of one and the same formula are congruent each other and each formula of A′ not containing the sign $\in$ is its only translation. Note also that the translation of a formula has the same free variables as the formula itself. The proof of the theorem will be done by proving that each formula deducible in A′ has a translation deducible in A (consequently, all its translations are deducible in A). The reasoning will be by induction.

First of all, we prove that all logical axioms of A′ have translations which are logical axioms of A. Obviously, each propositional logical axiom of A′ has a translation which is a propositional logical axiom of A. Consider now a logical axiom of A′ having the form $\forall\zeta\Phi' \Longrightarrow \Phi'(\mathcal{Z}/\zeta)$, where ζ is a variable, and $\mathcal{Z}$ is a expression of the same type as ζ. It is easy to observe the existence of a translation Φ of Φ' such that $\mathcal{Z}$ is free for ζ in Φ and $\Phi(\mathcal{Z}/\zeta)$ is a translation of $\Phi'(\mathcal{Z}/\zeta)$. Taking such a Φ and considering the formula $\forall\zeta\Phi \Longrightarrow \Phi(\mathcal{Z}/\zeta)$, we find again a translation which is a logical axiom of A. The case of a logical axiom of the form $\Phi'(\mathcal{Z}/\zeta) \Longrightarrow \exists\zeta\Phi'$ is similar.

The special axioms (0_0 − 17) do not contain the symbol $\in$, and hence they are translations of themselves. Hence these axioms of A′ again have translations which are axioms of A. As to the special axiom (18) of A′, we shall show the deducibility in A of the equivalence between this axiom and one of its translations; since (18) is an axiom of A too, the mentioned translation will turn out to be also deducible in A. To show the deducibility in A of such an equivalence, it is sufficient to apply the following remark to each subformula of (18) having the form $\mathcal{U} \geq_{s_0} \mathcal{V}$: if $\mathcal{U}$ and $\mathcal{V}$ are functional expressions not containing c_0 then the formula $\mathcal{U} \geq_{s_0} \mathcal{V}$ of A′ has a translation Θ in A such that $\mathcal{U} \geq_{s_0} \mathcal{V} \Longleftrightarrow \Theta$ is deducible in A. And to see the correctness of this remark, we take Θ to be the formula

$$\forall c_0 (\exists c_i (c_0 = c_i \,\&\, c_i \in s_0) \Longrightarrow \mathcal{U}c_0 \geq \mathcal{V}c_0),$$

where c_i is different from c_0 and c_i occurs neither in $\mathcal{U}$ nor in $\mathcal{V}$. Then the deducibility of $\mathcal{U} \geq_{s_0} \mathcal{V} \Longleftrightarrow \Theta$ in A is seen on the ground of the deducibility in A of the equivalences

(23) $$c_0 \in s_0 \Rightarrow \exists c_i (c_0 = c_i \,\&\, c_i \in s_0),$$

$$\forall c_0 (c_0 \in s_0 \Rightarrow \mathcal{U}c_0 \geq \mathcal{V}c_0) \Rightarrow (c_i \in s_0 \Rightarrow \mathcal{U}c_i \geq \mathcal{V}c_i),$$

$$c_0 = c_i \,\&\, \mathcal{U}c_i \geq \mathcal{V}c_i \Rightarrow \mathcal{U}c_0 \geq \mathcal{V}c_0 .$$

The last kind of axioms of A′ which have to be considered are the formulas of the form

(24) $$\exists s_0 \forall c_0 (c_0 \in s_0 \Leftrightarrow \Phi'),$$

where Φ' is a formula of A′ without free occurrences of s_0. Given such an axiom, we construct a translation of it deducible in A in the following way. We take a translation Φ of Φ' and a variable c_i different from c_0 and not occurring in Φ. From the fact that Φ is the translation of a formula and c_i is free for c_0 in Φ, the conclusion can be made that the implication

(25) $$c_0 = c_i \Rightarrow (\Phi \Leftrightarrow \Phi(c_i/c_0))$$

is deducible in A (this property of translations can be proved by induction on the construction of Φ', after proving the deducibility in A of each implication of the form

$$c_0 = c_i \Rightarrow \mathcal{Z} = \mathcal{Z}(c_i/c_0),$$

where $\mathcal{Z}$ is a functional expression). Consider now the following translation of (24):

$$\exists s_0 \forall c_0 (\exists c_i (c_0 = c_i \,\&\, c_i \in s_0) \Leftrightarrow \Phi).$$

Using the deducibility of (23) and (25) in A and the fact that (19) is an axiom of A, it is easy to show the deducibility in A of the above translation (one uses also the fact that

$$\forall c_0 (c_0 \in s_0 \Leftrightarrow \Phi) \Rightarrow (c_i \in s_0 \Leftrightarrow \Phi(c_i/c_0))$$

is deducible in A).

To complete the proof of the theorem, it remains to check that the inference rules of A′, whenever applied to formulas having translations deducible in A, always yield formulas with the same property. And no difficulties arise in checking this.∎

A denotation for the set $\mathcal{C}$ is also a thing which one could feel to be missing in the described systems, and especially in the system A′. The axioms (2), (9) and (10) of these systems illustrate a way for overcoming the lack of such a denotation. Namely, we can write $\exists c_i (\mathcal{Z} = c_i)$, with c_i not occurring in $\mathcal{Z}$, for expressing the statement that the value of $\mathfrak{Z}$ belongs to $\mathfrak{C}$. In particular, the notion of a normal element (cf. Definition II.1.2) can be formalized by using this way. Of course, there is no difficulty to enrich the system A' by introducing a denotation for $\mathfrak{C}$ together with a corresponding axiom (cf. Exercise 4).

For the sequel, a suitable notion of normal functional expression will be more usable then a straight-forward formalization of the notion of normal element. The definition is by induction:

(i) the empty string Λ is a *normal functional expression*;

(ii) whenever $\mathfrak{Z}$, $\mathfrak{U}$, $\mathfrak{V}$ are normal functional expressions, then the strings $\mathfrak{Z}c_i$ (for $i = 0, 1, 2, \ldots$), $\mathfrak{Z}$T, $\mathfrak{Z}$F and $\mathfrak{Z}(\mathfrak{U}, \mathfrak{V})$, are also *normal functional expressions*.

It is easy to see that the normal functional expressions are exactly those functional expressions which contain neither variables f_i nor L, R, $\supset$ or I. Clearly, the result of the concatenation of two normal functional expressions is again a normal functional expression. The next several propositions list the most useful properties of the normal expressions.

Proposition 6. If $\mathfrak{Z}$ is a normal functional expression, and the variable c_i does not occur in $\mathfrak{Z}$ then each formula of the kind $\exists c_i (\mathfrak{Z}c_j = c_i)$ is deducible in the system A.

Proof. Induction on the construction of $\mathfrak{Z}$. ∎

Corollary 4. If $\mathfrak{Z}$ is a normal functional expression, and $\mathfrak{S}$ is an iterative combinatory space, then the value of $\mathfrak{Z}$ in $\mathfrak{S}$ is a normal element for each valuation of the variables.

Of course, a direct proof of the above corollary is straight-forward.

Proposition 7. If $\mathfrak{U}$ and $\mathfrak{V}$ are normal functional expressions then $(\mathfrak{U}, \mathfrak{V})$ is also a normal functional expression, and the formulas $L(\mathfrak{U}, \mathfrak{V}) = \mathfrak{U}$, $R(\mathfrak{U}, \mathfrak{V}) = \mathfrak{V}$ are deducible in the system A.

Proof. Formalization of the proof of Proposition 1.4, using Proposition 6 in the places where the definition of the notion of a normal element was used. ∎

Corollary 5. If $\mathfrak{Z}$ is a normal functional expression then the formulas $L(f_0, \mathfrak{Z}) = f_0$, $R(\mathfrak{Z}, f_0) = f_0$ are deducible in the system A.

Proof. Formalization of the proof of Corollary 1.2. ∎

Proposition 8. If $\mathcal{Z}$ is a normal functional expression then the formulas

$$(f_1, f_2)\mathcal{Z} = (f_1\mathcal{Z}, f_2\mathcal{Z}),$$
$$(f_0 \supset f_1, f_2)\mathcal{Z} = (f_0\mathcal{Z} \supset f_1\mathcal{Z}, f_2\mathcal{Z}),$$

all formulas of the kind $c_i\mathcal{Z} = c_i$ and the formulas

$$(T\mathcal{Z} \supset f_1, f_2) = f_1, \quad (F\mathcal{Z} \supset f_1, f_2) = f_2$$

are deducible in the system A.

Proof. Formalization of the proofs of Propositions 1.5, 1.6 and 1.7. ∎

Proposition 9. If $\mathcal{Z}$ and $\mathcal{Y}$ are normal functional expressions then the formulas

$$f_0\mathcal{Z} = T\mathcal{Y} \Rightarrow [f_1, f_0]\mathcal{Z} = [f_1, f_0]f_1\mathcal{Z},$$
$$f_0\mathcal{Z} = F\mathcal{Y} \Rightarrow [f_1, f_0]\mathcal{Z} = \mathcal{Z}$$

are deducible in the system A.

Proof. Formalization of the proof of Proposition 3.4. ∎

We recommend to the reader to consider from the point of view of formalization of the proofs some more statements from the preceding sections (as an example for this, see Exercise 1 where the statement of Proposition 5.4 is written in the language of the system A).

Exercises

1. Show the deducibility in A of the formula

$$(f_0 \supset f_1, f_2) = R[R_* f_2 R, L]R[R_* R_* f_0 RR, L](f_0, L_*),$$

where L_* and R_* denote $(T,)$ and $(F,)$, respectively.

2. Let $\mathfrak{S} = \langle \mathcal{F}, I, \mathcal{C}, \Pi, L, R, \Sigma, T, F \rangle$ be an iterative combinatory space, M be a set and α be a surjection of M onto $\mathcal{C}$. Let an *M, α-valuation in $\mathfrak{S}$* be an arbitrary mapping v having the set of all variables as its domain and transforming, for all i in $\mathbb{N}$, the variables f_i, c_i and s_i into elements of $\mathcal{F}$, elements of M and subsets of M, respectively. Let the value $|\mathcal{Z}|$ of an arbitrary functional expression $\mathcal{Z}$ at a given M, α-valuation v in $\mathfrak{S}$ be an element of $\mathcal{F}$ defined recursively by means of the same equalities as in the definition of the ordinary valuations, except that the equality $|\mathcal{Z}c_i| = |\mathcal{Z}|v(c_i)$ is replaced by $|\mathcal{Z}c_i| = |\mathcal{Z}|\alpha(v(c_i))$. Let the truth definition for formulas be obtained from this definition in the same

way as in the case of ordinary valuations (of course, taking into account which are the admissible values of the variables c_i and s_i now). Show the correctness of the system A with respect to this semantic. Use this correctness to show the non-deducibility of the formula (22) in A.

3. Let the notion of translation of formulas of the system A' into formulas of the system A be defined in the same way as in the proof of Theorem 1. Show that, for each formula Φ' of A' and each translation Φ of this formula in A, the formula $\Phi' \Leftrightarrow \Phi$ is deducible in A', and Φ' is deducible in A' iff Φ is deducible in A.

4. Let A'_C be the system obtained from A' by means of the following modifications. We add to the alphabet of the system the letter C, and, for each functional expression $\mathfrak{X}$, using an atomic formula $\mathfrak{X} \in C$ is allowed (with the truth condition $|\mathfrak{X}| \in \mathcal{C}$). The additional axiom is $c_0 \in C$, and in the axioms $\forall \zeta \Phi \Longrightarrow \Phi(\mathfrak{X}/\zeta)$, $\Phi(\mathfrak{X}/\zeta) \Longrightarrow \exists \zeta \Phi$ we allow $\mathfrak{X}$ to be C in case ζ is a variable of the form s_i. Show that A'_C is a conservative extension of A', i.e. deducibility in A'_C is equivalent to deducibility in A' for formulas not containing C.

CHAPTER III

COMPUTABILITY IN ITERATIVE COMBINATORY SPACES

1. Explicit and fixed-point definability in partially ordered algebras

The notion of iterative combinatory space introduced in Chapter II encompasses as special cases some partially ordered semigroups of functions or function-like objects studied in Chapter I. In each of these semigroups, there was a corresponding notion of relative computability of an element of the semigroup in some set of its elements. All of these notions had similar definitions using the operations composition, combination and iteration in the considered semigroups. Since we have these operations in each iterative combinatory space, it is possible to give in the same spirit a general definition of relative computability in such a space, and this will be done in the next section. However, the corresponding general notion can be regarded as a special case of a certain other one, which is still more general and will be considered now. This will be done with the purpose of making the further exposition better motivated.

Some drill examples to the definitions in this section can be found in Exercises 1, 2, 3 after it.

We shall make use of the notion of *partially ordered algebra*. The term will mean any ordered pair $\langle \mathcal{F}, \mathcal{O} \rangle$, where $\mathcal{F}$ is some partially ordered non-empty set, and $\mathcal{O}$ is some set of monotonically increasing operations in $\mathcal{F}$. The notion of operation in $\mathcal{F}$ will be understood in the usual way, namely: each operation has a given arity which is a natural number, the n-ary operations are mappings of $\mathcal{F}^n$ into $\mathcal{F}$ when $n > 0$, and the 0-ary operations will be identified with elements of $\mathcal{F}$. For n-ary operations with $n > 0$, the monotonic increasing will be understood as in Section II.4, and all 0-ary operations will be considered monotonically increasing. If $\langle \mathcal{F}, \mathcal{O} \rangle$ is a partially ordered algebra then the set of the n-ary operations belonging to $\mathcal{O}$ will be denoted by $\mathcal{O}^{(n)}$.

Remark 1. Ordinary (non-ordered) algebras can be regarded as partially ordered algebras whose partial ordering reduces to the equality relation.

Definition 1. Let $\langle\mathcal{F}, \mathcal{O}\rangle$ be a partially ordered algebra. The set of the *explicitly definable elements* of $\mathcal{F}$ is introduced by means of the following two inductive clauses:

(i) all elements of $\mathcal{O}^{(0)}$ are considered *explicitly definable*;

(ii) whenever $\Omega\in\mathcal{O}^{(n)}$, $n>0$, and $\varphi_1, \ldots, \varphi_n$ are explicitly definable elements of $\mathcal{F}$, then $\Omega(\varphi_1, \ldots, \varphi_n)$ is also considered *explicitly definable*.

Of course, the partial ordering in $\mathcal{F}$ plays no role in the above definition, but, however, Remark 1 shows the harmlessness of our choice to study partially ordered algebras instead of ordinary ones.

The explicitly definable elements of $\mathcal{F}$ will be called also *explicitly definable operations of arity* 0. The notion of a explicitly definable operation of non-zero arity will be introduced in a similar way.

Definition 2. Let $\langle\mathcal{F}, \mathcal{O}\rangle$ be a partially ordered algebra, and l be a positive integer. The set of the *l-ary explicitly definable operations* in $\mathcal{F}$ is introduced by means of the following two inductive clauses:

(i) the operations $\lambda\tau_1\ldots\tau_l.\tau_i$, $i=1, \ldots, l$, and the operations

(1) $$\lambda\tau_1\ldots\tau_l.\,\omega,$$

where $\omega\in\mathcal{O}^{(0)}$, are considered *explicitly definable*;

(ii) whenever $\Omega\in\mathcal{O}^{(n)}$, $n>0$, and $\Phi_1, \ldots, \Phi_n$ are l-ary explicitly definable operations in $\mathcal{F}$, the operation

(2) $$\lambda\tau_1\ldots\tau_l.\,\Omega(\Phi_1(\tau_1, \ldots, \tau_l), \ldots, \Phi_n(\tau_1, \ldots, \tau_l))$$

is also considered *explicitly definable*.

Remark 2. An immediate corollary of the given definitions is that all operations belonging to $\mathcal{O}$ are explicitly definable. An easy induction shows that, for each explicitly definable element ω of $\mathcal{F}$, the corresponding operation (1) is also explicitly definable, and, for each n-ary explicitly definable operation Ω with $n>0$ and each n-tuple $\Phi_1, \ldots, \Phi_n$ of l-ary explicitly definable operations, the corresponding operation (2) is explicitly definable too.

Another statement easily provable by induction is that all explicitly definable operations in $\mathcal{F}$ are monotonically increasing.

A partially ordered algebra $\langle\mathcal{F}, \mathcal{O}\rangle$ can be extended by adding some explicitly definable operations to $\mathcal{O}$. Remark 2 shows that such an extension preserves the set of the explicitly definable operations.

Remark 3. The notion introduced by means of Definition 2 can be reduced to the one introduced by means of Definition 1. To show this, suppose a partially ordered algebra $\langle\mathcal{F}, \mathcal{O}\rangle$ and a positive integer l are given. Let $\mathcal{F}'$ consist of all monotonically increasing l-ary operations in $\mathcal{F}$, and let $\mathcal{F}'$ be supplied with the natural partial ordering in it. For each ω belonging to $\mathcal{O}^{(0)}$, let ω' be the corresponding element (1) of $\mathcal{F}'$. For each positive integer n and each Ω belonging to $\mathcal{O}^{(n)}$, let Ω' be the n-ary operation in $\mathcal{F}'$ such that, for any n-tuple $\Phi_1, \ldots, \Phi_n$ of elements of $\mathcal{F}'$, $\Omega'(\Phi_1, \ldots, \Phi_n)$ is the corresponding element (2) of $\mathcal{F}'$. Consider now the partially ordered algebra $\langle\mathcal{F}', \mathcal{O}'\rangle$, where $\mathcal{O}'$ consists of the elements $\lambda\tau_1\ldots\tau_l.\tau_i$ of $\mathcal{F}'$, $i=1, \ldots, l$, and of all elements ω' and operations Ω' corresponding to elements of $\mathcal{O}$. Then the l-ary explicitly definable operations in the algebra $\langle\mathcal{F}, \mathcal{O}\rangle$ are exactly the explicitly definable elements of the algebra $\langle\mathcal{F}', \mathcal{O}'\rangle$.

Among the explicitly definable operations certain very special ones will be singled out by means of the following definition.

Definition 3. Let $\langle\mathcal{F}, \mathcal{O}\rangle$ be a partially ordered algebra, and let l be a positive integer. An l-ary operation in $\mathcal{F}$ will be called *simple* iff this operation has some of the two forms described in clause (i) of Definition 2 or the form

$$\lambda\tau_1\ldots\tau_l.\,\Omega(\tau_{i_1}, \ldots, \tau_{i_n}), \tag{3}$$

where n is some positive integer, Ω belongs to $\mathcal{O}^{(n)}$ and $i_1, \ldots, i_n$ belong to $\{1, \ldots, l\}$.

The above definition will be used a bit later in the formulation of the next definition. The following property can be easily verified.

Proposition 1. Let $\langle\mathcal{F}, \mathcal{O}\rangle$ be a partially ordered algebra, l and m be a positive integers, Φ be an l-ary simple operation in $\mathcal{F}$ and let $j_1, \ldots, j_l$ be natural numbers from the set $\{1, \ldots, m\}$. Then the operation

$\lambda\theta_1\ldots\theta_m.\Phi(\theta_{j_1}, \ldots, \theta_{j_l})$ is also simple.

In many cases, much more operations than the explicitly definable ones can be defined by using least fixed points. Some operations defined in such a way will be called fixed-point definable (the more precise term should be "least-fixed-point definable", but it is somewhat long). Here is the rigorous definition of the notion of fixed-point definable operation of arity 0 (fixed-point definable element) in a partially ordered algebra.

Definition 4. Let $\langle\mathcal{F}, \mathcal{O}\rangle$ be a partially ordered algebra, and φ be an element of $\mathcal{F}$. The element φ is called *fixed-point definable* iff, for some positive integer l, there is a l-tuple $\Gamma_1, \ldots, \Gamma_l$ of simple l-ary operations in $\mathcal{F}$ such that the system of inequalities

(3) $$\tau_i \geq \Gamma_i(\tau_1, \ldots, \tau_l), \quad i=1, \ldots, l,$$

has a least solution $\langle\tau_1, \ldots, \tau_l\rangle$ in $\mathcal{F}^l$, and the component τ_1 of this solution is equal to φ.[65]

A parameterization of the above definition leads to the definition of a fixed-point definable operation in a partially ordered algebra.

Definition 5. Let $\langle\mathcal{F}, \mathcal{O}\rangle$ be a partially ordered algebra, n be a positive integer, and Φ be an n-ary operation in $\mathcal{F}$. The operation Φ is called *fixed-point definable* iff, for some positive integer l, there is a l-tuple $\Gamma_1, \ldots, \Gamma_l$ of simple $l+n$-ary operations in $\mathcal{F}$ such that, for each choice of $\theta_1, \ldots, \theta_n$ in $\mathcal{F}$, the system of inequalities

(4) $$\tau_i \geq \Gamma_i(\theta_1, \ldots, \theta_n, \tau_1, \ldots, \tau_l), \quad i=1, \ldots, l,$$

has a least solution $\langle\tau_1, \ldots, \tau_l\rangle$ in $\mathcal{F}^l$, and the component τ_1 of this solution is equal to $\Phi(\theta_1, \ldots, \theta_n)$.

Proposition 2. If $\langle\mathcal{F}, \mathcal{O}\rangle$ is a partially ordered algebra then all operations from $\mathcal{O}$ and all operations of the

---

[65] For the definition of the notion of least solution of such a system, cf. Section II.4. Note that, by Proposition II.4.1, if the system of inequalities (1) has a least solution, then it is the least solution also of the corresponding system of equations

$$\tau_i = \Gamma_i(\tau_1, \ldots, \tau_l), \quad i=1, \ldots, l.$$

Instead of "the component τ_1" one could equivalently write "some component" (due to the property from Proposition 1).

form $\lambda\theta_1\dots\theta_n.\theta_i$, $j=1, \dots, n$, are fixed-point definable.

Proof. If $\Omega \in \mathcal{O}^{(n)}$ and $n>0$, then, for each choice of $\theta_1, \dots, \theta_n$ in $\mathcal{F}$, $\Omega(\theta_1, \dots, \theta_n)$ is the least solution τ of the inequality $\tau \geq \Omega(\theta_1, \dots, \theta_n)$. The fixed-point definability of the elements of $\mathcal{O}^{(0)}$ and of the operations of the form $\lambda\theta_1\dots\theta_n.\theta_i$ is seen in a similar way.∎

Proposition 3. Let $<\mathcal{F}, \mathcal{O}>$ be a partially ordered algebra. If φ is a fixed-point definable element of $\mathcal{F}$ then, for each positive integer m, the operation $\lambda\theta_1\dots\theta_m.\varphi$ is also fixed-point definable. If Φ is a fixed-point definable n-ary operation in $\mathcal{F}$ with $n>0$ then, for each choice of the positive integer m and of the natural numbers $i_1, \dots, i_n$ belonging to $\{1, \dots, m\}$, the operation

$$\lambda\theta_1\dots\theta_m.\Phi(\theta_{i_1}, \dots, \theta_{i_n})$$

is also fixed-point definable.

Proof. Application of Definitions 4, 5 and Proposition 1.∎

Corollary 1. In any partially ordered algebra, all simple operations are fixed-point definable.

Remark 4. Somewhat later, the much stronger statement will be proved that all explicitly definable operations are fixed-point definable (there is no difficulty to prove it immediately, but we shall obtain it as a corollary from another result).

Proposition 4. In any partially ordered algebra, all fixed-point definable operations are monotonically increasing.

Proof. Application of Proposition II.4.2.∎

By Propositions 2 and 4, if we replace the set $\mathcal{O}$ of a partially ordered algebra $<\mathcal{F}, \mathcal{O}>$ by the set of all fixed-point definable operations in this algebra then we shall get another partially ordered algebra which is an enrichment of the given one.

Definition 6. Let $<\mathcal{F}, \mathcal{O}>$ be a partially ordered algebra, and let $\overline{\mathcal{O}}$ be the set of all fixed-point definable operations in $\mathcal{F}$. Then the partially ordered algebra $<\mathcal{F},\overline{\mathcal{O}}>$ will be called *the fixed-point enrichment of* $<\mathcal{F}, \mathcal{O}>$.

A natural question arising in connection with the above definition is what will happen when one applies formation of fixed-point enrichment twice. We shall show that the second application will produce nothing new. First we shall prove a slightly more precise result.

Theorem 1. Let $\langle\mathcal{F}, \mathcal{O}\rangle$ be a partially ordered algebra, and let $\langle\mathcal{F}, \bar{\mathcal{O}}\rangle$ be its fixed-point enrichment. Let l be a positive integer, n be a non-negative integer, and $\Gamma_1, \ldots, \Gamma_l$ belong to $\bar{\mathcal{O}}^{(l+n)}$. Then there are a natural number m and a $l+m$-tuple $\Gamma'_1, \ldots, \Gamma'_{l+m}$ of $l+m+n$-ary simple operations in $\langle\mathcal{F}, \mathcal{O}\rangle$ such that, for all $\theta_1, \ldots, \theta_n$ in $\mathcal{F}$, the system of inequalities (4) has a least solution $\langle\tau_1, \ldots, \tau_l\rangle$ in $\mathcal{F}^l$ iff the system

(5) $\quad \tau_i \geq \Gamma'_i(\theta_1, \ldots, \theta_n, \tau_1, \ldots, \tau_{l+m}), \quad i=1, \ldots, l+m,$

has a least solution $\langle\tau_1, \ldots, \tau_{l+m}\rangle$ in $\mathcal{F}^{l+m}$, and, if $\langle\varphi_1, \ldots, \varphi_{l+m}\rangle$ is the least solution of (5) in $\mathcal{F}^{l+m}$, then $\langle\varphi_1, \ldots, \varphi_l\rangle$ is the least solution of (4) in $\mathcal{F}^l$.

Proof. By Definitions 5 and 6, for each i from the set $\{1, \ldots, l\}$ a system of inequalities

(6) $$\xi_{i,j} \geq B_{i,j}(\theta_1, \ldots, \theta_n, \tau_1, \ldots, \tau_l, \xi_{i,1}, \ldots, \xi_{i,k_i}), \quad j=1, \ldots, k_i,$$

can be chosen, with $B_{i,1}, \ldots, B_{i,k_i}$ simple in $\langle\mathcal{F}, \mathcal{O}\rangle$, such that, for every fixed $\theta_1, \ldots, \theta_n, \tau_1, \ldots, \tau_l$ in $\mathcal{F}$, the system (6) has a least solution $\langle\xi_{i,1}, \ldots, \xi_{i,k_i}\rangle$ in $\mathcal{F}^{k_i}$, the component $\xi_{i,1}$ of this solution being equal to $\Gamma_i(\theta_1, \ldots, \theta_n, \tau_1, \ldots, \tau_l)$. We assume that the systems (6) corresponding to different numbers i have disjoint lists of unknowns $\xi_{i,j}$. Let $\tau_{l+1}, \ldots, \tau_{l+m}$ be all these unknowns, taken in the following order:

$$\xi_{1,1}, \ldots, \xi_{1,k_1}, \ldots, \xi_{l,1}, \ldots, \xi_{l,k_l}$$

(hence $m=k_1+\ldots+k_l$). We take (5) to be the system consisting of the inequalities

$$\tau_i \geq \xi_{i,1}, \quad i=1, \ldots, l,$$

and of all inequalities of all systems (6), written consecutively. Now we have to show that (4) is consentient with (5) with respect to least solutions in the sense described in the theorem. This can be done by applicating Theorem II.4.1 l times. Namely, we start by eliminating the unknowns $\xi_{l,j}$, $j=1, \ldots, k_l$, making use of the inequalities (6) with $i=l$.

The new system obtained through this elimination does not contain more the inequalities just mentioned, and the only change in the other inequalities is that

$$\tau_l \geq \xi_{l,1}$$

becomes

$$\tau_l \geq \Gamma_l(\theta_1, \ldots, \theta_n, \tau_1, \ldots, \tau_l).$$

By Theorem II.4.1, this new system has a least solution iff the system (5) has a least solution, and if $\langle \varphi_1, \ldots, \varphi_{l+m} \rangle$ is the least solution of (5) then the least solution of the new system can be obtained by deleting the last k_l members of $\langle \varphi_1, \ldots, \varphi_{l+m} \rangle$. The next step is the elimination of the unknowns $\xi_{l-1,j}$, $j=1, \ldots, k_{l-1}$, making use of the inequalities (6) with $i=l-1$. Then these inequalities drop out of the system and the inequality

$$\tau_{l-1} \geq \xi_{l-1,1}$$

becomes

$$\tau_{l-1} \geq \Gamma_{l-1}(\tau_1, \ldots, \tau_l, \theta_1, \ldots, \theta_n).$$

The new system of inequalities obtained thus is again consentient with (5) with respect to least solutions. Going on in the same manner, we consecutively eliminate the unknowns $\xi_{l-2,j}$, the unknowns $\xi_{l-3,j}$ and so on, and finally obtain the needed conclusion about the system (4).■

Corollary 2. Let $\langle \mathcal{F}, \mathcal{O} \rangle$ be a partially ordered algebra, and let $\langle \mathcal{F}, \overline{\mathcal{O}} \rangle$ be its fixed-point enrichment. Then the partially ordered algebras $\langle \mathcal{F}, \mathcal{O} \rangle$ and $\langle \mathcal{F}, \overline{\mathcal{O}} \rangle$ have one and the same set of fixed-point definable operations.

Proof. Since all fixed-point definable operations of $\langle \mathcal{F}, \mathcal{O} \rangle$ belong to $\overline{\mathcal{O}}$, by Proposition 2 (applied to $\langle \mathcal{F}, \overline{\mathcal{O}} \rangle$), all these operations are fixed-point definable in $\langle \mathcal{F}, \overline{\mathcal{O}} \rangle$.[66] Suppose now an arbitrary fixed-point definable operation in $\langle \mathcal{F}, \overline{\mathcal{O}} \rangle$ is given. Then this operation can be defined by means of a system of the form (3) or (4), with operations Γ_i simple in $\langle \mathcal{F}, \overline{\mathcal{O}} \rangle$. By Proposition 3, all these Γ_i belong to $\overline{\mathcal{O}}$, and therefore Theorem 1 can be applied. The new system of inequalities obtained according the theorem defines the same operation, and this shows the fixed-point definability of the operation in the partially ordered al-

---

[66] The same conclusion can be obtained also directly from Definitions 3, 5 and the inclusion $\mathcal{O} \subseteq \overline{\mathcal{O}}$ asserted in Proposition 2.

gebra $\langle\mathcal{F}, \mathcal{O}\rangle$. ∎

Corollary 3. Let $\langle\mathcal{F}, \mathcal{O}\rangle$ be a partially ordered algebra, and let $\langle\mathcal{F}, \overline{\mathcal{O}}\rangle$ be its fixed-point enrichment. Then the set $\overline{\mathcal{O}}$ is closed under substitution.

Proof. Let Φ_0 belongs to $\overline{\mathcal{O}}^{(m)}$, and $\Phi_1, \ldots, \Phi_m$ belong to $\overline{\mathcal{O}}^{(n)}$. We shall prove the fixed-point definability of the n-ary operation Φ in $\mathcal{F}$ defined by the equality

$$\Phi(\theta_1, \ldots, \theta_n) = \Phi_0(\Phi_1(\theta_1, \ldots, \theta_n), \ldots, \Phi_m(\theta_1, \ldots, \theta_n)),$$

(we neglect the small changes needed for the case of $m=0$ or $n=0$). For that purpose, consider the system of inequalities

$$\tau_0 \geq \Phi_0(\tau_1, \ldots, \tau_m),$$
$$\tau_i \geq \Phi_i(\theta_1, \ldots, \theta_n), \quad i=1, \ldots, n.$$

Since its least solution $\langle\tau_0, \tau_1, \ldots, \tau_m\rangle$ has first component $\Phi(\theta_1, \ldots, \theta_n)$, the operation Φ turns out to be fixed-point definable in $\langle\mathcal{F}, \overline{\mathcal{O}}\rangle$, and hence $\Phi \in \overline{\mathcal{O}}$. ∎

Corollary 4. In any partially ordered algebra, all explicitly definable operations are fixed-point definable.

Proof. Application of Definitions 1,2, Proposition 2 and the above corollary. ∎

The introduced notions and the proved results can be used in arbitrary partially ordered algebras, including such ones where not every system of inequalities of the form considered in the definition of fixed-point definability has a least solution. However, a special attention is deserved by the partially ordered algebras where all such systems have least solutions.

Definition 7. Let $\langle\mathcal{F}, \mathcal{O}\rangle$ be a partially ordered algebra. This algebra will be called *fixed-point precomplete* iff for each positive integer l, each natural number n, each l-tuple $\Gamma_1, \ldots, \Gamma_l$ of simple $l+n$-ary operations in $\mathcal{F}$ and each choice of $\theta_1, \ldots, \theta_n$ in $\mathcal{F}$, the corresponding system of inequalities (4) has a least solution $\langle\tau_1, \ldots, \tau_l\rangle$ in $\mathcal{F}^l$.

Definition 8. A partially ordered algebra will be called *fixed-point complete* iff it is fixed-point precomplete and all fixed-point definable operations in this algebra are explicitly definable.

The condition from the definition of precompleteness is equivalent to certain stronger conditions.

Proposition 5. If the word "simple" in the condition from Definition 7 is replaced by "explicitly definable" or by "fixed-point definable" then equivalent conditions arise.

Proof. Application of Theorem 1 and Corollary 4. ∎

An obvious necessary condition for the fixed-point precompleteness of a partially ordered algebra is the existence of a least element in this algebra (the inequality $\tau \geq \tau$ must have a least solution in a fixed-point precomplete partially ordered algebra). Of course, this condition is far from being sufficient. Certain sufficient conditions will be given in the next two propositions.

Proposition 6. Let $\langle\mathcal{F}, \mathcal{O}\rangle$ be a partially ordered algebra having the following three properties:

(i) there is a least element in $\mathcal{F}$;

(ii) each monotonically increasing infinite sequence of elements of $\mathcal{F}$ has a least upper bound;

(iii) the operations of $\mathcal{O}$ are continuous with respect to least upper bounds of monotonically increasing infinite sequences[67].

Then $\langle\mathcal{F}, \mathcal{O}\rangle$ is fixed-point precomplete.

Proof. If a system of the form (4) is given with all Γ_i simple in the partially ordered algebra $\langle\mathcal{F}, \mathcal{O}\rangle$ then, for any fixed $\theta_1, \ldots, \theta_n$ in $\mathcal{F}$, the Knaster–Tarski–Kleene Theorem (Theorem II.4.3) can be applied to the mapping Γ of

---

[67] In the case of operations with more than one argument, this can be understood in the sense of continuity with respect to each one of the arguments. It is easy to prove that such a continuity implies continuity with respect to all arguments taken together. For example, if Φ is a binary operation which is continuous with respect to each one of its both arguments, $\{\varphi_k\}_{k=0}^{\infty}$, $\{\psi_k\}_{k=0}^{\infty}$ are monotonically increasing sequences of elements of $\mathcal{F}$, and $\varphi=\sup\{\varphi_k\}_{k=0}^{\infty}$, $\psi=\sup\{\psi_k\}_{k=0}^{\infty}$, then $\varphi\psi=\sup\{\Phi(\varphi_k, \psi_k)\}_{k=0}^{\infty}$, since $\varphi\psi$ is obviously an upper bound of the last sequence, and whenever α is an arbitrary upper bound of it, then

$$\Phi(\varphi_i, \psi_j) \leq \alpha$$

for all $i, j \in \mathbb{N}$, and hence

$$\Phi(\varphi, \psi)=\sup\{\Phi(\varphi, \psi_j)\}_{j=0}^{\infty}=\sup\{\sup\{\Phi(\varphi_i, \psi_j)\}_{i=0}^{\infty}\}_{j=0}^{\infty} \leq \alpha.$$

$\mathcal{F}^l$ into itself defined by

$$\Gamma(<\tau_1, \ldots, \tau_l>) = <\tau_1', \ldots, \tau_l'>,$$

where

$$\tau_i' = \Gamma_i(\theta_1, \ldots, \theta_n, \tau_1, \ldots, \tau_l), \quad i=1, \ldots, l. \blacksquare$$

Proposition 7. Let $<\mathcal{F}, \mathcal{O}>$ be a partially ordered algebra, and let each chain in $\mathcal{F}$ (including the empty one) has a least upper bound. Then $<\mathcal{F}, \mathcal{O}>$ is fixed-point precomplete.

Proof. The same as the proof of Proposition 6, but using the Knaster–Tarski–Platek Theorem (Theorem II.4.4) instead of the Knaster–Tarski–Kleene one. ■

An way for obtaining fixed-point complete partially ordered algebras is the following one.

Proposition 8. The fixed-point enrichment of any fixed-point precomplete partially ordered algebra is fixed-point complete.

Proof. Application of Definition 6, Corollary 2 and Remark 2. ■

Of course, the fixed-point complete partially ordered algebras obtained according to Proposition 7 have infinitely many primitive operations. Fixed-point complete partially ordered algebras with finitely many primitive operations must be considered more interesting. One of the main results in this book will be to show the fixed-point completeness of certain partially ordered algebras corresponding naturally to iterative combinatory spaces and having finitely many primitive operations. This result will be formulated and proved further in this chapter.

It is appropriate to mention here also two very interesting other classes of fixed-point complete partially ordered algebras, which, too, are closely connected with the theory of computability. The first of these classes consists of the already mentioned L. Ivanov's iterative operative spaces (cf. Ivanov [1980, 1980 a, 1980 b, 1983, 1984, 1984 a, 1990], and especially Ivanov [1986]). Ivanov's theory can be successfully applied to the study of the iterative combinatory spaces and to other subjects in the theory of computability, in particular to the recursive functions with finite type arguments (Kleene [1959]). The other class has been introduced and studied by J. Zashev (cf. Zashev [1983, 1984, 1984 a, 1985, 1986, 1987, 1990]). In the structures from this class, the main role is played by an operation, which corresponds not to composition, but to application. The lack of an assumption about associativity of this operation creates considerable technical complications and requires a rather different approach, but the scope of the theory becomes large enough to encompass in a natural way also such structures as, for example, the Plotkin-Scott model of the λ-calculus (cf. Scott [1975]).

If there are no operations of arity 0 in $\mathcal{O}$ then the partially ordered algebra $\langle\mathcal{F}, \mathcal{O}\rangle$ cannot be fixed-point complete, since no explicitly definable elements of $\mathcal{F}$ will exist in this case, and it would be not possible a least element of $\mathcal{F}$ to exist and to be explicitly definable. Here is a necessary and sufficient condition for a given partially ordered algebra to be fixed-point complete.

Proposition 9. Let $\langle\mathcal{F}, \mathcal{O}\rangle$ be a partially ordered algebra. Then the following two conditions are equivalent:

(i) $\langle\mathcal{F}, \mathcal{O}\rangle$ is fixed-point complete;

(ii) for each natural number n (including $n=0$) and each explicitly definable $n+1$-ary operation Γ in $\langle\mathcal{F}, \mathcal{O}\rangle$, there is an explicitly definable n-ary operation Δ such that

$$\Delta(\theta_1, \ldots, \theta_n) = \mu\tau.\Gamma(\theta_1, \ldots, \theta_n, \tau)$$

for all $\theta_1, \ldots, \theta_n$ in $\mathcal{F}$.

Proof. The implication from (i) to (ii) follows from Definition 7 and Proposition 5. To prove the converse implication, one assumes (ii) and proves by induction on l that, for each n and each l-tuple $\Gamma_1, \ldots, \Gamma_l$ of explicitly definable $l+n$-ary operations in $\mathcal{F}$, there are explicitly definable n-ary operations $\Delta_1, \ldots, \Delta_l$ in $\mathcal{F}$ such that, for any choice of $\theta_1, \ldots, \theta_n$ in $\mathcal{F}$, the l-tuple

$$\langle\Delta_1(\theta_1, \ldots, \theta_n), \ldots, \Delta_l(\theta_1, \ldots, \theta_n)\rangle$$

is the least solution of the system (4) corresponding to the given $\Gamma_1, \ldots, \Gamma_l$. The induction step is by elimination based on Theorem II.4.1. ∎

Exercises

1. Let $\mathcal{F}$ be a distributive lattice with a greatest element $\mathbb{1}$ and a least element $\mathbf{0}$. Let $\mathcal{O}$ consist of the constants $\mathbb{1}, \mathbf{0}$ and of the binary operations $\wedge, \vee$ of the lattice $\mathcal{F}$. Give a description of the explicitly definable operations in $\langle\mathcal{F}, \mathcal{O}\rangle$ and show that $\langle\mathcal{F}, \mathcal{O}\rangle$ is fixed-point complete. Show also that $\langle\mathcal{F}, \mathcal{O}\rangle$ will be no more fixed-point complete if we remove the constant $\mathbf{0}$ from the set $\mathcal{O}$.

2. Let $\mathcal{F}$ be the set of the real numbers, partially ordered by the equality relation. Let $\mathcal{O}$ consist of the constant 1 and the binary operation of subtraction. Show

that the partially ordered algebra $\langle\mathcal{F}, \mathcal{O}\rangle$ has the integers as its explicitly definable elements and all rational numbers as its fixed-point definable elements. Give descriptions of all explicitly definable operations and of all fixed-point definable ones in $\langle\mathcal{F}, \mathcal{O}\rangle$.

3. Let $\mathcal{F}$ be the set of the real numbers, linearly ordered in the usual way. Let $\mathcal{O}$ consist of the constant 1, the binary operation of addition and all operations $\lambda\tau.\frac{\tau}{n}$, $n=2, 3, 4, \ldots$ Give a description of the explicitly definable operations in $\langle\mathcal{F}, \mathcal{O}\rangle$ and show that all fixed-point definable operations are explicitly definable in this case. Show that the set of the fixed-point definable operations remains the same if we take $\lambda\tau.\frac{\tau}{n}$ only with $n=2$.

4. Let $\langle\mathcal{F}, \mathcal{O}\rangle$ be such a partially ordered algebra that either $\mathcal{F}$ has no least element or the range of some operation belonging to $\mathcal{O}$ contains the least element of $\mathcal{F}$. Prove that in this case the operations Γ_i in Definitions 4 and 5 can be supposed to be only of the forms (1) and (3) from Definitions 2 and 3.

2. Computable elements and mappings in iterative combinatory spaces

From now on, until the end of this section it will be supposed that an iterative combinatory space $\mathfrak{S}=\langle\mathcal{F}, I, \mathcal{C}, \Pi, L, R, \Sigma, T, F\rangle$ is given.

Definition 1. Let $\mathcal{B}$ be a subset of $\mathcal{F}$. An element of $\mathcal{F}$ is called *$\mathfrak{S}$-computable in $\mathcal{B}$* iff this element can be generated from elements of the set $\{L, R, T, F\}\cup\mathcal{B}$ by means of the operations composition, combination and iteration in $\mathfrak{S}$ (sometimes we shall simply say "computable" instead of "$\mathfrak{S}$-computable", since the space $\mathfrak{S}$ will be fixed or the context will make clear which it is).[68] The set of all elements of $\mathcal{F}$, which are $\mathfrak{S}$-computable in $\mathcal{B}$, will be denoted by $\mathrm{COMP}_{\mathfrak{S}}(\mathcal{B})$.

By its definition, the introduced relative computability is transitive: if $\mathcal{B}\subseteq\mathrm{COMP}_{\mathfrak{S}}(\mathcal{B}')$ (in particular, if $\mathcal{B}\subseteq\mathcal{B}'$) then $\mathrm{COMP}_{\mathfrak{S}}(\mathcal{B})\subseteq\mathrm{COMP}_{\mathfrak{S}}(\mathcal{B}')$. By Example II.3.1 and by Defini-

---

[68] In our previous publications on iterative combinatory spaces, we used the term "recursive" instead of "computable".

tion II.3.3, the element I and the zero of $\mathfrak{S}$ are computable in each subset of $\mathcal{F}$. Propositions II.5.1 and II.5.4 show that the operation Σ preserves the computability in $\mathcal{B}$ (hence including Σ as an additional generating operation in the above definition would not enlarge the set $\mathrm{COMP}_{\mathfrak{S}}(\mathcal{B})$). Proposition II.5.5 shows that the iteration in the above definition can be replaced by $\mathfrak{S}_*$-iteration after enlarging $\{L, R, T, F\}$ to $\{I, L, R, T, F\}$.

We define also the notion of computability in $\mathcal{B}$ for a mapping of $\mathcal{F}^n$ into $\mathcal{F}$.

Definition 2. Let $\mathcal{B} \subseteq \mathcal{F}$, and let Γ be a mapping of $\mathcal{F}^n$ into $\mathcal{F}$, where n is some positive integer. Then Γ is called *$\mathfrak{S}$-computable in $\mathcal{B}$* (*computable in $\mathcal{B}$*, for short) iff, for arbitrary $\theta_1, \ldots, \theta_n$ in $\mathcal{F}$, there is an explicit expression for $\Gamma(\theta_1, \ldots, \theta_n)$ through $L, R, T, F, \theta_1, \ldots, \theta_n$ and elements of $\mathcal{B}$ by means of composition, combination and iteration in $\mathfrak{S}$, the form of the expression not depending on the concrete choice of $\theta_1, \ldots, \theta_n$.

Of course, a precise formulation of this definition can be given by using induction. Again including Σ as an additional operation does not enlarge the scope of the introduced notion.

The above notions of computability generalize the notions of $\mathfrak{A}$-computability introduced in Chapter I and thus enable reformulating some results from Chapter I as statements about computability in the corresponding combinatory spaces. In particular, such basic notions from the theory of computability as partial recursiveness and recursive enumerability turn out to be special cases of the general notion introduced in this section. Programmability in a FP-system (in the sense of Backus [1978]) is also a special case of this notion. The same will be shown further also for the notions of prime and search computability (the easier part of the proof is already carried out in Section I.7). Without giving such reformulations explicitly in the present moment, we shall have them in mind when developing the general theory. Some other computability notions from the literature also can be shown to be special cases of the introduced notion. This has been proved, for example, for the Friedman-Shepherdson computability by means of recursively enumerable definitional schemes (Friedman [1971], Shepherdson [1975]). Namely, as shown in Soskov [1987], this kind of computability can be characterized in the same way as search computability is characterized in Proposition I.7.2 and its conversion in Section 5, but with $\mathbb{N}^2$ instead of $(B^*)^2$.[69] Also the Kleene-recursiveness of functions with finite type arguments (Kleene [1959]) can be studied by means of suitable combinatory spaces. Namely, the algebraic approach to this notion in Ivanov [1984, 1986] by means of operative spaces can be modified in a way allowing certain non-symmetric combinatory spaces to be used instead of operative spaces (cf. Ivanov [1984, p. 50], as well as some of the exercises to Chapters 27 and 28 in Ivanov [1986]). To finish with this review of notions captured by our general definition, let us point also at some more exotic computability notions, such as the notions considered in Section I.8 (cf. also Exercises 8, 13, 16, 17, 18 after the present section, as well as the study of computable random functions presented in Section 4 of the Appendix).

Now we shall give an example generalizing Example I.2.1.

Example 1. The mapping Σ of $\mathcal{F}^3$ into $\mathcal{F}$ is $\mathfrak{S}$-computable in $\emptyset$ (by Propositions II.5.1 and II.5.4).

In order to become able to apply the considerations from Section 1, we note that the elements of $\mathrm{COMP}_{\mathfrak{S}}(\mathcal{B})$ are exactly the explicitly definable elements of the partially ordered algebra which arises when we consider the partially ordered semigroup $\mathcal{F}$ enriched by the operations combination and iteration of $\mathfrak{S}$ and the constants from the set $\{L, R, T, F\} \cup \mathcal{B}$ (the fact that both mentioned operations are monotonically increasing is known from Chapter II). Of course, the mappings of $\mathcal{F}^n$ in $\mathcal{F}$ $\mathfrak{S}$-computable in $\mathcal{B}$ are the n-ary operations explicitly definable in the same algebra. We shall denote this algebra by $\mathfrak{C}(\mathfrak{S}, \mathcal{B})$.

Remark 1.2 immediately implies the following important property of the computable mappings:

Proposition 1. For each subset $\mathcal{B}$ of $\mathcal{F}$, all mappings $\mathfrak{S}$-computable in $\mathcal{B}$ are monotonically increasing.

In the ordinary theory of computability (i.e. in the theory of recursive functions on $\mathbb{N}$), a certain role is played by such subclasses of the class of all computable functions as, for example, the class of the primitive recursive functions or the class of functions elementary in Kalmár's or Skolem's sense. A subset of $\mathrm{COMP}_{\mathfrak{S}}(\mathcal{B})$ with a similar role will be introduced also in the theory of the itera-

---

[69] Cf. Subsection (IV) of Section 5. The set $\mathbb{N}$ is considered a subset of B^* in virtue of the identification of the natural numbers with certain elements of $B^* \setminus B$ (cf. Subsection (I) of Section I.7).

tive combinatory spaces.

Definition 3. Let $\mathcal{B} \subseteq \mathcal{F}$. An element of $\mathcal{F}$ is called *$\mathfrak{S}$-elementary in* $\mathcal{B}$ (*elementary in* $\mathcal{B}$, for short) iff this element can be generated from elements of the set $\{I, L, R, T, F\} \cup \mathcal{B}$ by means of the operations composition, combination and branching in $\mathfrak{S}$. The set of all elements of $\mathcal{F}$, which are $\mathfrak{S}$-elementary in $\mathcal{B}$, will be denoted by $\mathrm{ELEM}_{\mathfrak{S}}(\mathcal{B})$.

Example 2. The elements L_* and R_* of $\mathcal{F}$ are $\mathfrak{S}$-elementary in $\emptyset$. Consequently, so is the element $\bar{m}$ of $\mathcal{F}$ for each natural number m.

Of course, the inclusion $\mathrm{ELEM}_{\mathfrak{S}}(\mathcal{B}) \subseteq \mathrm{COMP}_{\mathfrak{S}}(\mathcal{B})$ is seen on the basis of Example 1.

Remark 1. Exercise II.1.12 shows (in the notation used there) that $\mathrm{ELEM}_{\mathfrak{S}}(\mathcal{B})$ consists of those elements of $\mathcal{F}$ which can be generated from elements of the set $\{(I, I), R, K_1, K_2, St((I, I)), St((R, L)), St(St(I)), St(L), St(R), St(T), St(F)\} \cup St(\mathcal{B})$ by means of multiplication and branching. As seen from Exercise II.2.9, branching can be replaced by the operation Π_* in the above statement.

For the case of mappings of $\mathcal{F}^n$ into $\mathcal{F}$, the definition corresponding to Definition 3 looks as follows.

Definition 4. Let $\mathcal{B} \subseteq \mathcal{F}$, and let Γ be a mapping of $\mathcal{F}^n$ into $\mathcal{F}$, where n is some positive integer. Then Γ is called *$\mathfrak{S}$-elementary in* $\mathcal{B}$ (*elementary in* $\mathcal{B}$, for short) iff, for arbitrary $\theta_1, \ldots, \theta_n$ in $\mathcal{F}$, there is an explicit expression for $\Gamma(\theta_1, \ldots, \theta_n)$ through $I, L, R, T, F, \theta_1, \ldots, \theta_n$ and elements of $\mathcal{B}$ by means of composition, combination and branching in $\mathfrak{S}$, the form of the expression not depending on the concrete choice of $\theta_1, \ldots, \theta_n$.

Example 3. For each positive integer n, the mapping Π_* of $\mathcal{F}^n$ into $\mathcal{F}$ is $\mathfrak{S}$-elementary in $\emptyset$.

The mappings $\mathfrak{S}$-elementary in $\mathcal{B}$ are $\mathfrak{S}$-computable, again on the basis of Example 1.

Application of results from Section 1 is again possible after introducing another partially ordered algebra. This time we have to consider $\mathcal{F}$ enriched by the operations combination and branching of $\mathfrak{S}$ and the constants from the set $\{I, L, R, T, F\} \cup \mathcal{B}$. This partially ordered algebra will be denoted by $\mathfrak{E}(\mathfrak{S}, \mathcal{B})$. The elements of $\mathrm{ELEM}_{\mathfrak{S}}(\mathcal{B})$ can be characterized as the explicitly definable elements of $\mathfrak{E}(\mathfrak{S}, \mathcal{B})$, and the mappings of $\mathcal{F}^n$ into $\mathcal{F}$ elementary in $\mathcal{B}$ as the n-ary operations explicitly definable in this algebra.

Now an important point is that the operation of iteration is fixed-point definable in the partially ordered algebra $\mathfrak{C}(\mathfrak{S}, \mathcal{B})$ (this is clear from the equality

$$[\sigma, \chi] = \mu\tau.(\chi \to \tau\sigma, I),$$

since the operation $\mu\sigma\chi\tau.(\chi \to \tau\sigma, I)$ is explicitly definable in $\mathfrak{C}(\mathfrak{S}, \mathcal{B})$). Thus all primitive operations of $\mathfrak{C}(\mathfrak{S}, \mathcal{B})$ are fixed-point definable in $\mathfrak{C}(\mathfrak{S}, \mathcal{B})$, and hence, by Corollaries 1.4 and 1.2, all operations explicitly definable in the first of these two algebras are fixed-point definable in the second one. Hence the following holds:

Proposition 2. For each subset $\mathcal{B}$ of $\mathcal{F}$, all elements of $\mathrm{COMP}_{\mathfrak{S}}(\mathcal{B})$ and all mappings $\mathfrak{S}$-computable in $\mathcal{B}$ are fixed-point definable in the partially ordered algebra $\mathfrak{C}(\mathfrak{S}, \mathcal{B})$.

An important problem which naturally arises at this moment is whether the converse is true, i.e. whether all fixed-point definable operations in $\mathfrak{C}(\mathfrak{S}, \mathcal{B})$ are $\mathfrak{S}$-computable in $\mathcal{B}$. An affirmative answer to this question will be given further in this chapter.

According to the definitions given in Section 1, the fixed-point definability in the partially ordered algebra $\mathfrak{C}(\mathfrak{S}, \mathcal{B})$ means, roughly speaking, definability via the least solution of a system of the form 1.(3) or 1.(4) with mappings $\Gamma_1, \ldots, \Gamma_l$ which are simple with respect to $\mathfrak{C}(\mathfrak{S}, \mathcal{B})$. Taking into account the list of the primitive operations of this partially ordered algebra, we see that the simple operations in it are the ones having some of the following forms:

$$\lambda\psi_1 \cdots \psi_m . \psi_i ,$$
$$\lambda\psi_1 \cdots \psi_m . \alpha,$$
$$\lambda\psi_1 \cdots \psi_m . \psi_j\psi_i ,$$
$$\lambda\psi_1 \cdots \psi_m . (\psi_i, \psi_j),$$
$$\lambda\psi_1 \cdots \psi_m . (\psi_i \to \psi_j , \psi_k),$$

where i, j, k are fixed numbers from the set $\{1, \ldots, m\}$, and α is some fixed element of $\{I, L, R, T, F\} \cup \mathcal{B}$. So we see how the the systems 1.(3) and 1.(4) look. For example, in a system of the form 1.(3) each inequality has some of the forms

$$\tau_r \geq \alpha,$$
$$\tau_r \geq \tau_j\tau_i ,$$
$$\tau_r \geq (\tau_i, \tau_j),$$

$$\tau_r \geq (\tau_i \rightarrow \tau_j, \tau_k),$$
$$\tau_r \geq \tau_i.$$

In a system of the form 1.(4), the inequalities are of the same forms, except that parameters θ_s can occur in some places in the right-hand sides instead of some unknowns. A certain kind of canonization of such systems can be useful in some cases (for example, the systems can be assumed to contain no inequalities of the form $\tau_r \geq \tau_i$). For the time being, we shall not touch this subject in more detail.

Example 4. Let Γ be the mapping of $\mathcal{F}^2$ into $\mathcal{F}$ defined by

$$\Gamma(\theta_1, \theta_2) = [\theta_1, (L \rightarrow F, T)\theta_2].$$

Then, for each θ_1, θ_2 in $\mathcal{F}$, $\Gamma(\theta_1, \theta_2)$ is the component τ_1 of the least solution $\langle \tau_1, \tau_2, \tau_3, \tau_4, \tau_5, \tau_6, \tau_7, \tau_8 \rangle$ of the system of inequalities

$$\tau_1 \geq (\tau_2 \rightarrow \tau_3, \tau_4),$$
$$\tau_2 \geq \tau_5 \theta_2,$$
$$\tau_3 \geq \tau_1 \theta_1,$$
$$\tau_4 \geq I,$$
$$\tau_5 \geq (\tau_6 \rightarrow \tau_7, \tau_8),$$
$$\tau_6 \geq L,$$
$$\tau_7 \geq F,$$
$$\tau_8 \geq T.$$

The $\mathfrak{S}$-computability can be characterized by using functional expressions of the formal system A introduced in Section II.6.

Definition 5. Let $\mathcal{B} \subseteq \mathcal{F}$, $\varphi \in \mathcal{F}$, $\mathcal{Z}$ be a functional expression of the system A, and let a valuation of the variables of A in $\mathfrak{S}$ be given. It will be said that $\mathcal{Z}$ *expresses* φ *through* $\mathcal{B}$ *at the given valuation* iff all variables occurring in $\mathcal{Z}$ have values belonging to $\mathcal{B}$ at this valuation, and the value of $\mathcal{Z}$ at the same valuation is equal to φ.

The truth of the following two propositions is obvious.

Proposition 3. If some functional expression of A expresses an element φ of $\mathcal{F}$ through a subset $\mathcal{B}$ of $\mathcal{F}$ at some valuation then $\varphi \in \mathrm{COMP}_{\mathfrak{S}}(\mathcal{B})$.

Proposition 4. Let $\mathcal{B}$ be a subset of $\mathcal{F}$, φ be an element of $\mathrm{COMP}_{\mathfrak{S}}(\mathcal{B})$, and let a valuation of the variables of

A in $\mathfrak{S}$ be given such that each element of $\mathcal{B}$ is the value of some variable f_i at the given valuation. Then there is a functional expression $\mathcal{Z}$ of A such that $\mathcal{Z}$ expresses φ through $\mathcal{B}$ at the given valuation, and no variables c_i occur in $\mathcal{Z}$.

A similar definition can be given for expressibility of mappings of $\mathcal{F}^n$ into $\mathcal{F}$ can be given, and similar propositions will be valid.

Another obvious proposition will be formulated, namely the following one.

Proposition 5. Let $\mathcal{B}$ be a subset of $\mathcal{F}$, and let $\varphi_1, \ldots, \varphi_n$ be some elements of $\mathcal{F}$. Then the following statements hold:

(i) an element of $\mathcal{F}$ is $\mathfrak{S}$-computable in $\mathcal{B}\cup\{\varphi_1, \ldots, \varphi_n\}$ iff this element can be represented in the form $\Gamma(\varphi_1, \ldots, \varphi_n)$, where Γ is some mapping of $\mathcal{F}^n$ into $\mathcal{F}$ $\mathfrak{S}$-computable in $\mathcal{B}$;

(ii) a mapping of $\mathcal{F}^l$ into $\mathcal{F}$ is $\mathfrak{S}$-computable in $\mathcal{B}\cup\{\varphi_1, \ldots, \varphi_n\}$ iff this mapping can be represented in the form $\lambda\tau_1\ldots\tau_l.\Gamma(\varphi_1, \ldots, \varphi_n, \tau_1, \ldots, \tau_l)$, where Γ is some mapping of $\mathcal{F}^{n+l}$ into $\mathcal{F}$ $\mathfrak{S}$-computable in $\mathcal{B}$.

Proposition 6. Let $\mathcal{B}$ be a subset of $\mathcal{F}$. Then each element of $\mathcal{F}$ or mapping of $\mathcal{F}^l$ into $\mathcal{F}$ $\mathfrak{S}$-computable in $\mathcal{B}$ is also $\mathfrak{S}$-computable in some finite subset of $\mathcal{B}$.

Exercises

1. Let $\mathfrak{S}$ be an iterative combinatory space. Using Exercise II.3.8, show the existence of an element φ of $\mathrm{COMP}_{\mathfrak{S}}(\emptyset)$ such that $\varphi\bar{n}=\overline{2n}$ for all n in $\mathbb{N}$. Write the corresponding system of inequalities of the form 1.(3).

2. Let $\mathfrak{S}=\langle\mathcal{F}, I, \mathcal{C}, \Pi, L, R, \Sigma, T, F\rangle$ be an iterative combinatory space, l be a positive integer, and $K=\mathcal{F}^l$. Using the denotations from Exercise II.1.40 and the result from Exercise II.3.9, consider the iterative combinatory space

$$\mathfrak{S}^K=\langle\mathcal{F}', I', \mathcal{C}', \Pi', L', R', \Sigma', T', F'\rangle.$$

Generalizing the denotation $\mathcal{C}'$, let us adopt that, for each subset $\mathcal{B}$ of $\mathcal{F}$, $\mathcal{B}'$ denotes the set of all constant mappings of K into $\mathcal{B}$. Let $\Xi_1, \ldots, \Xi_l$ be the projection mappings from K into $\mathcal{F}$, i.e.

$$\Xi_i(\psi_1, \ldots, \psi_l) = \psi_i, \quad i = 1, \ldots, l,$$

for all $\psi_1, \ldots, \psi_l$ in $\mathcal{F}$. Let Γ be an arbitrary mapping of $\mathcal{F}^l$ into $\mathcal{F}$. Prove that Γ is $\mathfrak{S}$-computable in $\mathcal{B}$ in the sense of Definition 2 iff Γ is an element of $\mathcal{F}'$ $\mathfrak{S}^K$-computable in $\mathcal{B}' \cup \{\Xi_1, \ldots, \Xi_l\}$, and Γ is $\mathfrak{S}$-elementary in $\mathcal{B}$ in the sense of Definition 4 iff Γ is an element of $\mathcal{F}'$ $\mathfrak{S}^K$-elementary in $\mathcal{B}' \cup \{\Xi_1, \ldots, \Xi_l\}$.

3. (The First Recursion Theorem for left-linear mappings) Let $\mathfrak{S} = \langle \mathcal{F}, I, \mathcal{C}, \Pi, L, R, \Sigma, T, F \rangle$ be an iterative combinatory space, l be a positive integer, and $B_1, \ldots, B_l$ be left-homogeneous mappings of $\mathcal{F}^{l+1}$ into $\mathcal{F}$ which are $\mathfrak{S}$-computable in a given subset $\mathcal{B}$ of $\mathcal{F}$. Let the mappings $\Gamma_1, \ldots, \Gamma_l$ of $\mathcal{F}^l$ into $\mathcal{F}$ be defined by means of the equalities

$$\Gamma_i(\tau_1, \ldots, \tau_l) = B_i(I, \tau_1, \ldots, \tau_l), \quad i = 1, \ldots, l.$$

Prove the following statements[70]:

(i) the system of inequalities

$$\tau_i \geq \Gamma_i(\tau_1, \ldots, \tau_l), \quad i = 1, \ldots, l,$$

has a least solution $\langle \tau_1, \ldots, \tau_l \rangle$ in $\mathcal{F}^l$;

(ii) the components $\tau_1, \ldots, \tau_l$ of the mentioned least solution are $\mathfrak{S}$-recursive in $\mathcal{B}$;

(iii) the least solution of the above system of inequalities is also the least solution of the system of equations

$$\tau_i = \Gamma_i(\tau_1, \ldots, \tau_l), \quad i = 1, \ldots, l.$$

Hint. Use Theorem II.6.2 and Proposition II.4.1.

4. (Compare with Remark 1) Let $\mathfrak{S} = \langle \mathcal{F}, I, \mathcal{C}, \Pi, L, R, \Sigma, T, F \rangle$ be an iterative combinatory space. In the notation used in Exercise 1.12, prove that $\mathrm{COMP}_{\mathfrak{S}}(\mathcal{B})$ consists of those elements of $\mathcal{F}$ which can be generated from el-

---

[70] The statements in this exercise are true for arbitrary mappings $\Gamma_1, \ldots, \Gamma_l$ of $\mathcal{F}^l$ into $\mathcal{F}$ which are $\mathfrak{S}$-computable in $\mathcal{B}$, but such a generalization is much more difficult to be proved, and it will be the central result in the present book.

ements of the set $\{(I, I), R, K_1, K_2, St((I, I)), St((R, L)), St(St(I)), St(L), St(R), St(T), St(F)\} \cup St(\mathcal{B})$ by means of multiplication and iteration. Show that iteration can be replaced by $\mathfrak{S}_*$-iteration in this statement. Prove that a mapping Γ of $\mathcal{F}^n$ into $\mathcal{F}$ is $\mathfrak{S}$-computable in $\mathcal{B}$ iff this mapping is representable in the form

$$\Gamma(\theta_1, \ldots, \theta_n) = \Gamma'(St(\theta_1), \ldots, St(\theta_n)),$$

where Γ' is explicitly definable in the algebra obtained from the semigroup $\mathcal{F}$ through its enrichment by iteration or $\mathfrak{S}_*$-iteration and by the constants from the above set.

Hint. Use Exercises II.1.7, II.1.10, II.3.5 and Propositions II.5.1, II.5.5.

5. Let $\mathfrak{S} = \langle \mathcal{F}, I_M^{\sim}, \mathcal{C}, \Pi, L^{\sim}, R^{\sim}, \Sigma, T^{\sim}, F^{\sim} \rangle$ be the combinatory space from Exercise II.4.11 (the combinatory space of the $\mathbb{L}$-fuzzy relations corresponding to a given computational structure), but under the extra assumption that $\langle M, J, L, R, T, F, H \rangle$ is a standard computational structure on the natural numbers in the sense of Section I.3 (hence $M = \mathbb{N}$). Prove that, for each unary partial recursive function f, the corresponding $f^{\sim}$ is $\mathfrak{S}$-computable in the set $\{S^{\sim}, P^{\sim}\}$, where $S = \lambda u.\, u+1$, $P = \lambda u.\, u \dot{-} 1$.

Hint. Use Theorem I.3.1.

6. Let $\mathfrak{S}$ and S be such as in the previous exercise. Show that $(\mathbb{N}^2)^{\sim}$ is $\mathfrak{S}$-computable in $\{S^{\sim}, (\mathbb{N} \times \{0, 1\})^{\sim}\}$.

Hint. See the exercise to Section I.6.

7. Let $\mathfrak{S}$ be such as in Exercise 5, and let, in addition, the lattice $\mathbb{L}$ be a linearly ordered set. Let $\mathcal{H}$ be the set of all φ from $\mathcal{F}$ such that for any fixed l in $\mathbb{L} \setminus \{\mathbb{1}\}$ the set $\{\langle u, v \rangle \in \mathbb{N}^2 : \varphi(u, v) > l\}$ is recursively enumerable. Prove that, for each recursively enumerable subset f of $\mathbb{N}^2$, the corresponding $f^{\sim}$ belongs to $\mathcal{H}$, and, whenever $\mathcal{B} \subseteq \mathcal{H}$, then $\mathrm{COMP}_{\mathfrak{S}}(\mathcal{B}) \subseteq \mathcal{H}$.

8. In the situation from Exercises 5 and 7, suppose furthermore that the lattice $\mathbb{L}$ has finitely many elements. For each l in $\mathbb{L}$, let ε_l be the element of $\mathcal{F}$ defined in the following way:

$$\varepsilon_l(u, v) = \begin{cases} l & \text{if } u = v, \\ \mathbb{0} & \text{if } u \neq v. \end{cases}$$

Let $\mathcal{B}$ be the subset of $\mathcal{F}$ consisting of $S^{\sim}$, $P^{\sim}$, $(\mathbb{N}^2)^{\sim}$ and all elements ε_l of $\mathcal{F}$ with $\mathbb{0} < l < \mathbb{1}$. Prove the equality $\mathrm{COMP}_{\mathfrak{S}}(\mathcal{B})=\mathcal{H}$.

Hint. To prove the inclusion $\mathcal{H}\subseteq\mathrm{COMP}_{\mathfrak{S}}(\mathcal{B})$, suppose φ is an arbitrary element of $\mathcal{H}$. Let the elements of $\mathbb{L}$ be $l_1, l_2, \ldots, l_k$, where $\mathbb{1}=l_1>l_2>\ldots>l_k=0$. For $i=1, 2, \ldots, k-1$, choose some three-argument primitive recursive function h_i such that

$$\varphi(u, v)>l_{i+1} \Longleftrightarrow \exists w\in\mathbb{N}\,(h_i(u, v, w)=0)$$

for all natural numbers u and v, and set

$$\chi_i=(\lambda t.\, h_i(L(t), LR(t), R^2(t)))^{\sim},$$

$$\rho_i=\varepsilon_{l_i}(LR)^{\sim}.$$

Prove the equality $\varphi=\psi\Pi(I_{\mathbb{N}}{}^{\sim}, \Pi((\mathbb{N}^2)^{\sim}, (\mathbb{N}^2)^{\sim}))$, where

$$\psi=\Sigma(\chi_1, \Sigma(\chi_2, \ldots \Sigma(\chi_{k-2}, \Sigma(\chi_{k-1}, \varnothing^{\sim}, \rho_{k-1}), \rho_{k-2}), \ldots \rho_2), \rho_1).$$

9. Let $\mathfrak{S}=\langle\mathcal{F}, I_M{}^{\sim}, \mathcal{C}, \Pi, L^{\sim}, R^{\sim}, \Sigma, T^{\sim}, F^{\sim}\rangle$ be the combinatory space from Exercise II.4.13 (a combinatory space relevant to $\forall$-definedness), but under the extra assumption that $\langle M, J, L, R, T, F, H\rangle$ is a standard computational structure on the natural numbers. Let $S=\lambda u.\, u+1$, $P=\lambda u.\, u\dot{-}1$. Prove that, for each unary partial recursive function f, the corresponding $f^{\sim}$ is $\mathfrak{S}$-computable in the set $\{S^{\sim}, P^{\sim}\}$.

10. Let $\mathfrak{S}$, S and P be such as in the previous exercise. Prove that, for each recursively enumerable subset f of $\mathbb{N}^2$, the corresponding $f^{\sim}$ is $\mathfrak{S}$-computable in the set $\{S^{\sim}, P^{\sim}, (\mathbb{N}^2)^{\sim}\}$.

Hint. Take a three-argument primitive recursive function g such that

$$\langle u, v\rangle\in f \Longleftrightarrow \exists w\in\mathbb{N}\,(g(u, v, w)=0)$$

for all natural numbers u, v. Take also an one-argument partial recursive function f_0 such that $f_0\subseteq f$, and the first components of all pairs from f belong to $\mathrm{dom} f_0$. Set $h=\lambda t.\, g(L(t), LR(t), R^2(t))$ and prove the equality

$$f^{\sim}=\Sigma(h^{\sim}, f_0{}^{\sim}L^{\sim}, (LR)^{\sim})\Pi(I^{\sim}, \Pi((\mathbb{N}^2)^{\sim}, (\mathbb{N}^2)^{\sim})).$$

11. Let $\mathfrak{S}$, S and P be such as in Exercise 9, and let A be a subset of $\mathbb{N}$ belonging to the class Π^1_1 of the ana-

lytical hierarchy[71]. Prove that the element $\langle I_{\mathbb{N}}, A\rangle$ of $\mathcal{F}$ is $\mathfrak{S}$-computable in $\{S^{\sim}, P^{\sim}, (\mathbb{N}^2)^{\sim}\}$.

Hint. Take two-argument primitive recursive functions g_1 and g_2 with the following properties:

(i) a natural number u belongs to A iff each infinite sequence of natural numbers has some finite initial segment with a sequence number s satisfying the condition

$$g_1(s, u) = 0;$$ [72]

(ii) whenever s is the sequence number of a finite sequence $\langle s_0, s_1, \ldots, s_{n-1}\rangle$ of natural numbers, and t is an arbitrary natural number, then $g_2(s, t)$ is the sequence number of the sequence $\langle s_0, s_1, \ldots, s_{n-1}, t\rangle$.

Then set

$$h_i = \lambda t.\, g_i(L(t), R(t)), \quad i = 1, 2,$$
$$\iota = [\Pi(h_2{}^{\sim}\Pi(L^{\sim}, (\mathbb{N}^2)^{\sim}), R^{\sim}), h_1{}^{\sim}]$$

and prove the equality

$$\langle I_{\mathbb{N}}, A\rangle = \Sigma((\mathbb{N}^2)^{\sim}, I_{\mathbb{N}}{}^{\sim}, R^{\sim}\iota\Pi(F^{\sim}, I^{\sim})).$$

12. Let $\mathfrak{S}$ be such as in Exercise 9. Let $\mathcal{H}_0$ be the set of all elements $\langle f, A\rangle$ of $\mathcal{F}$ such that f is recursively enumerable, A is a Π^1_1-set, and $A \subseteq \operatorname{dom} f$. Prove that, whenever $\mathcal{B} \subseteq \mathcal{H}_0$, then $\mathrm{COMP}_{\mathfrak{S}}(\mathcal{B}) \subseteq \mathcal{H}_0$.

Hint. Use Exercises II.4.17 and II.4.18 to show that $\mathcal{H}_0$ is closed under iteration.

13. (Non-deterministic computability with unbounded non-determinism; cf. Skordev [1980, Chapter IV, Section 1.2, Example 9], and also Skordev [1987]) In the situation from Exercises 9 and 12, let $\mathcal{B}$ be the subset of $\mathcal{F}$ consisting of $S^{\sim}$, $P^{\sim}$ and $(\mathbb{N}^2)^{\sim}$. Prove the equality $\mathrm{COMP}_{\mathfrak{S}}(\mathcal{B}) = \mathcal{H}_0$.

---

[71] For the definition of this class, cf. for example Rogers [1967, § 16.1].

[72] Here and in the next condition (ii) a sufficiently good effective enumeration of the set of all finite sequences of natural numbers is supposed to be fixed, the sequence number of the empty sequence being equal to 0. The existence of a primitive recursive function g_1 with the property (i) follows from the assumption that A is a Π^1_1-set (cf. Rogers [1967, § 16.1, Corollary V]).

Hint. To prove the inclusion $\mathcal{H}_0 \subseteq \mathrm{COMP}_{\mathfrak{S}}(\mathcal{B})$, use the validity of $\langle f, A\rangle = f^{\sim}\langle I_{\mathbb{N}}, A\rangle$ for all $\langle f, A\rangle$ in $\mathcal{H}_0$.

14. Let $\mathfrak{S}$ and P be such as in Exercise 9. Let $\mathcal{E} = \langle e, \mathbb{N}\rangle$, where $e = \{\langle u, v\rangle \in \mathbb{N}^2 : u \geq v\}$. Prove that $\mathcal{E}$ is $\mathfrak{S}$-computable in the set $\{P^{\sim}, (\mathbb{N}\times\{0, 1\})^{\sim}\}$.

Hint. Prove the equality

$$\mathcal{E} = [P^{\sim}, \Sigma(I_{\mathbb{N}}{}^{\sim}, (\mathbb{N}\times\{0, 1\})^{\sim}, F^{\sim})].$$

15. Let $\mathfrak{S}$, S, P be such as in Exercise 9. Let $\mathcal{H}_0^b$ be the set of all elements $\langle f, A\rangle$ of $\mathcal{F}$ which have the following properties: (i) both f and A are recursively enumerable, and $A \subseteq \mathrm{dom} f$; (ii) for each u in A, the set $\{v : \langle u, v\rangle \in f\}$ is finite, and (iii) there is a partial recursive function which transforms each u from A into the cardinality of $\{v : \langle u, v\rangle \in f\}$. Prove that, whenever $\mathcal{B} \subseteq \mathcal{H}_0^b$, then $\mathrm{COMP}_{\mathfrak{S}}(\mathcal{B}) \subseteq \mathcal{H}_0^b$. Use this result to conclude that $(\mathbb{N}^2)^{\sim}$ is not $\mathfrak{S}$-computable in $\{S^{\sim}, P^{\sim}, (\mathbb{N}\times\{0, 1\})^{\sim}\}$ (compare with Exercise 6).

Hint. Use the fact that condition (iii) can be replaced by the requirement to exist an algorithm producing a list of the elements of $\{v : \langle u, v\rangle \in f\}$ for any given u in A.

16. (Non-deterministic computability with bounded non-determinism; Pazova [1978], cf. also Skordev [1987]). In the situation from Exercises 9 and 15, let $\mathcal{B}$ be the subset of $\mathcal{F}$ consisting of $S^{\sim}$, $P^{\sim}$ and $(\mathbb{N}\times\{0, 1\})^{\sim}$. Prove the equality $\mathrm{COMP}_{\mathfrak{S}}(\mathcal{B}) = \mathcal{H}_0^b$.

Hint. To prove the inclusion $\mathcal{H}_0^b \subseteq \mathrm{COMP}_{\mathfrak{S}}(\mathcal{B})$, suppose $\varphi = \langle f, A\rangle$ is an arbitrary element of $\mathcal{H}_0^b$. Take two-argument primitive recursive functions g_1, g_2 and one-argument primitive recursive function h such that the following equivalences hold for all natural numbers u, v, w:

$$\langle u, v\rangle \in f \iff \exists s \in \mathbb{N}\,(g_1(u, s) = 0 \;\&\; h(s) = v),$$

$$u \in A \;\&\; \mathrm{card}\{v : \langle u, v\rangle \in f\} = t \iff \exists s \in \mathbb{N}\,(g_2(u, s) = 0 \;\&\; h(s) = t).$$

Take also a two-argument partial recursive function g_3 such that

$$\{v : \langle u, v\rangle \in f\} = \{g_3(u, i) : i < t\},$$

whenever $u \in A \;\&\; \mathrm{card}\{v : \langle u, v\rangle \in f\} = t$. Consider one-argument functions h_1, h_2, h_3 corresponding to g_1, g_2, g_3 as in the hints to Exercises 10 and 11. Set

$$\iota=[\Pi(L^{\sim}, (SR)^{\sim}), \Sigma(h_2^{\sim}, \Sigma((\mathbb{N}\times\{0,1\})^{\sim}, T^{\sim}, h_1^{\sim}), F^{\sim})$$

and prove the equality

$$\varphi=\Sigma(h_2^{\sim}, h^{\sim}, h_3^{\sim}\Pi(L^{\sim}, \varepsilon(Ph)^{\sim}))\iota\Pi(I_{\mathbb{N}}^{\sim}, F^{\sim}),$$

where ε is the element of $\mathcal{F}$ defined in Exercise 14.

17. (Non-deterministic computability with unbounded non-determinism and possible unproductive termination) Let $\mathfrak{S}$, S, P be such as in Exercise 9. Let $\mathcal{H}$ be the set of all elements φ of $\mathcal{F}$ such that the first component of φ is recursively enumerable, and the second one is a Π_1^1-set. Let $\mathcal{B}$ be the subset of $\mathcal{F}$ consisting of $S^{\sim}$, $P^{\sim}$, $(\mathbb{N}^2)^{\sim}$ and $\langle\emptyset, \mathbb{N}\rangle$. Prove the equality $\mathrm{COMP}_{\mathfrak{S}}(\mathcal{B})=\mathcal{H}$.

Hint. Prove that $\langle f, \mathbb{N}\rangle\in\mathrm{COMP}_{\mathfrak{S}}(\mathcal{B})$ for each recursively enumerable binary relation f, and use the validity of $\langle f, A\rangle=\langle f, \mathbb{N}\rangle\langle I_{\mathbb{N}}, A\rangle$ for all $\langle f, A\rangle$ in $\mathcal{H}$.

18. (Non-deterministic computability with bounded non-determinism and possible unproductive termination) Let $\mathfrak{S}$, S, P be such as in Exercise 9. Let $\mathcal{H}^b$ be the set of all elements $\langle f, A\rangle$ of $\mathcal{F}$ such that both f and A are recursively enumerable and the conditions (ii) and (iii) from Exercise 15 are satisfied. Let $\mathcal{B}$ be the subset of $\mathcal{F}$ consisting of $S^{\sim}$, $P^{\sim}$, $(\mathbb{N}\times\{0,1\})^{\sim}$ and $\langle\emptyset, \mathbb{N}\rangle$. Prove the equality $\mathrm{COMP}_{\mathfrak{S}}(\mathcal{B})=\mathcal{H}^b$.

Hint. Prove that $\langle\{\langle u, v\rangle\in\mathbb{N}^2 : u>v\}, \mathbb{N}\rangle\in\mathrm{COMP}_{\mathfrak{S}}(\mathcal{B})$.

19. Let $\mathfrak{S}=\langle\mathcal{F}, I, \mathcal{C}, \Pi, L, R, \Sigma, T, F\rangle$ be an iterative combinatory space, $\mathcal{B}$ be subset of $\mathcal{F}$, and Γ be a mapping of $\mathcal{F}^l$ into $\mathcal{F}$ $\mathfrak{S}$-computable in $\mathcal{B}$. Prove that a mapping Γ' of $\mathcal{F}^l$ into $\mathcal{F}$ exists, also $\mathfrak{S}$-computable in $\mathcal{B}$, such that, for all $\theta_1, \ldots, \theta_l$ in $\mathcal{F}$ and all z in $\mathcal{C}$, the equality

$$\Gamma(\theta_1(z, I), \ldots, \theta_l(z, I))=\Gamma'\theta_1, \ldots, \theta_l)(z, I)$$

holds. Prove a similar result for mappings $\mathfrak{S}$-elementary in $\mathcal{B}$.

Hint. Use induction on the construction of Γ. For the case of iteration, apply Corollary II.3.1.

3. Representation of the partial recursive functions in iterative combinatory spaces

We first recall the representation of the natural numbers from Section II.2. Namely, if $\mathfrak{S} = \langle \mathcal{F}, I, \mathcal{C}, \Pi, L, R, \Sigma, T, F \rangle$ is an iterative combinatory space then an arbitrary natural number n is represented by the element $\bar{n} = R_*{}^n L_*$ of $\mathcal{F}$, where $L_* = (T, I)$, $R_* = (F, I)$. In order to introduce a representation of k-tuples of natural numbers by means of elements of $\mathcal{F}$, we first define an extension of the operation Π allowing its application to an arbitrary non-zero number of elements of $\mathcal{F}$. Namely, an element $(\varphi_1, \ldots, \varphi_k)$ is defined for each k-tuple $\langle \varphi_1, \ldots, \varphi_k \rangle$ of elements of $\mathcal{F}$, where $k \geq 1$, by setting

$$(\varphi) = \varphi, \quad (\varphi_0, \varphi_1, \ldots, \varphi_k) = \Pi(\varphi_0, (\varphi_1, \ldots, \varphi_k))$$

(the denotation (φ, ψ) used until now is obviously a particular case of the denotation just introduced). The representation of a k-tuple $\langle n_1, \ldots, n_k \rangle$ of natural numbers will be accomplished by means of the element $(\overline{n_1}, \ldots, \overline{n_k})$ of $\mathcal{F}$. Note that this element is normal (in the sense of Definition II.1.2) for every choice of the natural numbers $n_1, \ldots, n_k$.

Remark 1. Another way to represent $\langle n_1, n_2, \ldots, n_k \rangle$ is used in Ivanov [1986]. Following that way, we ought to represent the above k-tuple by the element $\overline{n_k} \ldots \overline{n_2}\,\overline{n_1}$ of $\mathcal{F}$.

Now we shall fix the representation of the natural numbers and of k-tuples of natural numbers by means of corresponding functional expressions of the formal system A from Section II.7.

Definition 1. For each natural number n, a functional expression n^* is defined in the following way: 0^* is $(\mathrm{T},)$ and $(n+1)^*$ is $(\mathrm{F},)\, n^*$ for each natural number n.

Obviously, for each natural number n the expression n^* does not contain variables, and, if some iterative combinatory space $\mathfrak{S}$ is given, then the value of n^* in $\mathfrak{S}$ is equal to $\bar{n}$ (independently from the choice of the valuation of the variables). We note also that all expressions n^* are normal in the sense of Section II.7, and the following formulas are deducible in the system A:

$$\mathrm{L}\,0^* = \mathrm{T}, \quad \mathrm{R}\,0^* = \Lambda,$$

$$\mathrm{L}(n+1)^* = \mathrm{F}n^*, \quad \mathrm{R}(n+1)^* = n^*$$

(by Propositions II.7.6 and II.7.1).

To fix the representation of k-tuples of natural numbers, it is natural to define the formal counterpart of the extension of Π defined above. We adopt the convention that $(\mathcal{U})$ denotes $\mathcal{U}$, and $(\mathcal{U}_0, \mathcal{U}_1, \ldots, \mathcal{U}_k)$ denotes $(\mathcal{U}_0, (\mathcal{U}_1, \ldots, \mathcal{U}_k))$, whatever the functional expressions $\mathcal{U}, \mathcal{U}_0, \mathcal{U}_1, \ldots, \mathcal{U}_k$ are. Obviously, $(\mathcal{U}_1, \ldots, \mathcal{U}_k)$ is a normal functional expression, whenever $\mathcal{U}_1, \ldots, \mathcal{U}_k$ are normal functional expressions. Having this denotations at our disposal, we agree to use the functional expression $(n_1^*, \ldots, n_k^*)$ as a representation of the k-tuple $\langle n_1, \ldots, n_k\rangle$ of natural numbers. This is again a normal functional expression not containing variables.

Now a notion of representation will be introduced for (possibly partial) functions of one or more arguments in $\mathbb{N}$ (in the sequel, such functions will be called "functions in $\mathbb{N}$" or simply "functions").

Definition 2. Let f be a function of k arguments in $\mathbb{N}$. A functional expression $\mathcal{U}$ will be called to *represent* f iff $\mathcal{U}$ contains no variables and, whenever the equality

(1) $$f(n_1, \ldots, n_k) = m$$

holds, then the formula

(2) $$\mathcal{U}(n_1^*, \ldots, n_k^*) = m^*$$

is deducible in the system **A**.

Remark 2. A possible defect of Definition 3 is connected with the fact that there is a function with not uniquely determined number of arguments, namely the empty function. Fortunately, the defect is not actually present, since each functional expression without variables turns out to represent the empty function independently of the choice of the value of k for the application of Definition 3.

Remark 3. Obviously, whenever a functional expression $\mathcal{U}$ represents a function f, then $\mathcal{U}$ represents also all restrictions of f. This can be considered also a defect, but it is not a logical one, and it is compensated to some extent by certain advantages of the definition. In any case, there is no problem to impose additional requirements when it proves to be useful.

Here are several simple examples to the definition.

Example 1. The empty expression Λ represents the function $I_{\mathbb{N}}$ (by Proposition II.7.1; the using of this proposition will be no more explicitly mentioned).

Example 2. The functional expression (F,) represents the function $\lambda u.\, u+1$, and the functional expression R represents the partial function $\lambda u.\, u-1$ (not defined at 0).

Example 3. The functional expression $(L\supset, R)$ represents the total function $\lambda u.\, u \dot{-} 1$ (by Proposition II.7.7 and Corollary II.7.1; the using of the last corollary will be no more explicitly mentioned).

Example 4. The functional expression [, T] represents no function whose domain is non-empty. This can be seen by using the correctness of A.

A necessary condition for the representability of a function by some functional expression is the existence of a partial recursive extension of this function. It is so since, for every functional expression $\mathcal{U}$ and any positive integer k, the set of those $k+1$-tuples $\langle n_1, \ldots, n_k, m\rangle$ of natural numbers, for which the formula (2) is deducible, can be shown to be the graph of some partial recursive function (this follows from the correctness of the system A and the fact that its notion of deduction is decidable). The main result in this section will be that the converse statement is also true.

We first introduce an abbreviation. For each functional expression $\mathcal{U}$ and each positive integer k, we shall denote by $\mathcal{U}^k$ the functional expression $\mathcal{U}\mathcal{U}\ldots\mathcal{U}$, with k repetitions of $\mathcal{U}$. By $\mathcal{U}^0$, the empty string will be denoted. Thus n^* can be written as $(F,)^n(T,)$ for each natural number n.

Proposition 1. Whenever $\mathcal{U}_1, \ldots, \mathcal{U}_k$ are normal functional expressions, and $0 \le i < k$, then the formula

$$R^i(\mathcal{U}_1, \ldots, \mathcal{U}_k) = (\mathcal{U}_{i+1}, \ldots, \mathcal{U}_k)$$

is deducible in the system A.

Proof. Induction on i using Proposition II.7.6. ∎

Corollary 1. Whenever $\mathcal{U}_1, \ldots, \mathcal{U}_k$ are normal functional expressions, and $k \ge 1$, then the formula

$$R^{k-1}(\mathcal{U}_1, \ldots, \mathcal{U}_k) = \mathcal{U}_k$$

is deducible in the system A.

Proposition 2. Whenever $\mathcal{U}_1, \ldots, \mathcal{U}_k$ are normal functional expressions, and $1 \leq i < k$, then the formula

$$\mathrm{L}\ \mathrm{R}^{i-1}(\mathcal{U}_1, \ldots, \mathcal{U}_k) = \mathcal{U}_i$$

is deducible in the system A.

Proof. Application of Propositions 1 and II.7.6.∎

Proposition 3. The functional expression $[\mathrm{R}, (\mathrm{L} \supset \mathrm{F}, \mathrm{T})]$ represents the function $\lambda n.0$.

Proof. Let us denote the mentioned functional expression by $\mathcal{U}$. By Proposition II.7.7, the formula

$$(\mathrm{L} \supset \mathrm{F}, \mathrm{T})\, n^* = (\mathrm{L}\, n^* \supset \mathrm{F}\, n^*, \mathrm{T}\, n^*)$$

is deducible in A for each natural number n, and hence the formula $(\mathrm{L} \supset \mathrm{F}, \mathrm{T})\, 0^* = \mathrm{F}\, 0^*$ and all formulas

$$(\mathrm{L} \supset \mathrm{F}, \mathrm{T})\, (n+1)^* = \mathrm{T}\, (n+1)^*$$

are deducible. Therefore, by Proposition II.7.8, the formula $\mathcal{U} 0^* = 0^*$ and all formulas $\mathcal{U}(n+1)^* = \mathcal{U}\mathrm{R}(n+1)^*$ are deducible in A. We conclude that $\mathcal{U} 0^* = 0^*$ and all formulas $\mathcal{U}(n+1)^* = \mathcal{U} n^*$ are deducible, and this enables proving by induction that $\mathcal{U} n^* = 0^*$ is deducible for each natural number n.∎

Proposition 4. Whenever $\mathcal{U}_1, \ldots, \mathcal{U}_l, \mathcal{V}_1, \ldots, \mathcal{V}_l$ $(l \geq 1)$ are functional expressions, $\mathcal{X}$ is a normal functional expression, and the formulas

$$\mathcal{U}_j \mathcal{X} = \mathcal{V}_j, \quad j = 1, \ldots, l,$$

are deducible in the system A, then so is also the formula

$$(\mathcal{U}_1, \ldots, \mathcal{U}_l)\mathcal{X} = (\mathcal{V}_1, \ldots, \mathcal{V}_l).$$

Proof. Induction on l using Propositions II.7.1, II.7.7 and Corollary II.7.2.∎

Proposition 5. Let f_0 be a function of l arguments in $\mathbb{N}$, and $f_1, \ldots, f_l$ be functions of k arguments in $\mathbb{N}$. Let the functional expressions $\mathcal{U}_0, \mathcal{U}_1, \ldots, \mathcal{U}_l$ represent the functions $f_0, f_1, \ldots, f_l$, respectively. Then the functional expression $\mathcal{U}_0(\mathcal{U}_1, \ldots, \mathcal{U}_l)$ represents the function

$$\lambda n_1 \ldots n_k \,.\, f_0(f_1(n_1, \ldots, n_k), \ldots, f_l(n_1, \ldots, n_k)).$$

Proof. Denote by f the last function. Suppose $n_1, \ldots, n_k, m$ are natural numbers satisfying the condition (1). Then there are natural numbers $m_1, \ldots, m_l$ such that the following equalities hold:

$$f_j(n_1, \ldots, n_k) = m_j, \quad j = 1, \ldots, l,$$

$$f_0(m_1, \ldots, m_l) = m.$$

It follows from this equalities that the following formulas are deducible in the system A:

$$\mathcal{U}_j(n_1^*, \ldots, n_k^*) = m_j^*, \quad j = 1, \ldots, l,$$

(3) $$\mathcal{U}_0(m_1^*, \ldots, m_l^*) = m^*.$$

By Proposition 4, the deducibility of the first l of them implies the deducibility of the formula

$$(\mathcal{U}_1, \ldots, \mathcal{U}_l)(n_1^*, \ldots, n_k^*) = (m_1^*, \ldots, m_l^*),$$

and from here, taking into account the deducibility of (3), we conclude that

$$\mathcal{U}_0(\mathcal{U}_1, \ldots, \mathcal{U}_l)(n_1^*, \ldots, n_k^*) = m^*$$

is also deducible.∎

Proposition 6. Let f, g, h be total functions of k, $k+2$, $k+1$ arguments, respectively, in $\mathbb{N}$, and let, for all i and $n_1, \ldots, n_k$ in $\mathbb{N}$ the following equalities hold:

$$h(0, n_1, \ldots, n_k) = f(n_1, \ldots, n_k),$$

$$h(i+1, n_1, \ldots, n_k) = g(h(i, n_1, \ldots, n_k), i, n_1, \ldots, n_k).$$

Let the functional expressions $\mathcal{U}$ and $\mathcal{V}$ represent f and g, respectively, and let $\mathcal{W}$ be the functional expression

$$\mathrm{LR}[(\mathrm{RL}, \mathcal{Y}\mathrm{R}), (\mathrm{L} \supset \mathrm{F}, \mathrm{T})\,\mathrm{L}]\,(\mathrm{L}, \mathcal{U}\mathrm{R}, \mathcal{X}\mathrm{L}, \mathrm{R}),$$

where $\mathcal{X}$ is a functional expression representing the function $\lambda i.0$, and $\mathcal{Y}$ is the functional expression

$$(\mathcal{V}, (\mathrm{F},)\,\mathrm{LR}, \mathrm{R}^2).$$

Then $\mathcal{W}$ represents h.

Proof. Let $n_1, \ldots, n_k$ be some given natural numbers, and let, for each natural number i,

$$m_i = h(i, n_1, \ldots, n_k).$$

The statement to be proved is that all formulas

$$\mathcal{W}(i^*, n_1^*, \ldots, n_k^*) = m_i^*$$

are deducible in the system A. Let i_0 be some fixed natural number. We shall prove the deducibility of the above formula for $i = i_0$.

First of all, we note the deducibility in A of the formula

$$(\mathrm{L}, \mathcal{U}\mathrm{R}, \mathcal{X}\mathrm{L}, \mathrm{R})(i_0^*, n_1^*, \ldots, n_k^*) = (i_0^*, m_0^*, 0^*, n_1^*, \ldots, n_k^*)$$

and of all formulas

$$\mathcal{Y}(m_i^*, i^*, n_1^*, \ldots, n_k^*) = (m_{i+1}^*, (i+1)^*, n_1^*, \ldots, n_k^*).$$

The deducibility of the mentioned formulas follows from Propositions 1, 2, 4 and the equalities

$$f(n_1, \ldots, n_k) = m_0,$$

$$g(m_i, i, n_1, \ldots, n_k) = m_{i+1}.$$

Let $\mathcal{Z}$ be the functional expression

$$[(RL, \mathcal{Y} R), (L \supset F, T) L]$$

occurring in $\mathcal{W}$. Now we shall prove that, for $i = 0, 1, \ldots, i_0$, the following formula is deducible in A:

$$\mathcal{Z}((i_0 - i)^*, m_i^*, i^*, n_1^*, \ldots, n_k^*) = (0^*, m_{i_0}^*, i_0^*, n_1^*, \ldots, n_k^*)$$

This will be done by means of induction going downwards from i_0 to 0. The induction makes use of the deducibility of the formula

$$(L \supset F, T) L (0^*, m_{i_0}^*, i_0^*, n_1^*, \ldots, n_k^*) = F 0^*$$

and of the formulas

$$(L \supset F, T) L ((i_0 - i)^*, m_i^*, i^*, n_1^*, \ldots, n_k^*) = T (i_0 - i)^*,$$

where $i < i_0$. Using their deducibility and Proposition II.7.8, we observe the deducibility of the formula

$$\mathcal{Z}(0^*, m_{i_0}^*, i_0^*, n_1^*, \ldots, n_k^*) = (0^*, m_{i_0}^*, i_0^*, n_1^*, \ldots, n_k^*)$$

and of the formulas

$$\mathcal{Z}((i_0 - i)^*, m_i^*, i^*, n_1^*, \ldots, n_k^*) = \mathcal{Z}(RL, \mathcal{Y} R)((i_0 - i)^*, m_i^*, i^*, n_1^*, \ldots, n_k^*)$$

where $i < i_0$. The deducibility of the first formula gives the induction base. By the property of $\mathcal{Y}$ indicated at the beginning, the deducibility of the other ones implies the deducibility of the formulas

$$\mathcal{Z}((i_0 - i)^*, m_i^*, i^*, n_1^*, \ldots, n_k^*) = \mathcal{Z}((i_0 - (i+1))^*, m_{i+1}^*, (i+1)^*, n_1^*, \ldots, n_k^*),$$

where $i < i_0$, and this makes the induction step possible.

Applying the proved statement for $i = 0$, we get the deducibility of the formula

$$\mathcal{Z}(i_0^*, m_0^*, 0^*, n_1^*, \ldots, n_k^*) = (0^*, m_{i_0}^*, i_0^*, n_1^*, \ldots, n_k^*).$$

It follows from here that the formula

$$L R \mathcal{Z}(i_0^*, m_0^*, 0^*, n_1^*, \ldots, n_k^*) = m_{i_0}^*$$

is also deducible. Combining this with the property of $(L, \mathcal{U}R, \mathcal{X}L, R)$ indicated at the beginning, we get the needed deducibility of the formula

$$\mathcal{W}(i_0{}^*, n_1{}^*, \ldots, n_k{}^*) = m_{i_0}{}^*. \blacksquare$$

Proposition 7. Each primitive recursive function can be represented by some functional expression.

Proof. Application of Example 2, Corollary 1 and Propositions 2, 3, 5 and 6. ■

In the case of partial functions, the notion of representation used above has a certain drawback mentioned in Remark 3. In connection with this one more definition will be given.

Definition 3. Let f be a function of k arguments in $\mathbb{N}$. A functional expression $\mathcal{U}$ will be called to *represent f strongly* iff $\mathcal{U}$ represents f and, whenever $n_1, \ldots, n_k$ are such natural numbers that $\langle n_1, \ldots, n_k \rangle \notin \operatorname{dom} f$, then, for each choice of an iterative combinatory space $\mathfrak{S} = \langle \mathcal{F}, I, \mathcal{C}, \Pi, L, R, \Sigma, T, F \rangle$, the value of $\mathcal{U}(n_1, \ldots, n_k)^*$ in $\mathfrak{S}$ is the least element of $\mathcal{F}$.

A remark similar to Remark 2 can be made in connection of this definition, and unfortunately this time the arising problem is more unpleasant. In order to obviate the difficulty, we shall further assume that functions are given together with the information about the number of their arguments.

Of course, if the function f in the above definition is a total one then the additional requirement is trivially fulfilled. Here is an example with a non-total function f.

Example 5. The functional expression $[\ , T]$ represents strongly the empty function (does not matter of how many arguments).

Proposition 8. Let f be a total function of $k+1$ arguments in $\mathbb{N}$, and let h be the function of k arguments defined by means of the equality

$$h(n_1, \ldots, n_k) \simeq \mu i [f(i, n_1, \ldots, n_k) = 0].$$

Let the functional expression $\mathcal{U}$ represents f, and let $\mathcal{W}$ be the functional expression

$$L[((F,)L, R), (L \supset F, T)\mathcal{U}](\mathcal{X}L,),$$

where $\mathcal{X}$ is a functional expression representing the function $\lambda n.0$. Then $\mathcal{W}$ represents h strongly.

Proof. We first observe the deducibility of the formulas

(4) $(\mathcal{K}L,)(n_1^*, \ldots, n_k^*) = (0^*, n_1^*, \ldots, n_k^*)$,

(5) $((F,)L, R)(i^*, n_1^*, \ldots, n_k^*) = ((i+1)^*, n_1^*, \ldots, n_k^*)$

for every choice of the natural numbers $n_1, \ldots, n_k, i$.

Suppose now that natural numbers $n_1, \ldots, n_k$ and l are given such that $h(n_1, \ldots, n_k) = l$. We have to prove the deducibility of the formula $\mathcal{W}(n_1^*, \ldots, n_k^*) = l^*$. To do this, we set

$$m_i = f(i, n_1, \ldots, n_k), \quad i = 0, 1, \ldots, l,$$

and note that $m_i > 0$ for $i = 0, 1, \ldots, l-1$, but $m_l = 0$. Since $\mathcal{U}$ represents f, all formulas

$$\mathcal{U}(i^*, n_1^*, \ldots, n_k^*) = m_i^*, \quad i = 0, 1, \ldots, l,$$

are deducible. Using this, we observe the deducibility of the formulas

$$(L \supset F, T)\,\mathcal{U}(i^*, n_1^*, \ldots, n_k^*) = T\,m_i^*, \quad i = 0, 1, \ldots, l-1,$$

and of the formula

$$(L \supset F, T)\,\mathcal{U}(l^*, n_1^*, \ldots, n_k^*) = F\,0^*.$$

Let $\mathcal{Z}$ be the functional expression

$$[((F,)L, R), (L \supset F, T)\,\mathcal{U}]$$

occurring in $\mathcal{W}$. From Proposition II.7.8 and the deducibility of the above mentioned formulas, we conclude that following formulas are also deducible:

(6) $\mathcal{Z}(i^*, n_1^*, \ldots, n_k^*) = \mathcal{Z}((F,)L, R)(i^*, n_1^*, \ldots, n_k^*)$,
$i = 0, 1, \ldots, l-1$,

$$\mathcal{Z}(l^*, n_1^*, \ldots, n_k^*) = (l^*, n_1^*, \ldots, n_k^*).$$

Making use of the deducibility of the formulas (5) and (6), we see the deducibility of the formulas

$$\mathcal{Z}(i^*, n_1^*, \ldots, n_k^*) = \mathcal{Z}((i+1)^*, n_1^*, \ldots, n_k^*),$$
$$i = 0, 1, \ldots, l-1.$$

Now an induction going downwards from l to 0 proves that

$$\mathcal{Z}(i^*, n_1^*, \ldots, n_k^*) = (l^*, n_1^*, \ldots, n_k^*)$$

is a deducible formula for each natural number not greater than l. In particular, the formula

$$\mathcal{Z}(0^*, n_1^*, \ldots, n_k^*) = (l^*, n_1^*, \ldots, n_k^*)$$

is deducible. Taking into account the deducibility of the formula (4) and the formula

$$L(l^*, n_1^*, \ldots, n_k^*) = l^*,$$

we finally get the needed conclusion that the formula

$$\mathcal{W}(n_1^*, \ldots, n_k^*) = l^*$$

is deducible.

Now suppose that natural numbers $n_1, \ldots, n_k$ are given such that $\langle n_1, \ldots, n_k \rangle \notin \mathrm{dom}\, h$. Let $\mathfrak{S} = \langle \mathcal{F}, I, \mathcal{C}, \Pi, L, R, \Sigma, T, F \rangle$ be an arbitrary iterative combinatory space, and let o be its least element. We have to prove that the value of the functional expression $\mathcal{W}(n_1^*, \ldots, n_k^*)$ in $\mathfrak{S}$ is equal to o. In other words, the element

$$\varphi = L[\,|((F,)\,L, R)|, (L \to F, T)\,|\mathcal{U}|\,]\,|(\mathcal{X}L,)|\,(\overline{n_1}, \ldots, \overline{n_k})$$

of $\mathcal{F}$ has to be shown equal to o. By the deducibility of the formulas (4), (5) and the correctness of the system A, the equalities

$$|(\mathcal{X}L,)|\,(\overline{n_1}, \ldots, \overline{n_k}) = (\overline{0}, \overline{n_1}, \ldots, \overline{n_k})$$
$$|((F,)\,L, R)|\,(\overline{i}, \overline{n_1}, \ldots, \overline{n_k}) = (\overline{i+1}, \overline{n_1}, \ldots, \overline{n_k})$$

hold. Let x be an arbitrary fixed element of $\mathcal{C}$, and let $\mathcal{A}$ be the subset of $\mathcal{C}$ consisting of all elements of the form $(\overline{i}, \overline{n_1}, \ldots, \overline{n_k})x$, where $i \in \mathbb{N}$. The second of the above equalities shows that $\mathcal{A}$ is invariant with respect to the element $\sigma = |((F,)\,L, R)|$ of $\mathcal{F}$. Let χ be the element $(L \to F, T)\,|\mathcal{U}|$ of $\mathcal{F}$. We shall check now the inequality

(7) $$o \underset{\mathcal{A}}{\geq} (\chi \to o\sigma, L).$$

Let y be an arbitrary element of $\mathcal{A}$, i.e.

$$y = (\overline{i}, \overline{n_1}, \ldots, \overline{n_k})x$$

for some natural number i. Then

$$(\chi \to o\sigma, L)y = (\chi y \to o\sigma y, Ly).$$

We have

$$\chi y = (L \to F, T)\,|\mathcal{U}|\,(\overline{i}, \overline{n_1}, \ldots, \overline{n_k})x,$$

and for some positive integer m the formula

$$\mathcal{U}(i^*, n_1^*, \ldots, n_k^*) = m^*$$

is deducible since $f(i, n_1, \ldots, n_k) > 0$. For this m, making use of the correctness of the system A, we get

$$\chi y = (L \to F, T)\,\overline{m}\,x = T\overline{m},$$

and from here we conclude that

$$(\chi \to o\sigma, L)y = (T\overline{m} \to o\sigma y, Ly) = o\sigma y.$$

But σy is a normal element of the given combinatory space, since

$$\sigma y = (\overline{i+1}, \overline{n_1}, \ldots, \overline{n_k})x.$$

Therefore, the equality $o\sigma y = o$ holds, and, consequently,

$$oy = (\chi \to o\sigma, L)y.$$

Thus the validity of (7) is established, and, using the definition of iteration and the fact that $\mathcal{A}$ is invariant with respect to σ, we conclude that

$$o \underset{\mathcal{A}}{\geq} L[\sigma, \chi],$$

i.e. $o = L[\sigma, \chi]y$ for all y in $\mathcal{A}$. In particular,

$$o = L[\sigma, \chi](\overline{0}, \overline{n_1}, \ldots, \overline{n_k})x = \varphi x.$$

Since this is true for all x in $\mathcal{C}$, we see that $\varphi = o$. ∎

We have now everything needed for the proof of a representation theorem for all partial recursive functions.

Theorem 1. For each partial recursive function, there is some functional expression which represents this function strongly.

Proof. Let h be a partial recursive function of k arguments. By the Kleene Normal Form Theorem (Kleene [1952, § 63]), there are a $k+1$-argument primitive recursive function f and a one-argument primitive recursive function g, such that for all natural numbers $n_1, \ldots, n_k$ the equality

$$h(n_1, \ldots, n_k) \simeq g(\mu i[f(i, n_1, \ldots, n_k) = 0])$$

holds. Define the k-argument partial recursive function h_0 by means of the equality

$$h_0(n_1, \ldots, n_k) \simeq \mu i[f(i, n_1, \ldots, n_k) = 0].$$

By Propositions 7 and 8, there are some functional expression $\mathcal{W}_0$ strongly representing h_0 and some functional expression $\mathcal{V}$ representing g. Let $\mathcal{W}$ be the functional expression $\mathcal{V}\mathcal{W}_0$. Since

$$h(n_1, \ldots, n_k) \simeq g(h_0(n_1, \ldots, n_k))$$

for all $n_1, \ldots, n_k$, an application of Proposition 5 shows that $\mathcal{W}$ represents h. It remains only to show that the representation is a strong one.

Suppose that natural numbers $n_1, \ldots, n_k$ are given such that $\langle n_1, \ldots, n_k \rangle \notin \mathrm{dom}\, h$. Let $\mathfrak{S} = \langle \mathcal{F}, I, \mathcal{C}, \Pi, L, R, \Sigma, T, F \rangle$ be an arbitrary iterative combinatory space, and let o be its least element. We have to prove that the value of the functional expression $\mathcal{W}\langle n_1, \ldots, n_k \rangle^*$ in $\mathfrak{S}$ is equal to o. But this is clear, since $\langle n_1, \ldots, n_k \rangle \notin \mathrm{dom}\, h_0$, and therefore

$$|\mathcal{W}|(\overline{n_1}, \ldots, \overline{n_k})| = |\mathcal{V}|\,|\mathcal{W}_0|(\overline{n_1}, \ldots, \overline{n_k})| = |\mathcal{V}|\,o = o.$$ ∎

Corollary 2. Let $\mathfrak{S} = \langle \mathcal{F}, I, \mathcal{C}, \Pi, L, R, \Sigma, T, F\rangle$ be an iterative combinatory space, and let f be a partial recursive function of k arguments. Then there is an element φ of $\mathrm{COMP}_{\mathfrak{S}}(\emptyset)$ such that, for all natural numbers $n_1, \ldots, n_k, m$:

(i) if the equality (1) holds, then $\varphi(\overline{n_1}, \ldots, \overline{n_k}) = \overline{m}$ in $\mathfrak{S}$;

(ii) if $\langle n_1, \ldots, n_k\rangle \notin \mathrm{dom} f$, then $\varphi(\overline{n_1}, \ldots, \overline{n_k})$ is the least element of $\mathcal{F}$.

Note that a direct proof (without using functional expressions) of this corollary can be obtained by an appropriate modification of the proof of Theorem 1.

The representability of the partial recursive functions in the system A can be used to obtain certain negative results concerning this system and, more generally, the problems of decidability and axiomatizability of the theory of iterative combinatory spaces. Here are some results of this kind. The sign $\vdash$ in their formulations and proofs means deducibility in the system A (with an exception in Remark 5 below).

Theorem 2. A functional expression $\mathcal{U}$ of the system A with the following properties can be found:

(i) $\mathcal{U}$ does not contain variables;

(ii) the set $\{n \in \mathbb{N} : \vdash \mathcal{U} n^* = 0^*\}$ is not recursive (hence the recursive unsolvability of the problem of deciding whether a given formula is deducible in A);

(iii) there is a natural number n such that neither of the formulas $\mathcal{U} n^* = 0^*$ and $\neg(\mathcal{U} n^* = 0^*)$ is deducible in A, but the second of them is true in all iterative combinatory spaces (hence the syntactic and the semantic incompleteness of the system A);

(iv) whenever an iterative combinatory spaces is given, the non-recursive set from (ii) coincides with the set of the natural numbers n such that $\mathcal{U} n^* = 0^*$ is true in the given combinatory space (hence the recursive unsolvability of the problem of deciding whether a given formula is true in all combinatory spaces from a given non-empty class of iterative combinatory spaces).

Proof. Let E be a recursively enumerable subset of $\mathbb{N}$ such that $\mathbb{N} \setminus E$ is not recursively enumerable. Take f to be the restriction of the constant function $\lambda n.0$ to the set E, and take $\mathcal{U}$ to be a functional expression which represents f strongly. Then the formula $\mathcal{U}n^* = 0^*$ will be deducible for all n in E, and this formula will be non-deducible for all n in $\mathbb{N} \setminus E$ (by the correctness of A and the fact that the value of 0^* in an iterative combinatory space is different from the least element of the space). Hence the complement of the set $\{n \in \mathbb{N} : \vdash \mathcal{U}n^* = 0^*\}$ is not recursively enumerable. Since the system A is consistent, the set $\{n \in \mathbb{N} : \vdash \neg(\mathcal{U}n^* = 0^*)\}$ is a subset of this complement. But this subset is recursively enumerable due to the fact that A has a decidable notion of deduction. Hence there is some natural number n belonging to the difference of these two sets, and for such an n no one of the formulas $\mathcal{U}n^* = 0^*$ and $\neg(\mathcal{U}n^* = 0^*)$ could be deducible. It will be shown now that the second of these two formulas is true in all iterative combinatory spaces. To show this, we note than the concerned n does not belong to $\mathrm{dom} f$, since otherwise the first one of the formulas would be deducible. Therefore, if $\mathfrak{S} = \langle \mathcal{F}, I, \mathcal{C}, \Pi, L, R, \Sigma, T, F \rangle$ is an iterative combinatory space, then the value of $\mathcal{U}n^*$ in $\mathfrak{S}$ will be the least element of $\mathcal{F}$, and hence the second of the formulas will be true in $\mathfrak{S}$. By the correctness of the system A, the proof of (iv) can be reduced to showing that if for some natural number n the formula $\mathcal{U}n^* = 0^*$ is true in some iterative combinatory space then, for the same n, this formula is deducible in A. And it is really so, since such a natural number n is obliged to be in $\mathrm{dom} f$ (otherwise the value of $\mathcal{U}n^*$ would be the least element of the space) ∎

Remark 4. The functional expression $\mathcal{U}$ used in the above proof can be (in principle) effectively found. As to the natural number n with the property (iii), to assure such a possibility also for its construction, it is sufficient to choose E to be a creative set.

Remark 5. One could try to replace the system A with some other formal system in order to avoid the negative results in paragraphs (i) and (ii) of Theorem 2. Unfortunately, there are no reasons to be optimistic in this respect, since very few properties of A have been used in the proof of Theorem 2, namely: the strong representability of the partial recursive functions, the correctness of the system, its consistency (which follows from the correctness) and the fact that the system has a decidable notion of deduction. These properties will be preserved at least in the case when A is replaced by some correct extension of it having a decidable notion of deduction (in particular, the properties are preserved if we replace A by the system A' from Section II.7). But even if these conditions are weakened to some extent, the unpleasant situation remains. For example, the recursive unsolvability from (ii) remains if the formulas of the kind $\mathcal{U} = \mathcal{V}$ of A are among the formulas of the considered system, simple (not necessarily strong) representability of the partial recursive functions is present, the notion of deduction of the system is decidable, and instead of correctness and consistency the following weaker assumption is fulfilled: there is no functional expression $\mathcal{Z}$ such that both formulas $\mathcal{Z}=0^*$ and $\mathcal{Z}=1^*$ are deducible. Indeed, suppose that a system S with these properties is given. Then take an one-argument partial recursive function f which has no recursive extension and no values different from 0 and 1. Let $\mathcal{U}$ be a functional expression representing this function. We assert that $\{n\in\mathbb{N} : \vdash \mathcal{U}n^*=0^*\}$ is a non-recursive set again (the sign $\vdash$ meaning deducibility in S this time). Otherwise, we could obtain a recursive extension g of f by setting $g(n)=0$ in the case when $\mathcal{U}n^*=0^*$ is deducible, and $g(n)=1$ in the opposite case.

The formulation of the next theorem does not concern the deductive machinery of the system A and uses only a small part of its syntax and semantics (however, the proof of the theorem makes use of the information in Theorem 1 about the deductive machinery of A).

Theorem 3. There are functional expressions $\mathcal{U}$ and $\mathcal{V}$ with the following properties:

(i) $\mathcal{U}$ and $\mathcal{V}$ do not contain variables;

(ii) for every choice of an iterative combinatory space, the set $\{n\in\mathbb{N} : |\mathcal{U}n^*|=|\mathcal{V}|\}$ is not recursively enumerable, and this set does not depend on the choice of the combinatory space.

Proof. Let $\mathcal{U}$ be chosen in the same way, as in the proof of Theorem 2, and let $\mathcal{V}$ be the functional expression [, T]. Then the set described in (ii) always coincides with the set $\mathbb{N}\setminus E$ from the same proof.∎

Theorem 3 can be used for showing the non-axiomatizability of any given non-empty class of iterative combinatory spaces. Let $\mathfrak{K}$ be such a class of combinatory spaces. We shall show that no formal system S is possible with the following properties: the system S has a decidable notion of deduction, the equalities between functional expressions are among the formulas of S, and an equality between functional expressions without variables is deducible in S iff this equality is true in each combinatory spaces of the class $\mathfrak{K}$. Suppose S is a formal system with this properties. Then, taking functional expressions $\mathcal{U}$ and $\mathcal{V}$ with the properties from Theorem 3, we can form the set of all natural numbers n such that the formula $\mathcal{U}n^*=\mathcal{V}$ is deducible in S. This set must be recursively enumerable due to the assumption that S has a decidable notion of deduction. On the other hand, according to the last assumption about S, this set must coincide with the set from paragraph (ii) of Theorem 3, and this is a contradiction.

We should like to end this section by mentioning that some other representations of number theoretic functions in an iterative combinatory space can be obtained by application of the representation theorems from Georgieva [1984] and Ivanov [1980, 1986] to the companion operative space of the given combinatory space.

Exercises

1. Let A_0 be the formal system defined in the following way. The language of A_0 is contained in the language of A and consists again of functional expressions and formulas. Functional expressions of A_0 are those functional expressions of A which contain neither variables nor the symbol $\supset$. Formulas of A_0 are those formulas of A which have the form $\mathcal{U}=\mathcal{X}$, where $\mathcal{U}$, $\mathcal{X}$ are functional expressions of A_0, and $\mathcal{X}$ is normal (in the sense of Section II.7). The system has the following axioms and rules of inference, where $\mathcal{U}$, $\mathcal{V}$, $\mathcal{X}$, $\mathcal{Y}$, $\mathcal{Z}$ are functional expressions of A_0, and $\mathcal{X}$, $\mathcal{Y}$, $\mathcal{Z}$ are normal:

$$\mathcal{X}=\mathcal{X} \qquad L(\mathcal{X}, \mathcal{Y})=\mathcal{X} \qquad R(\mathcal{X}, \mathcal{Y})=\mathcal{Y}$$

$$\frac{\mathcal{U}=\mathcal{X} \quad \mathcal{V}\mathcal{X}=\mathcal{Y}}{\mathcal{V}\mathcal{U}=\mathcal{Y}} \qquad \frac{\mathcal{U}\mathcal{X}=\mathcal{Y} \quad \mathcal{V}\mathcal{X}=\mathcal{Z}}{(\mathcal{U}, \mathcal{V})\mathcal{X}=(\mathcal{Y}, \mathcal{Z})}$$

$$\frac{\mathcal{V}\mathcal{X}=T\mathcal{Z} \quad [\mathcal{U}, \mathcal{V}]\mathcal{U}\mathcal{X}=\mathcal{Y}}{[\mathcal{U}, \mathcal{V}]\mathcal{X}=\mathcal{Y}} \qquad \frac{\mathcal{V}\mathcal{X}=F\mathcal{Z}}{[\mathcal{U}, \mathcal{V}]\mathcal{X}=\mathcal{X}}$$

Show that all formulas deducible in A_0 are deducible also in A. Adopting the same definitions for the notions of representation and strong representation of a function as in the case of the system A, but with A replaced by A_0, prove that all partial recursive functions are strongly representable in A_0.

Hint. Make use of the deducibility in A_0 of the formulas $(n+1)^*=(Fn^*, n^*)$. Use the functional expression $R[(F, F), L](L, TR)$ instead of $(L\supset F, T)$.

2. (Soskov [1979, p. 40]) Let M be the set B^* from

Moschovakis [1969][79] in the special case of $B=\mathbb{N}$ (of course, the identification of some elements of $B^*\setminus B$ with natural numbers must be declined in this case). Let $\mathfrak{U}$ be the corresponding computational structure $\mathfrak{M}_B$, P be the mapping $\lambda u.\, u\dot{-}1$ of $\mathbb{N}$ into $\mathbb{N}$, and Z be the mapping of $\mathbb{N}$ into M defined as follows: $Z(u)=\langle O, O\rangle$ for all u in $\mathbb{N}\setminus\{0\}$, $Z(0)=O$ (pay attention to the difference between "0" and "O"!). Let $\mathcal{F}=\mathcal{F}_{\mathrm{p}}(M)$, k be a positive integer, and f be a partial mapping of $\mathbb{N}^k$ into $\mathbb{N}$. Prove that the following two conditions are equivalent:

(i) there is an element φ of $\mathcal{F}$ such that φ is $\mathfrak{U}$-computable in $\{P, Z\}$ and, for all natural numbers $n_1, \ldots, n_k$, the equality

$$f(n_1, \ldots, n_k)\simeq\varphi(\langle n_1, \langle n_2, \ldots, \langle n_{k-1}, n_k\rangle\rangle \ldots\rangle\rangle)$$

holds;

(ii) f is partial recursive, and, for all $n_1, \ldots, n_k$ satisfying the condition $\langle n_1, \ldots, n_k\rangle\in\operatorname{dom} f$, the inequality $f(n_1, \ldots, n_k)\le\max\{n_1, \ldots, n_k\}$ holds.

Hint. To prove the implication from (ii) to (i), apply Corollary 2 to represent f in the iterative combinatory space $\mathfrak{S}_{\mathrm{p}}(\mathfrak{U})$ and, in addition, construct $\mathfrak{U}$-computable in $\{P, Z\}$ elements α, β, γ of $\mathcal{F}_{\mathrm{p}}(M)$ such that, for all $n_1, \ldots, n_k, m, n$ in $\mathbb{N}$,

$$\alpha(\langle n_1, \langle n_2, \ldots, \langle n_{k-1}, n_k\rangle\rangle \ldots\rangle\rangle)=(\overline{n_1}, \ldots, \overline{n_k})(n_1),$$
$$\beta(\langle n_1, \langle n_2, \ldots, \langle n_{k-1}, n_k\rangle\rangle \ldots\rangle\rangle)=\max\{n_1, \ldots, n_k\},$$
$$\gamma(\langle\overline{m}(n_1), n\rangle)=\min\{m, n\}$$

(where $\overline{i}$ is the element of $\mathcal{F}$ representing i in $\mathfrak{S}_{\mathrm{p}}(\mathfrak{U})$).

3. Prove the existence of a combinatory space $\langle\mathcal{F}, I, \mathcal{C}, \Pi, L, R, \Sigma, T, F\rangle$ which is not iterative, but, however, for all σ, χ, ρ in $\mathcal{F}$, the element $\mu\tau.(\chi\rightarrow\tau\sigma, \rho)$ exists, and the equality

$$\mu\tau.(\chi\rightarrow\tau\sigma, \rho)=\rho\,\mu\tau.(\chi\rightarrow\tau\sigma, I)$$

holds.

Hint. Use the completeness theorem for the predicate calculus and the semantic incompleteness of the system A with respect to the class of the iterative combinatory

[79] Cf. also Section I.7 of the present book.

spaces.

4. The First Recursion Theorem for iterative combinatory spaces

In the whole section (including the exercises), an iterative combinatory space

$$\mathfrak{S} = \langle \mathcal{F}, I, \mathcal{C}, \Pi, L, R, \Sigma, T, F \rangle$$

and a subset $\mathcal{B}$ of $\mathcal{F}$ will be supposed to be given. In Proposition 2.2, the fact was established that all elements of $\mathcal{F}$ and all mappings in $\mathcal{F}$ $\mathfrak{S}$-computable in $\mathcal{B}$ are fixed-point definable in the corresponding partially ordered algebra $\mathfrak{C}(\mathfrak{S}, \mathcal{B})$. The problem was noted whether the converse is true, and giving an affirmative answer was promised. In the present section a theorem will be proved giving this answer. The theorem will be called "First Recursion Theorem", since it is similar to the First Recursion Theorem from the ordinary recursion theory dealing with partial recursive functions in $\mathbb{N}$. The logical relationship between the two theorems will be discussed in the next section on the basis of the considerations from Chapter I.

The simplest case of the First Recursion Theorem can be formulated as follows: whenever Γ is a mapping of $\mathcal{F}$ into $\mathcal{F}$, and Γ is $\mathfrak{S}$-computable in $\mathcal{B}$, then the element $\mu\tau.\Gamma(\tau)$ exists and this element is also $\mathfrak{S}$-computable in $\mathcal{B}$. This is a particular case of the following more general formulation:

Theorem 1 (Non-parameterized version of the First Recursion Theorem). Let l be a positive integer, and $\Gamma_1, \dots, \Gamma_l$ be mappings of $\mathcal{F}^l$ into $\mathcal{F}$ which are $\mathfrak{S}$-computable in $\mathcal{B}$. Then the system of inequalities

(1) $$\tau_r \geq \Gamma_r(\tau_1, \dots, \tau_l), \quad r = 1, \dots, l,$$

has a least solution $\langle \tau_1, \dots, \tau_l \rangle$, and the components of this solution are $\mathfrak{S}$-computable in $\mathcal{B}$.

If we allow the right-hand sides of the inequalities in (1) to contain parameters, then we obtain a stronger version of the theorem.

Theorem 2 (Parameterized version of the First Recursion Theorem). Let l and n be positive integers, and $\Gamma_1, \dots, \Gamma_l$ be mappings of $\mathcal{F}^{l+n}$ into $\mathcal{F}$ which are $\mathfrak{S}$-computable in $\mathcal{B}$. Then mappings $\Delta_1, \dots, \Delta_l$ of $\mathcal{F}^n$ into $\mathcal{F}$ exist such that, for any choice of $\theta_1, \dots, \theta_n$ in $\mathcal{F}$, the l-tuple

$$\langle \Delta_1(\theta_1, \ldots, \theta_n), \ldots, \Delta_l(\theta_1, \ldots, \theta_n) \rangle$$

is the least solution $\langle \tau_1, \ldots, \tau_l \rangle$ of the system of inequalities

(2) $\tau_r \geq \Gamma_r(\theta_1, \ldots, \theta_n, \tau_1, \ldots, \tau_l)$, $r = 1, \ldots, l$,

and $\Delta_1, \ldots, \Delta_l$ are $\mathfrak{S}$-computable in $\mathfrak{B}$.

Remark 1. Theorem 2 is stronger than Theorem 1, since there is a simple way to obtain Theorem 1 from it (namely by introducing fictitious parameters and replacing them by suitable elements of $\mathcal{F}$), and no simple way is seen for the converse reduction. We shall restrict ourselves to proving Theorem 2.

Remark 2. The case of arbitrary positive integer l is in a sense not essentially more general than the case of $l = 1$, since there is an easy way to reduce the first case to the second one (see Proposition II.4.3). Such a reduction has been used in Skordev [1980].[74] However, the proof which will be given here (following the papers Skordev [1984, 1989]) will not use a reduction of this sort.

Remark 3. The case of arbitrary positive integer n in Theorem 2 can be reduced to the case of $n = 1$ by considering the mappings $\Gamma'_1, \ldots, \Gamma'_l$ of $\mathcal{F}^{l+1}$ into $\mathcal{F}$ defined by means of the equalities

$$\Gamma'_r(\theta, \tau_1, \ldots, \tau_l) =$$
$$\Gamma_r(\theta\overline{0}, \theta\overline{1}, \ldots, \theta\overline{n-2}, \theta R_*^{n-1}, \tau_1, \ldots, \tau_l), \quad r = 1, \ldots, l.$$

Indeed, suppose that $\Delta_1', \ldots, \Delta_l'$ are mappings of $\mathcal{F}$ into $\mathcal{F}$ $\mathfrak{S}$-computable in $\mathfrak{B}$ and such that, for every fixed θ in $\mathcal{F}$, the l-tuple $\langle \Delta'_1(\theta), \ldots, \Delta'_l(\theta) \rangle$ is the least solution $\langle \tau_1, \ldots, \tau_l \rangle$ of the system of inequalities

$$\tau_r \geq \Gamma'_r(\theta, \tau_1, \ldots, \tau_l), \quad r = 1, \ldots, l.$$

Since the system of inequalities (2) is equivalent to the system

$$\tau_r \geq \Gamma'_r(\Pi_*(\theta_1, \ldots, \theta_n), \tau_1, \ldots, \tau_l), \quad r = 1, \ldots, l,$$

we can obtain the needed $\Delta_1, \ldots, \Delta_l$ by setting

$$\Delta_i(\theta_1, \ldots, \theta_n) = \Delta'_i(\Pi_*(\theta_1, \ldots, \theta_n)), \quad i = 1, \ldots, l.$$

---

[74] The proof given there is based on a previously proved normal form theorem for the considered mappings, and therefore the case of a single inequality turned out to be easier then.

This reduction will be used in the proof of Theorem 2 for reducing the amount of writing.

Proof of Theorem 2. Making use of the above remark, we shall restrict ourselves to the case of $n=1$. Thus we suppose that $\Gamma_1, \ldots, \Gamma_l$ are mappings of $\mathcal{F}^{n+1}$ into $\mathcal{F}$ which are $\mathfrak{S}$-computable in $\mathcal{B}$, and we shall prove the existence of mappings $\Delta_1, \ldots, \Delta_l$ of $\mathcal{F}$ into $\mathcal{F}$, also $\mathfrak{S}$-computable in $\mathcal{B}$, such that, for any fixed θ in $\mathcal{F}$, the l-tuple $\langle\Delta_1(\theta), \ldots, \Delta_l(\theta)\rangle$ is the least solution $\langle\tau_1, \ldots, \tau_l\rangle$ of the system of inequalities

(3) $\qquad \tau_r \geq \Gamma_r(\theta, \tau_1, \ldots, \tau_l), \quad r=1, \ldots, l.$

Without loss of generality, we can suppose that $\Gamma_1, \ldots, \Gamma_l$ are simple operations of the partially ordered algebra $\mathfrak{C}(\mathfrak{S}, \mathcal{B})$. In fact, by Proposition 2.2, the mappings $\Gamma_1, \ldots, \Gamma_l$ are fixed-point definable in the partially ordered algebra $\mathfrak{C}(\mathfrak{S}, \mathcal{B})$, and this enables applying Theorem 1.1 to the system (3). According to that theorem, there are a natural number m and an $l+m$-tuple $\Gamma'_1, \ldots, \Gamma'_{l+m}$ of $l+m+1$-ary simple operations of $\mathfrak{C}(\mathfrak{S}, \mathcal{B})$ such that, for each fixed θ in $\mathcal{F}$, if $\langle\varphi_1, \ldots, \varphi_{l+m}\rangle$ is the least solution $\langle\tau_1, \ldots, \tau_{l+m}\rangle$ of the system

(4) $\qquad \tau_r \geq \Gamma'_r(\theta, \tau_1, \ldots, \tau_{l+m}), \quad r=1, \ldots, l+m,$

then $\langle\varphi_1, \ldots, \varphi_l\rangle$ is the least solution $\langle\tau_1, \ldots, \tau_l\rangle$ of the system (3). Thus the system (3) can be replaced by the system (4) for the purpose of the present proof, and so the theorem is reduced to its particular case when the given $\Gamma_1, \ldots, \Gamma_l$ are simple operations of the partially ordered algebra $\mathfrak{C}(\mathfrak{S}, \mathcal{B})$. From now on, they will be assumed to be such operations, i.e. each Γ_r will be assumed to have one of the following forms:

$$\lambda\psi_0\psi_1\cdots\psi_l\cdot\psi_i,$$
$$\lambda\psi_0\psi_1\cdots\psi_l\cdot\alpha,$$
$$\lambda\psi_0\psi_1\cdots\psi_l\cdot\psi_j\psi_i,$$
$$\lambda\psi_0\psi_1\cdots\psi_l\cdot(\psi_i, \psi_j),$$
$$\lambda\psi_0\psi_1\cdots\psi_l\cdot(\psi_i\rightarrow\psi_j, \psi_k),$$

where i, j, k are fixed numbers from the set $\{0, 1, \ldots, l\}$, and α is some fixed element of $\{I, L, R, T, F\}\cup\mathcal{B}$.

One more reduction of the problem will be done, namely to consider only the case when the only mapping of the first form among $\Gamma_1, \ldots, \Gamma_l$ is $\lambda\psi_0\psi_1\cdots\psi_l\cdot\psi_0$, and all numbers i, j, k belong to the set $\{1, \ldots, l\}$, i.e. the case when each inequality of the system (3) has one of the following forms:

(5) $$\tau_r \geq \theta,$$

(6) $$\tau_r \geq \alpha,$$

(7) $$\tau_r \geq \tau_j\tau_i,$$

(8) $$\tau_r \geq (\tau_i, \tau_j),$$

(9) $$\tau_r \geq (\tau_i \rightarrow \tau_j, \tau_k),$$

where i, j, k are fixed numbers from the set $\{1, \ldots, l\}$, and α is some fixed element of $\{I, L, R, T, F\} \cup \mathcal{B}$. In fact, if it is not this case then, making use of Theorem II.4.1 (on elimination), we could replace the system (3) by another one obtained in the following way: we add the inequalities $\tau_{l+1} \geq I$, $\tau_{l+2} \geq \theta$, replace each inequality of the form $\tau_r \geq \tau_i$ by the inequality $\tau_r \geq \tau_i\tau_{l+1}$ and replace by τ_{l+2} each occurrence of θ in some inequality not of the form $\tau_r \geq \theta$. We shall further assume that each inequality of (3) has one of the forms (5) - (9).

The next steps in the proof have an intuitive explanation as the arrangement of some coding. Its purpose is to enable imitation of a sort of stack-implementation of a system of mutually recursive monadic procedures[75]. The components $\tau_1, \ldots, \tau_l$ of the least solution can be regarded as such a system of procedures, and the elements $\overline{1}, \ldots, \overline{l}$ of $\mathcal{F}$ will be used as their "codes" (the elements of $\{I, L, R, T, F\} \cup \mathcal{B}$ must be regarded as the basic primitive procedures, and θ as an external procedure which can be chosen arbitrarily). Since the operations Π and Σ correspond to certain non-monadic operations on monadic procedures, some additional "codes" will be used for certain, so to say, auxiliary monadic procedures having one more parameter besides their argument (such procedures arise from procedures with two parameters—arguments by means of so-called "curryfication"). Here is a more concrete description of these other "codes" and their intuitive meaning:

(i) if $1 \leq r \leq l$, and there is an inequality of the form (8) with this r in the system (3), then, for each x in $\mathcal{C}$, the element $\overline{l+r}(x, I)$ will be the "code" for the "proce-

---

[75] The term "monadic" is used here in the sense of "having one parameter-argument".

dure" $(I, \tau_j x)$, which arises at the representation of the inequality

$$\tau_r x \geq (\tau_i, \tau_j) x$$

into the form

$$\tau_r x \geq (I, \tau_j x) \tau_k x ;$$

(ii) if $1 \leq r \leq l$, and there is an inequality of the form (9) with this r in the system (3), then, for each x in $\mathcal{C}$, the element $\overline{l+r}(x, I)$ will be the "code" for the "procedure" $(I \rightarrow \tau_j x, \tau_k x)$, which arises at the representation of the inequality

$$\tau_r x \geq (\tau_i \rightarrow \tau_j, \tau_k) x$$

into the form

$$\tau_r x \geq (I \rightarrow \tau_j x, \tau_k x) \tau_i x ;$$

(iii) for each y in $\mathcal{C}$, the element $\overline{2l+1}(y, I)$ of $\mathcal{F}$ will be the "code" for the "procedure" (y, I), which arises at the representation of elements of the form $(I, \tau_j x) y$ into the form $(y, I) \tau_j x$.

Note that all "codes" listed above are normal elements of $\mathcal{F}$.

The "stacks contents" will be represented by means of products of the form $\eta_1 \ldots \eta_p \overline{0} c$, where c is some fixed element of $\mathcal{C}$, and $\eta_1, \ldots, \eta_p$ are the "codes" of some "procedures" subject to execution in this order (the product $\overline{0} c$ representing the contents of an "empty stack"). Of course, the object domain where the "procedures" act is represented by means of the set $\mathcal{C}$. The task of "application" of the above sequence of "procedures" to an "object" x will be represented by means of the element

$$(x, \eta_1 \ldots \eta_p \overline{0} c),$$

which obviously belongs to $\mathcal{C}$ (since $\eta_1 \ldots \eta_p \overline{0} c \in \mathcal{C}$). The implementation of the given "system of procedures" will consist in a step by step transformation of such "task representations" one into another. We are going now to construct a mapping E of $\mathcal{F}$ into $\mathcal{F}$, $\mathfrak{S}$-elementary in $\mathcal{B}$, such that, for any fixed θ in $\mathcal{F}$, the element $E(\theta)$ of $\mathcal{F}$ "performs" the mentioned transformation.

The conditions which must be satisfied by $E(\theta)$ are the following ones:

1) If $1 \leq r \leq l$, and there is an inequality of the form (5) or (6) with this r in the system (3), then

$$E(\theta)(x, \overline{r} z) = (\theta x, z)$$

or

$$E(\theta)(x, \overline{r}z) = (\alpha x, z),$$

respectively, for all x and z in $\mathscr{C}$.

2) If $1 \le r \le l$, and there is an inequality of the form (7) with this r in the system (3), then

$$E(\theta)(x, \overline{r}z) = (x, \overline{l}\,\overline{j}z)$$

for all x and z in $\mathscr{C}$.

3) If $1 \le r \le l$, and there is an inequality of the form (8) or (9) with this r in the system (3), then for all x, y, z in $\mathscr{C}$,

$$E(\theta)(x, \overline{r}z) = (x, \overline{l}\,\overline{l+r}(x, I)z)$$

in both cases,

$$E(\theta)(y, \overline{l+r}(x, I)z) = (x, \overline{j}\,\overline{2l+1}(y, I)z)$$

if the inequality has the form (8), and

$$E(\theta)(y, \overline{l+r}(x, I)z) = (y \rightarrow (x, \overline{j}z), (x, \overline{k}z))$$

in the other case.

4) For all x, y, z in $\mathscr{C}$, the equalities

$$E(\theta)(x, \overline{2l+1}(y, I)z) = ((y, x), z),$$

$$E(\theta)(x, \overline{0}z) = (x, \overline{0}z)$$

hold.

From the already presented intuitive point of view, all equalities in 1)–4) except the second one in 4) correspond to rules of the mentioned step by step transformation of "task representations". The equality $E(\theta)(x, \overline{0}z) = (x, \overline{0}z)$ is needed for technical reasons, and it has the intuitive meaning that "tasks with an empty stack" are transformed into themselves.

Of course, the existence of such a mapping E must be proved before going on further. This can be done by using Proposition II.2.2. We recall that a binary operation Δ in $\mathscr{F}$ has been defined in Section II.2 by means of the equality

$$\Delta(\varphi, \psi) = (LR \rightarrow \varphi(L, R^2), \psi(L, R^2)),$$

and then the definition was extended by setting

$$\Delta(\varphi_0, \varphi_1, \ldots, \varphi_n) = \Delta(\varphi_0, \Delta(\varphi_1, \ldots, \Delta(\varphi_{n-1}, \varphi_n)\ldots)).$$

It is clear that, for each natural number n greater than 1, the restriction of the operation Δ to $\mathscr{F}^n$ is a mapping $\mathfrak{S}$-elementary in $\emptyset$. We recall also the equalities

$$\Delta(\varphi_0, \ldots, \varphi_n)(x, \bar{r}) = \varphi_r(x, I), \quad r = 0, 1, \ldots, n,$$
$$\Delta(\varphi_0, \ldots, \varphi_n)(x, (F, I)^n) = \varphi_n(x, I),$$

which have been proved in Proposition II.2. We note that these equalities imply immediately the equalities

$$\Delta(\varphi_0, \ldots, \varphi_n)(x, \bar{r}y) = \varphi_r(x, y), \quad r = 0, 1, \ldots, n,$$
$$\Delta(\varphi_0, \ldots, \varphi_n)(x, (F, I)^n y) = \varphi_n(x, y).$$

Having in mind all this, we see that it is sufficient to define the mapping E by means of the equality

$$E(\theta) = \Delta(E_0(\theta), E_1(\theta), \ldots, E_{2l+1}(\theta)),$$

where $E_0, E_1, \ldots, E_{2l+1}$ are $\mathfrak{S}$-elementary in $\mathfrak{B}$ and satisfy the following conditions:

1_0) If $1 \leq r \leq l$, and there is an inequality of the form (5) or (6) with this r in the system (3), then

$$E_r(\theta)(x, z) = (\theta x, z)$$

or

$$E_r(\theta)(x, z) = (\alpha x, z),$$

respectively, for all x and z in $\mathcal{C}$.

2_0) If $1 \leq r \leq l$, and there is an inequality of the form (7) with this r in the system (3), then

$$E_r(\theta)(x, z) = (x, \bar{l}\,\bar{j}z)$$

for all x and z in $\mathcal{C}$.

3_0) If $1 \leq r \leq l$, and there is an inequality of the form (8) or (9) with this r in the system (3), then, for all x, y, z in $\mathcal{C}$,

$$E_r(\theta)(x, z) = (x, \bar{l}\,\overline{l+r}(x, I)z)$$

in both cases,

$$E_{l+r}(\theta)(y, (x, I)z) = (x, \bar{j}\,\overline{2l+1}(y, I)z)$$

if the inequality has the form (8), and

$$E_{l+r}(\theta)(y, (x, I)z) = (y \to (x, \bar{j}z), (x, \bar{k}z))$$

in the other case.

4_0) For all x, y, z in $\mathcal{C}$, the equalities

$$E_{2l+1}(\theta)(x, \bar{0}(y, I)z) = ((y, x), z),$$
$$E_0(\theta)(x, z) = (x, \bar{0}z)$$

hold.

The existence of $E_0, E_1, \ldots, E_{2l+1}$ with the above properties is seen by means of their explicit construction. Namely we can set $E_r(\theta) = (\theta L, R)$ for the first case of 1_0) and $E_r(\theta) = (\alpha L, R)$ for the second case. To satisfy 2_0), we can set $E_r(\theta) = (L, \bar{i}\bar{j}R)$. Satisfying 3_0) can be achieved by setting $E_r(\theta) = (L, \bar{i}\,\overline{l+r})$ in both cases,

$$E_{l+r}(\theta) = (LR, \bar{j}\,\overline{2l+1}\,(L, R^2))$$

if the inequality has the form (8), and

$$E_{l+r}(\theta) = (L \to (L, \bar{j}R)R, (L, \bar{k}R)R)$$

in the other case. Lastly, it is convenient to set

$$E_{2l+1}(\theta) = ((LR^2, L), R^3), \quad E_0(\theta) = (L, \bar{0}R).$$

Note the possibility that no conditions are imposed on some E_s with $l < s \leq 2l$; in this case we can choose E_s arbitrarily (for example, we can set $E_s(\theta) = I$).

From now on, a mapping E with the listed properties will be supposed to be fixed. Using this mapping, we shall construct certain mappings $\Delta_1, \ldots, \Delta_l$ of $\mathcal{F}$ into $\mathcal{F}$, which will be $\mathfrak{S}$-computable in $\mathcal{B}$ on the basis of their construction, and it will be proved (by hard work) that, for any fixed θ in $\mathcal{F}$, the l-tuple $\langle \Delta_1(\theta), \ldots, \Delta_l(\theta) \rangle$ is the least solution $\langle \tau_1, \ldots, \tau_l \rangle$ of the system of inequalities (3).

The idea how to construct $\Delta_r(\theta)$ is straight-forward after the intuitive explanations given until now. If some "object" x is given, and $\tau_r = \Delta_r(\theta)$ has "to be executed at x", then it is natural to form the "task representation" $(x, \bar{r}\bar{0}x)$ (taking x in the role of c) and then to start an iteration of the step by step transformation performed by $E(\theta)$. The termination condition for this iteration must obviously be "obtaining a task representation with an empty stack". The "result of the execution" will be the "object" y in such a "terminal task representation" $(y, \bar{0}x)$. Writing all this in a formal way, we get the following definition of the mappings $\Delta_1, \ldots, \Delta_l$:

(10) $\quad \Delta_r(\theta) = L[E(\theta), (L \to F, T)R](I, r\bar{0}), \quad r = 1, \ldots, l.$

The $\mathfrak{S}$-computability of these mappings in $\mathcal{B}$ is evident, and the rest of the proof is devoted to the other statement about them.

From now on, the element θ of $\mathcal{F}$ will be considered as fixed, and, for the sake of brevity, we set

$$\varepsilon = \mathrm{E}(\theta),$$

$$\iota = [\mathrm{E}(\theta), (L \to F, T)R],$$

(11) $$\delta_r = \Delta_r(\theta), \quad r = 1, \ldots, l.$$

For the needs of the proof, we set also, for each c in $\mathcal{C}$,

$$\delta_r^c = L[\mathrm{E}(\theta), (L \to F, T)R](I, r\bar{0}c), \quad r = 1, \ldots, l.$$

We shall prove that $\langle \delta_1^c, \ldots, \delta_l^c \rangle$ is the least solution $\langle \tau_1, \ldots, \tau_l \rangle$ of the system of inequalities (3) (from here, using the arbitrariness in the choice of c, it would be easy to draw the same conclusion about $\langle \delta_1, \ldots, \delta_l \rangle$).

The elements $\bar{1}, \ldots, \bar{l}$ of $\mathcal{F}$ and all elements of the form $\bar{s}(z, I)$, mentioned in the paragraphs (i) - (iii) of the intuitive explanations at the beginning of the proof, will be called *coding elements*. As already noted, all these elements are normal.

For any given c in $\mathcal{C}$, let $\mathcal{P}_c$ be the set of all elements of $\mathcal{C}$ having the form $\eta_1 \ldots \eta_p \bar{0} c$, where p is a natural number (possibly 0), and $\eta_1, \ldots, \eta_p$ are coding elements. Let $\mathcal{T}_c$ (the set of all possible "task representations" corresponding to c) be the set of all elements of $\mathcal{C}$ having the form (x, y) for some x from $\mathcal{C}$ and some y from $\mathcal{P}_c$. From now on, except for the concluding part of the proof, an element c of $\mathcal{C}$ will be supposed to be fixed.

One more definition will be useful for the further exposition. Let y_1, y_2, z_1, z_2 be some elements of $\mathcal{C}$. It will be said that y_1, y_2 are *proportional* to z_1, z_2 iff

$$y_1 = \eta_1 \ldots \eta_p z_1, \quad y_2 = \eta_1 \ldots \eta_p z_2$$

for some p and some $\eta_1, \ldots, \eta_p$ such as in the definition of the set $\mathcal{P}_c$.

In view of the considerable length of the present proof, some statements in it will be formulated as lemmas accompanied with their own proofs.

Lemma 1. Let z_1, z_2 be given elements of $\mathcal{C}$, λ_1, λ_2 be given elements of $\mathcal{F}$. Whenever x, y_1, y_2 are elements of $\mathcal{C}$, y_1, y_2 being proportional to z_1, z_2, let the inequality

$$\lambda_1(x, y_1) \geq \lambda_2(x, y_2)$$

hold. Then, for any choice of the coding element η, the inequality

(12) $$\lambda_1 \varepsilon(x, \eta y_1) \geq \lambda_2 \varepsilon(x, \eta y_2)$$

holds under the same conditions on x, y_1, y_2.

Proof. Since the inequality $\lambda_1(x, y_1) \geq \lambda_2(x, y_2)$ can be written in the form $\lambda_1(I, y_1)x \geq \lambda_2(I, y_2)x$, it follows that $\lambda_1(I, y_1) \geq \lambda_2(I, y_2)$ whenever y_1, y_2 are elements of $\mathcal{E}$ proportional to z_1, z_2. Let η be an arbitrary coding element, x, y_1, y_2 be element of $\mathcal{E}$, y_1, y_2 being proportional to z_1, z_2. There are several possibilities for η, and they will be scrutinized separately.

At first, let us consider the case when $\eta = \overline{r}$ for some r from the set $\{1, \ldots, l\}$. If the corresponding inequality in (3) is $\tau_r \geq \theta$ then, by the condition 1) on E, the equalities

$$\lambda_t \varepsilon(x, \eta y_t) = \lambda_t(\theta x, y_t) = \lambda_t(I, y_t)\theta x, \quad t = 1, 2,$$

hold, and the validity of (12) follows. The case,when the inequality in (3) concerning τ_r is of the form (6), can be treated in the same way.

Consider now the case, when this inequality has the form $\tau_r \geq \tau_j \tau_i$. Then, by the condition 2) on E, the equalities

$$\lambda_t \varepsilon(x, \eta y_t) = \lambda_t(x, \overline{i}\,\overline{j} y_t), \quad t = 1, 2,$$

hold. Since $\overline{i}\,\overline{j} y_1, \overline{i}\,\overline{j} y_2$ are again proportional to z_1, z_2, the inequality (12) turns out to be valid also in this case.

Suppose now the inequality is of the form $\tau_r \geq (\tau_i, \tau_j)$ or of the form $\tau_r \geq (\tau_i \rightarrow \tau_j, \tau_k)$. Then, by the first clause of condition 3) on E, the equalities

$$\lambda_t \varepsilon(x, \eta y_t) = \lambda_t(x, \overline{i}\,\overline{l+r}(x, I) y_t), \quad t = 1, 2,$$

hold, and since $\overline{i}\,\overline{l+r}(x, I) y_1$, $\overline{i}\,\overline{l+r}(x, I) y_2$ are proportional to z_1, z_2, the validity of (12) is sure again. So we finished with the case when $\eta = \overline{r}$ for some r from the set $\{1, \ldots, l\}$.

Another possibility is that $\eta = \overline{l+r}(z, I)$ for some z in $\mathcal{E}$ and some r from the set $\{1, \ldots, l\}$. Two cases are possible now: the case, when an inequality of the form $\tau_r \geq (\tau_i, \tau_j)$ is present, and the case, when there is an inequality of the form $\tau_r \geq (\tau_i \rightarrow \tau_j, \tau_k)$. In both cases, the second clause of condition 3) is applied. In the first of the cases, we get the equalities

$$\lambda_t \varepsilon(x, \eta y_t) = \lambda_t(z, \overline{j}\,\overline{2l+1}(x, I)y_t), \quad t=1, 2,$$

and we use the fact that the elements $\overline{j}\,\overline{2l+1}(x, I)y_t$ are proportional to the elements z_t. In the second case, we have the equalities

$$\lambda_t \varepsilon(x, \eta y_t) = \lambda_t(x \to (z, \overline{j}y_t), (z, \overline{k}y_t)) = (x \to \lambda_t(z, \overline{j}y_t), \lambda_t(z, \overline{k}y_t)), \quad t=1, 2,$$

and we use the fact that both $\overline{j}y_1, \overline{j}y_2$ and $\overline{k}y_1, \overline{k}y_2$ are proportional to z_1, z_2.

The last possibility is that $\eta = \overline{2l+1}(z, I)$ for some z in $\mathscr{C}$. Then the first equality in condition 4) is applicable, and we get the equalities

$$\lambda_t \varepsilon(x, \eta y_t) = \lambda_t((z, x), y_t), \quad t=1, 2,$$

which immediately imply the validity of (12) in this case. ∎

Lemma 2. The set $\mathcal{T}_c$ is invariant with respect to ε.

Proof. Let λ_1 and λ_2 be arbitrary elements of $\mathscr{F}$ satisfying the inequality $\lambda_1 \underset{\mathcal{T}_c}{\geq} \lambda_2$. We have to prove the inequality $\lambda_1 \varepsilon \underset{\mathcal{T}_c}{\geq} \lambda_2 \varepsilon$. Consider an arbitrary element of $\mathcal{T}_c$. By the definition of the set $\mathcal{T}_c$, this element has the form (x, y), where $x \in \mathscr{C}$ and $y \in \mathscr{P}_c$. There are two possibilities: $y = \overline{0}c$ or $y = \eta y_0$, where η is some coding element, and y_0 is again an element of $\mathscr{P}_c$. In the first case, making use of the second equality in condition 4) on E, we observe that

$$\lambda_t \varepsilon(x, y) = \lambda_t(x, y), \quad t=1, 2,$$

and hence

$$\lambda_1 \varepsilon(x, y) \geq \lambda_2 \varepsilon(x, y).$$

The same inequality is true also in the second case, by Lemma 1, applied to $z_1 = z_2 = \overline{0}c$. ∎

Lemma 3. There is an element β of $\mathscr{F}$ such that

$$\beta(x, \eta z) = (\eta x, z)$$

for each coding element η and all x, z in $\mathscr{C}$.

Proof. We set $\beta = \Delta(\beta_0, \beta_1, \ldots, \beta_{2l+1})$, where Δ is the operation from Section II.2, and $\beta_0, \beta_1, \ldots, \beta_{2l+1}$ are elements of $\mathscr{F}$ satisfying the following conditions:

$$\beta_s(x, z)=(\bar{s}x, z), \quad s=1, \ldots, l,$$

$$\beta_s(x, (y, I)z)=(\bar{s}(y, I)x, z), \quad s=l+1, \ldots, 2l+1. \blacksquare$$

Remark 4. From the proof of the above lemma, it is seen that the element β can be chosen to be $\mathfrak{S}$-elementary in $\mathfrak{O}$, but we shall not make use of this fact.

Lemma 4. For each z_0 in $\mathcal{C}$, there is an element γ of $\mathcal{F}$ such that, whenever y, z are proportional to $\bar{0}c$, z_0, then $\gamma y=z$.

Proof. Let β be an element of $\mathcal{F}$ having the property from Lemma 3, and let

$$\rho=L[\beta, (L \to F, T)R].$$

It is easy to see that, for all x, z in $\mathcal{C}$ and all coding elements η, the following equalities hold:

$$\rho(x, \bar{0}z)=x, \quad \rho(x, \eta z)=\rho(\eta x, z).$$

Therefore

$$\rho(x, \eta_p \cdots \eta_1 \bar{0}c)=\eta_1 \cdots \eta_p x$$

for all x in $\mathcal{C}$ and each finite sequence $\eta_1, \ldots, \eta_p$ of coding elements. Making use of this, we check that, for each z_0 in $\mathcal{C}$, the corresponding element

$$\gamma=\rho(z_0, \rho(\bar{0}c, I))$$

has the needed property. $\blacksquare$

Lemma 5. Whenever x, y, z, z_0 belong to $\mathcal{C}$, and y, z are proportional to $\bar{0}c$, z_0, then

$$\iota(x, z) \geq \iota(I, z_0)L\iota(x, y).$$

Proof. Let z_0 be an arbitrary element of $\mathcal{C}$. We take an element γ with the property from Lemma 4 and note that, whenever x, y, z belong to $\mathcal{C}$, and y, z are proportional to $\bar{0}c$, z_0, then

$$\iota(x, z)=\iota(x, \gamma y)=\varkappa(x, y),$$

where $\varkappa=\iota(L, \gamma R)$. Making use of Lemma 1, we conclude that, for any choice of the coding element η, the equality

$$\iota\mathcal{E}(x, \eta z)=\varkappa\mathcal{E}(x, \eta y)$$

holds under the same conditions on x, y, z. Since

$$\iota\mathcal{E}(x, \eta z)=\iota(x, \eta z)=\varkappa(x, \eta y)$$

under the above conditions on η, x, y, z, we see that

$$\varkappa(x, \eta y)=\varkappa\mathcal{E}(x, \eta y)$$

whenever $x \in \mathcal{C}$, $y \in \mathcal{P}_c$ and η is a coding element. On the other hand, if we take merely $\bar{0}c$ and z_0 as y and z, respectively, in the equality $\varkappa(x, y) = \iota(x, z)$, we get

$$\varkappa(x, \bar{0}c) = \iota(x, z_0) = \iota(I, z_0)L(x, \bar{0}c).$$

Thus, for any y in $\mathcal{P}_c$ and all x in $\mathcal{C}$, the following equality holds:

$$\varkappa(x, y) = ((L \rightarrow F, T)R \rightarrow \varkappa\varepsilon, \iota(I, z_0)L))(x, y).$$

Hence we have the inequality

$$\varkappa \underset{\mathcal{T}_c}{\geq} ((L \rightarrow F, T)R \rightarrow \varkappa\varepsilon, \iota(I, z_0)L)).$$

From this inequality, Lemma 2 and the definition of iteration, the inequality

$$\varkappa \underset{\mathcal{T}_c}{\geq} \iota(I, z_0)L\iota$$

follows, i.e.

$$\varkappa(x, y) \geq \iota(I, z_0)L\iota(x, y)$$

for all x in $\mathcal{C}$ and all y in $\mathcal{P}_c$ holds. Since

$$\varkappa(x, y) = \iota(x, z)$$

whenever y, z are proportional to $\bar{0}c$, z_0, the proof of the lemma is thus completed. ∎

Lemma 6. For all x and z_0 in $\mathcal{C}$, the inequalities

$$\iota(x, \bar{r}z_0) \geq \iota(I, z_0)\delta_r^c x, \quad r = 1, \ldots, l,$$

hold.

Proof. Application of Lemma 5 to $y = \bar{r}\bar{0}c$, $z = \bar{r}z_0$. ∎

Lemma 7. For all x in $\mathcal{C}$, the inequalities

$$\iota(x, \bar{r}\bar{0}c) \geq (I, \bar{0}c)\delta_r^c x, \quad r = 1, \ldots, l,$$

hold.

Proof. We apply Lemma 6 to $z_0 = \bar{0}c$ and use the fact that $\iota(I, 0\bar{c}) = (I, 0c)$. ∎

Lemma 8. For all y in $\mathcal{C}$, the equality

$$L\iota(I, \overline{2l+1}(y, I)\bar{0}c) = (y, I)$$

holds.

Proof. For all x in $\mathcal{C}$, we have

$$L\iota(I, \overline{2l+1}(y, I)\bar{0}c)x = L\iota\varepsilon(x, \overline{2l+1}(y, I)\bar{0}c) =$$
$$L\iota((y, x), \bar{0}c) = L((y, x), \bar{0}c) = (y, x) = (y, I)x. \blacksquare$$

Lemma 9. The l-tuple $\langle\delta_1^c, \ldots, \delta_l^c\rangle$ is a solution of the system of inequalities (3).

Proof. Let r be some of the numbers $1, \ldots, l$. We shall prove the inequality

$$\delta_r^c \geq \Gamma_r(\theta, \delta_1^c, \ldots, \delta_l^c).$$

For all x in $\mathscr{C}$, we have

$$\delta_r^c x = L\iota(I, \overline{r}\,\overline{0}\,c)x = L\iota\varepsilon(x, \overline{r}\,\overline{0}\,c),$$

and therefore it is sufficient to prove that

$$L\iota\varepsilon(x, \overline{r}\,\overline{0}\,c) \geq \Gamma_r(\theta, \delta_1^c, \ldots, \delta_l^c)x$$

for all x in $\mathscr{C}$. The various possibilities concerning the form of the inequality

$$\tau_r \geq \Gamma_r(\theta, \tau_1, \ldots, \tau_l)$$

will be considered separately.

If the inequality is $\tau_r \geq \theta$ then we have

$$L\iota\varepsilon(x, \overline{r}\,\overline{0}\,c) = L\iota(\theta x, \overline{0}\,c) = L\iota(I, \overline{0}\,c)\theta x = \theta x = \Gamma_r(\theta, \delta_1^c, \ldots, \delta_l^c)x.$$

The case of $\tau_r \geq \alpha$ is quite the same - one has only to replace θ by α.

Let the inequality be $\tau_r \geq \tau_j\tau_i$. Then, applying Lemma 6, we get

$$L\iota\varepsilon(x, \overline{r}\,\overline{0}\,c) = L\iota(x, \overline{i}\,\overline{j}\,\overline{0}\,c) \geq L\iota(I, \overline{j}\,\overline{0}\,c)\delta_i^c x = \delta_j^c\delta_i^c x = \Gamma_r(\theta, \delta_1^c, \ldots, \delta_l^c)x.$$

Now suppose the inequality is $\tau_r \geq (\tau_i, \tau_j)$ or it is $\tau_r \geq (\tau_i \longrightarrow \tau_j, \tau_k)$. Then, again by application of Lemma 6, we get

$$L\iota\varepsilon(x, \overline{r}\,\overline{0}\,c) = L\iota(x, \overline{i}\,\overline{l+r}(x, I)\overline{0}\,c) \geq L\iota(I, \overline{l+r}(x, I)\overline{0}\,c)\delta_i^c x.$$

We note also that, for all y in $\mathscr{C}$, we have

$$L\iota(I, \overline{l+r}(x, I)\overline{0}\,c)y = L\iota\varepsilon(y, \overline{l+r}(x, I)\overline{0}\,c).$$

At this point, the reasoning branches.

Let us consider first the case when the inequality is $\tau_r \geq (\tau_i, \tau_j)$. Then, for all y in $\mathscr{C}$, we have

$$L\iota\varepsilon(y, \overline{l+r}(x, I)\overline{0}\,c) = L\iota(x, \overline{j}\,\overline{2l+1}(y, I)\overline{0}\,c) \geq L\iota(I, \overline{2l+1}(y, I)\overline{0}\,c)\delta_j^c x = (y, I)\delta_j^c x = (I, \delta_j^c x)y$$

(Lemmas 6 and 8 have been applied). Taking advantage of the arbitrariness in the choice of y, we conclude that

$$L\iota(I, \overline{l+r}(x, I)\overline{0}c) \geq (I, \delta_j^c x),$$

and hence

$$L\iota\varepsilon(x, \overline{r}\overline{0}c) \geq (I, \delta_j^c x)\delta_i^c x = (\delta_i^c, \delta_j^c)x = \Gamma_r(\theta, \delta_1^c, \ldots, \delta_l^c)x.$$

Now we go to the case when the inequality has the form $\tau_r \geq (\tau_i \rightarrow \tau_j, \tau_k)$. Then we have, for all y in $\mathscr{C}$,

$$L\iota\varepsilon(y, \overline{l+r}(x, I)\overline{0}c) = L\iota(y \rightarrow (x, \overline{j}\overline{0}c), (x, \overline{k}\overline{0}c)) = (y \rightarrow \delta_j^c x, \delta_k^c x) = (I \rightarrow \delta_j^c x, \delta_k^c x)y.$$

Therefore

$$L\iota(I, \overline{l+r}(x, I)\overline{0}c) \geq (I \rightarrow \delta_j^c x, \delta_k^c x),$$

and hence

$$L\iota\varepsilon(x, \overline{r}\overline{0}c) \geq (I \rightarrow \delta_j^c x, \delta_k^c x)\delta_i^c x = (\delta_i^c \rightarrow \delta_j^c, \delta_k^c)x = \Gamma_r(\theta, \delta_1^c, \ldots, \delta_l^c)x. \blacksquare$$

We are going now to prove that $\langle \delta_1^c, \ldots, \delta_l^c \rangle$ is the least solution of the system (3).

Lemma 10. Let $\langle \varphi_1, \ldots, \varphi_l \rangle$ be an arbitrary solution of the system (3). Then $\varphi_r \geq \delta_r^c$, $r = 1, \ldots, l$.

Proof. We define an element π of $\mathscr{F}$ such that

$$\pi(x, \overline{r}z) = (\varphi_r x, z), \quad r = 1, \ldots, l,$$

$$\pi(x, \overline{s}z) = \varepsilon(x, \overline{s}z), \quad s = l+1, \ldots, 2l+1,$$

for all x and z in $\mathscr{C}$.

Such an element can be constructed with the help of Proposition II.2.2 in a similar way as E(θ) was. An intuitive interpretation can be given to the element π in the same spirit as in the case of E(θ). This interpretation is the same as the one of E(θ), except that now the elements $\overline{1}, \ldots, \overline{l}$ of $\mathscr{F}$ play the role of "codes" of $\varphi_1, \ldots, \varphi_l$, respectively, and $\varphi_1, \ldots, \varphi_l$ are treated as primitive procedures. In other words, we give a new intuitive interpretation of a part of the "procedure denotations", and the element π "performs" the step by step transformation of "task representations" corresponding to this new interpretation. The formal treatment of the corresponding concept of "result of carrying out the tasks" can be done by using the element

$$\upsilon = L[\pi, (L \rightarrow F, T)R],$$

which can be regarded as "transforming the task representations into the results of carrying out the tasks".

We shall prove now that

(13) $$\upsilon(x, \eta z) \geq \upsilon\mathcal{E}(x, \eta z)$$

for all x, z in $\mathcal{C}$ and all coding elements η. To do this, we first show that

(14) $$\upsilon(x, \bar{r} z) = \upsilon(I, z)\varphi_r x, \quad \upsilon(I, \bar{r} z) = \upsilon(I, z)\varphi_r$$

for $r = 1, \ldots, l$ and all x, z in $\mathcal{C}$. In fact, we have

$$\upsilon(I, \bar{r} z)x = \upsilon(x, \bar{r} z) = \upsilon\pi(x, \bar{r} z) =$$
$$\upsilon(\varphi_r x, z) = \upsilon(I, z)\varphi_r x.$$

Now we go to the proof of (13) in the case of $\eta = \bar{r}$, where $1 \leq r \leq l$. In this case,

$$\upsilon(x, \eta z) = \upsilon(I, z)\varphi_r x \geq \upsilon(I, z)\Gamma_r(\theta, \varphi_1, \ldots, \varphi_l)x.$$

We shall prove the equality

(15) $$\upsilon\mathcal{E}(x, \eta z) = \upsilon(I, z)\Gamma_r(\theta, \varphi_1, \ldots, \varphi_l)x.$$

The proof will be by consideration of the various cases concerning the form of the inequality

$$\tau_r \geq \Gamma_r(\theta, \tau_1, \ldots, \tau_l).$$

If this inequality is $\tau_r \geq \theta$ then

$$\upsilon\mathcal{E}(x, \eta z) = \upsilon(\theta x, z) = \upsilon(I, z)\theta x =$$
$$\upsilon(I, z)\Gamma_r(\theta, \varphi_1, \ldots, \varphi_l)x.$$

The situation is completely similar also in the case of inequality of the form $\tau_r \geq \alpha$. If the inequality is $\tau_r \geq \tau_j\tau_i$ then

$$\upsilon\mathcal{E}(x, \eta z) = \upsilon(x, \bar{i}\,\bar{j} z) = \upsilon(I, \bar{j} z)\varphi_i x =$$
$$\upsilon(I, z)\varphi_j\varphi_i x = \upsilon(I, z)\Gamma_r(\theta, \varphi_1, \ldots, \varphi_l)x.$$

Let the inequality has the form $\tau_r \geq (\tau_i, \tau_j)$ or the form $\tau_r \geq (\tau_i \rightarrow \tau_j, \tau_k)$. Then

$$\upsilon\mathcal{E}(x, \eta z) = \upsilon(x, \bar{i}\,\overline{l+r}(x, I)z) = \upsilon(I, \overline{l+r}(x, I)z)\varphi_i x.$$

We note also that for all y in $\mathcal{C}$,

$$\upsilon(I, \overline{l+r}(x, I)z)y = \upsilon(y, \overline{l+r}(x, I)z) =$$
$$\upsilon\pi(y, \overline{l+r}(x, I)z) = \upsilon\mathcal{E}(y, \overline{l+r}(x, I)z).$$

Suppose first the inequality has the form $\tau_r \geq (\tau_i, \tau_j)$. Then

$$\upsilon\mathcal{E}(y, \overline{l+r}(x, I)z) = \upsilon(x, \bar{j}\,\overline{2l+1}(y, I)z) =$$
$$\upsilon(I, \overline{2l+1}(y, I)z)\varphi_j x.$$

But

$$\upsilon(I,\ \overline{2l{+}1}(y,\ I)z)w = \upsilon(w,\ \overline{2l{+}1}(y,\ I)z) = \upsilon\pi(w,\ \overline{2l{+}1}(y,\ I)z) = \upsilon\varepsilon(w,\ \overline{2l{+}1}(y,\ I)z) = \upsilon((y,\ w),\ z) = \upsilon(I,\ z)(y,\ I)w.$$

Therefore

$$\upsilon(I,\ \overline{2l{+}1}(y,\ I)z) = \upsilon(I,\ z)(y,\ I),$$

and hence

$$\upsilon(I,\ \overline{l{+}r}(x,\ I)z)y = \upsilon(I,\ z)(y,\ I)\varphi_j x = \upsilon(I,\ z)(I,\ \varphi_j x)y.$$

Thus

$$\upsilon(I,\ \overline{l{+}r}(x,\ I)z) = \upsilon(I,\ z)(I,\ \varphi_j x),$$

and, consequently,

$$\upsilon\varepsilon(x,\ \eta z) = \upsilon(I,\ z)(I,\ \varphi_j x)\varphi_i x = \upsilon(I,\ z)(\varphi_i,\ \varphi_j)x = \upsilon(I,\ z)\Gamma_r(\theta,\ \varphi_1,\ \ldots,\ \varphi_l)x.$$

Now suppose the inequality is $\tau_r \geq (\tau_i \to \tau_j,\ \tau_k)$. Then

$$\upsilon\varepsilon(y,\ \overline{l{+}r}(x,\ I)z) = \upsilon(y \to (x,\ \overline{j}z),\ (x,\ \overline{k}z)) = (y \to \upsilon(x,\ \overline{j}z),\ \upsilon(x,\ \overline{k}z)) = (y \to \upsilon(I,\ z)\varphi_j x,\ \upsilon(I,\ z)\varphi_k x) = \upsilon(I,\ z)(I \to \varphi_j x,\ \varphi_k x)y.$$

Hence

$$\upsilon(I,\ \overline{l{+}r}(x,\ I)z) = \upsilon(I,\ z)(I \to \varphi_j x,\ \varphi_k x),$$

and therefore

$$\upsilon\varepsilon(x,\ \eta z) = \upsilon(I,\ z)(I \to \varphi_j x,\ \varphi_k x)\varphi_i x = \upsilon(I,\ z)(\varphi_i \to \varphi_j,\ \varphi_k)x = \upsilon(I,\ z)\Gamma_r(\theta,\ \varphi_1,\ \ldots,\ \varphi_l)x.$$

Thus the equality (15) is established in all possible cases, and so the inequality (13) is proved under the assumption that $\eta = \overline{r}$ for some r from the set $\{1,\ \ldots,\ l\}$.

It remains to prove (13) for $\eta = \overline{l{+}r}(y,\ I)$, where $y \in \mathscr{C}$, $1 \leq r \leq l+1$. Then

$$\upsilon(x,\ \eta z) = \upsilon(x,\ \overline{l{+}r}(y,\ I)z) = \upsilon\pi(x,\ \overline{l{+}r}(y,\ I)z) = \upsilon\varepsilon(x,\ \overline{l{+}r}(y,\ I)z) = \upsilon\varepsilon(x,\ \eta z),$$

and hence (13) is valid again.

Making use of (13) we shall prove now the inequality

$$\upsilon \underset{\mathcal{T}_c}{\geq} ((L \to F,\ T)R \to \upsilon\varepsilon,\ \upsilon). \tag{16}$$

Let x be an arbitrary element of $\mathscr{C}$, and y be an arbitrary element of $\mathscr{F}_c$. We have to prove the inequality

(17) $\quad v(x, y) \geq ((L \rightarrow F, T)R \rightarrow v\varepsilon, v)(x, y)$.

There are two possibilities: $y=\bar{0}c$ or $y=\eta y_0$, where η is some coding element, and y_0 again belongs to $\mathcal{G}_c$. In the first case

$$((L \rightarrow F, T)R \rightarrow v\varepsilon, v)(x, y) = v(x, y),$$

and in the second one

$$((L \rightarrow F, T)R \rightarrow v\varepsilon, v)(x, y) = v\varepsilon(x, y) \leq v(x, y),$$

hence (17) holds in both cases, and thus (16) is established.

By Lemma 2 and the definition of iteration, the inequality (16) implies the inequality

$$v \underset{\mathcal{T}_c}{\geq} v\iota.$$

Let r be some of the numbers $1, \ldots, l$, and let $x \in \mathcal{C}$. Then $(x, \bar{r}\bar{0}c) \in \mathcal{T}_c$, and therefore, by the above inequality, we have

$$v(x, \bar{r}\bar{0}c) \geq v\iota(x, \bar{r}\bar{0}c).$$

Making use of (14) and of the definition of v, we get

$$v(x, \bar{r}\bar{0}c) = v(I, \bar{0}c)\varphi_r x = L(I, \bar{0}c)\varphi_r x = \varphi_r x.$$

On the other hand,

$$v\iota(x, \bar{r}\bar{0}c) = v(I, \bar{0}c)\delta_r^c x = \delta_r^c x.$$

Thus

$$\varphi_r x \geq \delta_r^c x, \quad r = 1, \ldots, l,$$

for all x in $\mathcal{C}$, and therefore

$$\varphi_r \geq \delta_r^c, \quad r = 1, \ldots, l. \blacksquare$$

Lemmas 9 and 10 show that $\langle \delta_1^c, \ldots, \delta_l^c \rangle$ is the least solution of the system (3), independently of the choice of c in $\mathcal{C}$. Hence $\langle \delta_1^c, \ldots, \delta_l^c \rangle$ does not depend on this choice. Making use of this, we shall now show that

$$\delta_r^c = \delta_r, \quad r = 1, \ldots, l,$$

where δ_r are the elements defined by (10)-(11). In fact, if $r \in \{1, \ldots, l\}$ then

$$\delta_r^c x = \delta_r^x x = L\iota(I, \bar{r}\bar{0}x)x = L\iota(I, \bar{r}\bar{0})x = \delta_r x$$

for all x in $\mathcal{C}$.

The proof of the First Recursion Theorem (in its parameterized version) is thus completed. ∎

Some applications of the First Recursion Theorem will be given in the next sections. We finish this section by mentioning explicitly a useful thing from the proof of the theorem. Namely, the components $\Delta_r(\theta)$ of the least solution of the system of inequalities (3) are defined by means of the equalities (10), where the mapping E is $\mathfrak{S}$-elementary in $\mathcal{B}$. This is a certain kind of normal form, and applications of this fact also will be given in the sequel.

Exercises

1. Show that, after a small modification of the proof of Theorem 2, the mapping E used there can be chosen to be of the form

$$\mathrm{E}(\theta) = (\pi \rightarrow \sigma, (\theta L, R^{m+2})),$$

where χ, σ are elements of $\mathcal{F}$, $\mathfrak{S}$-elementary in $\emptyset$, and m is some positive integer.

Hint. Without loss of generality, it can be assumed that there is only one inequality of the form $\tau_r \geq \theta$ in the system (3). Take m to be the corresponding r, and choose χ so that

$$\chi(x, \bar{i}) = \begin{cases} F & \text{if } i \neq m \text{ and } i \leq 2l+1, \\ T & \text{if } i = m. \end{cases}$$

2. Give a modified proof of the First Recursion Theorem avoiding the use of coding elements with the intuitive meaning described in (i) and (ii).

Hint. Show that inequalities of the form $\tau_r \geq (\alpha, \tau_j)$ with $\alpha \in \{I, R\}$ and of the form $\tau_r \geq (R \rightarrow \tau_j, \tau_k)$ can be used instead of using inequalities of the forms (8), (9).

3. (Cf. the translation operation in Ivanov [1986]) Prove the existence of a mapping T of $\mathcal{F}$ into $\mathcal{F}$ with the following properties:

(i) T is $\mathfrak{S}$-computable in $\emptyset$;

(ii) for all θ in $\mathcal{F}$ and all natural numbers n, the equalities

$$\mathrm{T}(\theta) L_* = L_* \theta, \quad \mathrm{T}(\theta) R_* = R_* \mathrm{T}(\theta),$$
$$\mathrm{T}(\theta)\bar{n} = \bar{n}\theta$$

hold.

5. Application of the First Recursion Theorem to some concrete iterative combinatory spaces

(I) The relationship to the First Recursion Theorem in the ordinary recursion theory. In this subsection, the relationship will be discussed of Theorem 4.1 and 4.2 to the Kleene First Recursion Theorem (Kleene [1952, § 66, Theorem XXVI]) in the ordinary recursion theory.

There are two kinds of iterative combinatory spaces closely related to ordinary recursion theory, namely the spaces $\mathfrak{S}_p(\mathfrak{U})$ and $\mathfrak{S}_m(\mathfrak{U})$ (from Examples II.1.2 and II.2.1, respectively) corresponding to any standard computational structure $\mathfrak{U}=\langle \mathbb{N}, J, L, R, T, F, H\rangle$ on the natural numbers. If $\mathcal{B}$ is some subset of the semigroup $\mathcal{F}$ of the combinatory space, i.e. a subset of $\mathcal{F}_p(\mathbb{N})$ or of $\mathcal{F}_m(\mathbb{N})$, respectively, then $\mathfrak{U}$-computability in $\mathcal{B}$ is equivalent to computability in $\mathcal{B}$ in the considered combinatory space. Therefore, having in mind the results from Sections I.3 and I.6, it is natural to take $\mathcal{B}=\{S, P\}$ in the case of $\mathfrak{S}_p(\mathfrak{U})$ and $\mathcal{B}=\{S, P, \mathbb{N}^2\}$ in the case of $\mathfrak{S}_m(\mathfrak{U})$, where S, P are the functions $\lambda u. u+1$ and $\lambda u. u \dot{-} 1$, respectively. In the first case, the elements of the combinatory space, which are computable in $\mathcal{B}$, are the unary partial recursive functions, and the mappings computable in $\mathcal{B}$ are the μ-recursive operators which transform unary functions into unary ones. In the second case, the mappings computable in $\mathcal{B}$ are the enumeration operators transforming binary relations into binary ones. We shall apply now Theorems 4.1 and 4.2 to these two cases, restricting ourselves to the case of $l=1$. In the statements obtained in this way, we shall consider unessential the restriction only to unary functions and binary relations, because of the effective one-to-one correspondence between $\mathbb{N}$ and $\mathbb{N}^k$ with $k>1$. For the sake a brevity, we shall consider only Theorem 4.2 regarding Theorem 4.1 as a particular case of it.

Let us consider first $\mathcal{F}=\mathcal{F}_p(\mathbb{N})$ with $\mathcal{B}=\{S, P\}$. Then we get the statement that, whenever Γ is a μ-recursive mapping of $\mathcal{F}^{n+1}$ into $\mathcal{F}$, then, for all $\theta_1, \dots, \theta_n$ in $\mathcal{F}$, the element $\mu\tau. \Gamma(\theta_1, \dots, \theta_n, \tau)$ of $\mathcal{F}$ exists, and this element is uniformly μ-recursive in $\theta_1, \dots, \theta_n$ (in the case of $n=0$, this means that $\mu\tau. \Gamma(\tau)$ exists and it is a partial recursive function). A comparison with the statement of the Kleene First Recursion Theorem shows that neither of both compared statements covers the other one. The main differences between these statements are the following ones: (i) the statement formulated above concerns μ-recursive operators, whereas the Kleene Recursion Theorem deals with arbitrary recursive operators, and (ii) in the conclusion of the above statement, μ-recursiveness in $\theta_1, \ldots, \theta_n$ is asserted, whereas in the conclusion of the Kleene Recursion Theorem only partial recursiveness in $\theta_1, \ldots, \theta_n$ is claimed. Thus the above statement has a stronger assumption and a stronger conclusion than the Kleene First Recursion Theorem, and hence this statement is a result different from the Kleene Theorem. Moreover, the statement is not directly obtainable by the usual proofs of the Kleene First Recursion Theorem (such as the proof in Kleene [1952, § 66] or Rogers [1967, § 11.5]).[76] The situation becomes simpler when $n=0$, since the conclusions of the compared statements are equivalent in this case. Therefore the Kleene First Recursion Theorem is stronger in the case of $n=0$.

Now consider $\mathcal{F}=\mathcal{F}_m(\mathbb{N})$ with $\mathcal{B}=\{S, P, \mathbb{N}^2\}$. Then we get the statement that, whenever Γ is an enumeration operator acting from $\mathcal{F}^{n+1}$ into $\mathcal{F}$, then, for all $\theta_1, \ldots, \theta_n$ in $\mathcal{F}$, the element $\mu\tau.\Gamma(\theta_1, \ldots, \theta_n, \tau)$ of $\mathcal{F}$ exists, and this element is uniformly enumeration reducible in $\theta_1, \ldots, \theta_n$ (in the case of $n=0$, this means that $\mu\tau.\Gamma(\tau)$ exists and it is a recursively enumerable relation[77]). The Kleene First Recursion Theorem follows easily from the above statement (cf. Rogers [1967, Theorem 11-XII]). In this sense our First Recursion Theorem is a generalization of the Kleene First Recursion Theorem.

We note also that some steps toward a more direct covering of the Kleene First Recursion Theorem by the abstract theory are undertaken in the paper Ivanov [1981], where some supplements to the theory in this direction are made.

Anyway, we hope that no serious objections could arise

---

[76] Unfortunately, we are not able to give a bibliographical reference concerning this statement. A direct proof of its validity is known to the author since 1968 or 1969, and then he presented the result in a seminar talk at Moscow University.

[77] This is a well-known fact (see, e.g. Rogers [1967, Theorem 11-XI])

against the name given by us to the considered result from the previous section.

(II) **Elimination of recursion in FP-systems.** In Section I.4, the programmability in a FP-system (in the sense of Backus [1978]) has been shown to be equivalent to $\mathfrak{U}$-computability in a certain subset $\mathcal{B}$ of $\mathcal{F}_p(M)$, where $\mathfrak{U}=\langle M, J, L, R, T, F, H\rangle$ is the computational structure corresponding to the given FP-system (cf. Example I.1.3), and $\mathcal{B}$ consists of all primitive functions of the system and of the constant functions corresponding to the various elements of M. However, this result was obtained under a strong restriction on the use of the FP-system, namely only programmability without recursion has been considered. The restriction is quite unpleasant, since recursion in the form of so-called definitions is allowed in the original FP-systems, and almost all interesting examples of programs in such a system use recursive definitions. Now we are able to show that using the mentioned form of recursion does not enlarge the class of the programmable functions.

A definition of the mentioned kind is an equality whose left-hand side is some non-primitive functional symbol, and whose right-hand side is some functional form, possibly containing the functional symbol from the left-hand side (in this case, the definition will be called to be recursive). To use a series of definitions is also allowed, and thus right-hand sides of equalities may contain several non-primitive symbols. Clearly, a functional form containing non-primitive functional symbols represents an operation in $\mathcal{F}_p(M)$; an inspection of the proofs of Lemmas I.1.2–I.1.7 shows that this operation is $\mathfrak{U}$-computable in $\mathcal{B}$. Backus is not completely explicit about the semantics of the recursive definitions, but his exposition does not contradict to the traditional least-fixed-point semantics, and we shall adopt it in the further considerations. So a series of definitions can be considered as defining the least solution of the corresponding system of equations.

But $\mathfrak{U}$-computability in $\mathcal{B}$ is equivalent to $\mathfrak{S}_p(M)$-computability in $\mathcal{B}$. Therefore an application of the First Recursion Theorem from Section 4 shows that the functions defined by a series of definitions of the FP-system are always $\mathfrak{S}_p(M)$-computable in $\mathcal{B}$, and therefore, by Theorem I.4.1, they are programmable without using definitions. Thus the use of definitions and, in particular, of recursive ones in the programs of an FP-system can be eliminated.

An analysis of the proof of the First Recursion Theorem given in Section 4 shows its constructive character in the sense that an algorithm exists which transforms any l-tuple of expressions for $\Gamma_1, \ldots, \Gamma_l$ into some l-tuple of expressions for the components of the corresponding least solution. As a consequence, the existence of an algorithm follows which transforms arbitrary programs in an FP-system into such ones which do not contain definitions. Of course, the algorithm will be quite bad from a practical point of view (complicated programs produced by the algorithm, much longer execution time of these programs in comparison with the execution time of the source ones). There are some particular cases when better results can be obtained by means of specific methods (certain such cases are indicated by Backus himself).

It is worth mentioning that also a result about the elimination of the essential recursions in the programming language LISP can be established in a similar way.

(III) Equivalence of prime and search computability on B^* with $\mathfrak{M}_B$-computability in corresponding subsets of $\mathcal{F}_m(B^*)$. In this subsection, the same assumptions will be adopted and the same denotations will be used as in Section I.7, where we started to study a possibility to characterize prime and search computability in the sense of Moschovakis (1969) in the terms of $\mathfrak{M}_B$-computability. We recall that the following results (Propositions I.7.1 and I.7.2) have been established for each subset A of B^* and each choice of $\psi_1, \ldots, \psi_l$ in $\mathcal{F}_m(B^*)$ ($\mathcal{C}_A$ denoting the set of all constant single-valued functions whose domain is B^* and whose values belong to A, and $\mathfrak{M}_B$ being the computational structure from Example I.1.2):

All elements of $\mathcal{F}_m(B^*)$, which are $\mathfrak{M}_B$-computable in $\mathcal{C}_A \cup \{\psi_1, \ldots, \psi_l\}$, belong to $PC(A, \psi_1, \ldots, \psi_l)$; all elements of $\mathcal{F}_m(B^*)$, which are $\mathfrak{M}_B$-computable in $\mathcal{C}_A \cup \{\psi_1, \ldots, \psi_l, (B^*)^2\}$, belong to $SC(A, \psi_1, \ldots, \psi_l)$.

It was mentioned there that also the converse statements are also true if only one-argument functions from the classes $PC(A, \psi_1, \ldots, \psi_l)$ and $SC(A, \psi_1, \ldots, \psi_l)$ are considered, but the proofs will be given further in the book. Now the time has arrived to prove these converse statements, and thus to obtain the following result:

Theorem 1. Let A be a subset of B^* and $\psi_1, \ldots, \psi_l$,

φ be elements of $\mathcal{F}_{\mathrm{m}}(B^*)$. Then

(i) $\varphi \in \mathrm{PC}(A, \psi_1, \ldots, \psi_l)$ iff φ is $\mathfrak{M}_B$ -computable in $\mathcal{C}_A \cup \{\psi_1, \ldots, \psi_l\}$;

(ii) $\varphi \in \mathrm{SC}(A, \psi_1, \ldots, \psi_l)$ iff φ is $\mathfrak{M}_B$-computable in $\mathcal{C}_A \cup \{\psi_1, \ldots, \psi_l, (B^*)^2\}$.

Proof. Of course, we have to prove only the implications from left to right. To do this, we firstly recall that the sets $\mathrm{PC}(A, \psi_1, \ldots, \psi_l)$ and $\mathrm{SC}(A, \psi_1, \ldots, \psi_l)$ are defined on the basis of two partial multiple-valued operations $\{e\}(q_1, \ldots, q_n)$ and $\{e\}_\nu(q_1, \ldots, q_n)$ from elements $e, q_1, \ldots, q_n$ of B^* into B^*, and these operations are introduced by means of somewhat complicated recursive definitions. We have, roughly speaking, to show that these recursive definition cannot take us out of the scope of the $\mathfrak{M}_B$-computability, and this will be done by a suitable application of the First Recursion Theorem to the combinatory space

$$\mathfrak{S} = \mathfrak{S}_{\mathrm{m}}(\mathfrak{M}_B).$$

Let $\mathcal{F} = \mathcal{F}_{\mathrm{m}}(B^*)$, $M = B^*$, $I = I_M$. We define two elements ω and ω_ν of $\mathcal{F}$ in the following way: $\langle p, r\rangle \in \omega$ iff there are a natural number n and elements $e, q_1, \ldots, q_n$ of M such that $p = \langle e, \langle\!\langle q_1, \ldots, q_n\rangle\!\rangle\rangle$ and r is a value of $\{e\}(q_1, \ldots, q_n)$, and similarly for ω_ν with the only difference that $\{e\}_\nu$ occurs instead of $\{e\}$. The most important part of the proof consists in proving that

$$\omega \in \mathrm{COMP}_{\mathfrak{S}}(\{\psi_1, \ldots, \psi_l\}), \tag{1}$$

$$\omega_\nu \in \mathrm{COMP}_{\mathfrak{S}}(\{\psi_1, \ldots, \psi_l, (B^*)^2\}). \tag{2}$$

This will be done in several steps.

First of all, we note that, according to the definition of the operation $\{e\}(q_1, \ldots, q_n)$, the relation ω is the least element τ of $\mathcal{F}$ which satisfies the following conditions for all e, g, h, q, r, s, t in M and all natural numbers j, k, m, n:

0) if $1 \le j \le l$ then

$$\tau(\langle\langle\!\langle 0, 1+m, j\rangle\!\rangle, \langle\!\langle s, t_1, \ldots, t_m\rangle\!\rangle\rangle) = \psi_j(s);$$

1) $\tau(\langle\langle\!\langle 1, n, r\rangle\!\rangle, \langle\!\langle q_1, \ldots, q_n\rangle\!\rangle\rangle) = r$;

2) $\tau(\langle\langle\!\langle 2, m+1\rangle\!\rangle, \langle\!\langle s, t_1, \ldots, t_m\rangle\!\rangle\rangle) = s$;

3) $\tau(\langle\langle\!\langle 3,\ m+2\rangle\!\rangle,\ \langle\!\langle s_1,\ s_2,\ t_1,\ \ldots,\ t_m\rangle\!\rangle\rangle) = \langle s_1,\ s_2\rangle$;

4_0) $\tau(\langle\langle\!\langle 4,\ m+1,\ 0\rangle\!\rangle,\ \langle\!\langle s,\ t_1,\ \ldots,\ t_m\rangle\!\rangle\rangle) = L(s)$;

4_1) $\tau(\langle\langle\!\langle 4,\ m+1,\ 1\rangle\!\rangle,\ \langle\!\langle s,\ t_1,\ \ldots,\ t_m\rangle\!\rangle\rangle) = R(s)$;

5) $\tau(\langle\langle\!\langle 5,\ m,\ g,\ h\rangle\!\rangle,\ \langle\!\langle t_1,\ \ldots,\ t_m\rangle\!\rangle\rangle) =$

$$\tau(\langle g,\ \langle\!\langle \tau(\langle h,\ \langle\!\langle t_1,\ \ldots,\ t_m\rangle\!\rangle\rangle),\ t_1,\ \ldots,\ t_m\rangle\!\rangle\rangle);$$

6_0) if $s \in B^{\circ}$ then

$\tau(\langle\langle\!\langle 6,\ m+1,\ g,\ h\rangle\!\rangle,\ \langle\!\langle s,\ t_1,\ \ldots,\ t_m\rangle\!\rangle\rangle) =$

$$\tau(\langle g,\ \langle\!\langle s,\ t_1,\ \ldots,\ t_m\rangle\!\rangle\rangle);$$

6_1) $\tau(\langle\langle\!\langle 6,\ m+1,\ g,\ h\rangle\!\rangle,\ \langle\!\langle \langle s_1,\ s_2\rangle,\ t_1,\ \ldots,\ t_m\rangle\!\rangle\rangle) =$

$$\tau(\langle h,\ \langle\!\langle \tau(\langle\langle\!\langle 6,\ m+1,\ g,\ h\rangle\!\rangle,\ \langle\!\langle s_1,\ t_1,\ \ldots,\ t_m\rangle\!\rangle\rangle),$$

$$\tau(\langle\langle\!\langle 6,\ m+1,\ g,\ h\rangle\!\rangle,\ \langle\!\langle s_2,\ t_1,\ \ldots,\ t_m\rangle\!\rangle\rangle),$$

$$s_1,\ s_2,\ t_1,\ \ldots,\ t_m\rangle\!\rangle\rangle);$$

7) whenever $k < n$, then

$\tau(\langle\langle\!\langle 7,\ n,\ k,\ g\rangle\!\rangle,\ \langle\!\langle q_1,\ldots,\ q_k,\ q_{k+1},\ q_{k+2},\ \ldots,\ q_n\rangle\!\rangle\rangle) =$

$$\tau(\langle g,\ \langle\!\langle q_{k+1},\ q_1,\ldots,\ q_k,\ q_{k+2},\ \ldots,\ q_n\rangle\!\rangle\rangle);$$

8) $\tau(\langle\langle\!\langle 8,\ k+m+1,\ k\rangle\!\rangle,\ \langle\!\langle e,\ s_1,\ \ldots,\ s_k,\ t_1,\ \ldots,\ t_m\rangle\!\rangle\rangle) =$

$$\tau(\langle e,\ \langle\!\langle s_1,\ \ldots,\ s_k\rangle\!\rangle\rangle).$$

As to the relation ω_ν, it is the least element τ of $\mathcal{F}$ which, for all e, g, h, q, r, s, t in M and all natural numbers j, k, m, n, satisfies the above conditions and the following additional condition:

9) $\tau(\langle\langle\!\langle 9,\ n,\ g\rangle\!\rangle,\ \langle\!\langle q_1,\ldots,\ q_n\rangle\!\rangle\rangle) =$

$$\{r\colon\ \tau(\langle g,\ \langle\!\langle r,\ q_1,\ldots,\ q_n\rangle\!\rangle\rangle) \ni 0\}.$$

The formulated characterizations of ω and ω_ν as least elements of $\mathcal{F}$ satisfying certain conditions remain valid if the following condition is added to the other ones:

*) each element of $\operatorname{dom}\tau$ has some of the forms indicated as arguments of τ in the left-hand sides of the equalities in the other conditions.

We shall show that each of the systems of conditions 1)–8), *) and 1)–9), *) is equivalent to a certain equality of the form $\tau = \Gamma(\tau)$, where Γ is a mapping of $\mathcal{F}$ into itself, $\mathfrak{S}$–computable in $\{\psi_1,\ \ldots,\ \psi_l\}$ in the case of the first system of conditions, and $\mathfrak{S}$–computable in $\{\psi_1,\ \ldots,\ \psi_l,\ M^2\}$ in the case of the second one. When this will be done, then an application of the First Recursion Theorem will immediately yield the validity of (1) and (2). We shall give the construction of Γ for the case of the second system of conditions, and the construction for the other case will be obtainable by means of an obvious simplification.

In order to construct the mapping Γ with the needed properties, we shall first define some elements of $\mathcal{F}$ representing certain subsets of M, and, in particular, certain subsets connected with the natural numbers (by saying "natural numbers", we mean the elements of M representing natural numbers). If K is a subset of M then the representing function of K is the total function on M having the value 1 at all points of K and the value 0 at all other points of M (1 and 0 considered as elements of M).

For each natural number j, let χ_j be the function representing the one-element set $\{j\}$. All functions χ_j are $\oslash$-computable in $\emptyset$, as it is seen from the equalities

$$\chi_0 = (I \to F,\ (L \to F,\ T)),$$
$$\chi_{j+1} = (I \to (\chi_j L,\ \chi_0 R),\ F).$$

Let $\chi_{\mathbf{N}}$ be the function representing the set of all natural numbers, and $\chi_<$ be the function representing the set of all elements of M having the form $\langle k, n\rangle$, where k and n are natural numbers, and $k < n$. We shall prove that $\chi_{\mathbf{N}}$ and $\chi_<$ belong to $\mathrm{COMP}_{\oslash}(\emptyset)$. This can be done directly, but we prefer to use the First Recursion Theorem and the fact that $\chi_{\mathbf{N}}$ and $\chi_<$ are the only solutions of the equations

$$\tau = (\chi_0 \to T,\ (I \to (\chi_0 R \to \tau L,\ F),\ F))$$

and

$$\tau = (I \to (\chi_{\mathbf{N}} L \to (\chi_{\mathbf{N}} R \to$$
$$(L \to (R \to \tau(L^2,\ LR),\ F),\ R),\ F),\ F),\ F),$$

respectively.

Let $\chi_=$ be the function representing the set of all elements of M having the form $\langle k, n\rangle$, where k and n are natural numbers, and $k = n$. This function also belongs to $\mathrm{COMP}_{\oslash}(\emptyset)$ due to the equality

$$\chi_= = (I \to (\chi_{\mathbf{N}} L \to (\chi_{\mathbf{N}} R \to$$
$$(\chi_< \to F,\ (\chi_<(R,\ L) \to F,\ T)),\ F),\ F),\ F).$$

The next element of $\mathcal{F}$ which will be considered is the function $\chi_{\langle\langle\,\rangle\rangle}$ representing the set of all elements of M having the form $\langle\langle u_1, \ldots, u_i\rangle\rangle$. The function $\chi_{\langle\langle\,\rangle\rangle}$ is the only solution of the equality

$$\tau=(I\to(\chi_N L\to(L\to(R\to\tau(L^2, R^2), F), \chi_0 R), F), F),$$

hence $\chi_{«»}\in \mathrm{COMP}(\emptyset)_{\circledS}$ too.

Now the existence of an element σ_1 of $\mathrm{COMP}_{\circledS}(\emptyset)$ with the following property will be proved: whenever k is a natural number, and an element $«q_1, \ldots, q_n»$ with $n\geq k$ is given, then

(3) $$\sigma_1(<k, «q_1, \ldots, q_n»>)=«q_1, \ldots, q_k».$$

In fact, by the First Recursion Theorem, the following equation has a solution τ belonging to $\mathrm{COMP}_{\circledS}(\emptyset)$

$$\tau=(L\to(L, (LR^2, R\tau(L^2, (L^2, R^2)R))), (F, F)).$$

If we denote by σ_1 such a solution τ then an induction on k shows the validity of the equality (3).

Also the existence of an element σ_2 of $\mathrm{COMP}_{\circledS}(\emptyset)$ with the following property will be proved: whenever k is a natural number, and an element $«q_1, \ldots, q_n»$ with $n>k$ is given, then

$$\sigma_2(<k, «q_1, \ldots, q_n»>)=«q_{k+1}, q_1, \ldots, q_k, q_{k+2}, \ldots, q_n».$$

In fact, it is sufficient to take as σ_2 an $\circledS$-computable in $\emptyset$ solution τ of the equation

$$\tau=(L\to(LR, (LR, (L, R^2))(LR^2, \tau(L^2, (L^2, R^2)R))), R).$$

Now we shall note the following common feature of the elements of M appearing as arguments in the left-hand sides of the equalities in conditions 0)-9): all these elements are ordered pairs of the form

(4) $$<«u_1, \ldots, u_i», «v_1, \ldots, v_j»>$$

with $i\geq 2$ and $u_2=j$. Let D be the set of all such elements of M. The elements of D are exactly those elements of M for which none of the functions I, $\chi_{«»}L$, $\chi_{«»}R$, L^3, $\chi_=(LR^2L, LR)$ has a value in B°. The elements τ of $\mathcal{F}$ with domains contained in D are exactly those which are representable in the form

$$\tau=(I\to(\chi_{«»}L\to(\chi_{«»}R\to(L^3\to(\chi_=(LR^2L, LR)\to\rho, \emptyset), \emptyset), \emptyset), \emptyset), \emptyset).$$

Therefore we shall try to express the system of conditions 1)-9), *) by an equality of the form

(5) $$\tau=(I\to(\chi_{«»}L\to(\chi_{«»}R\to(L^3\to(\chi_=(LR^2L, LR)\to \mathrm{P}(\tau), \emptyset), \emptyset), \emptyset), \emptyset), \emptyset),$$

where P is $\mathfrak{S}$-computable in $\{\psi_1, \ldots, \psi_l, M^2\}$. Of course, we have to bother about the behaviour of $P(\tau)$ only on the elements of the set D. If p is the element (4) of D then $u_1 = LRL(p)$ and therefore we shall look for a mapping P of the form

$$\begin{aligned}P(\tau) = (\chi_0 LRL \to P_0(\tau), (\chi_1 LRL \to P_1(\tau), (\chi_2 LRL \to P_2(\tau),\\ (\chi_3 LRL \to P_3(\tau), (\chi_4 LRL \to P_4(\tau), (\chi_5 LRL \to P_5(\tau),\\ (\chi_6 LRL \to P_6(\tau), (\chi_7 LRL \to P_7(\tau), (\chi_8 LRL \to P_8(\tau),\\ (\chi_9 LRL \to P_9(\tau), \emptyset))))))))))\end{aligned}$$

with $P_0, \ldots, P_9$ $\mathfrak{S}$-computable in $\{\psi_1, \ldots, \psi_l, M^2\}$. It is not difficult to check that the equation (5) will be surely equivalent to the system of conditions 1)-9), *) if we define $P_0(\tau), \ldots, P_9(\tau)$ in the following way:[78]

$$\begin{aligned}P_0(\tau) = (\chi_3 L^2 \to (LR \to (\chi_1 LR^3 L \to \psi_1 LR^2,\\ (\chi_2 LR^3 L \to \psi_2 LR^2, \ldots, (\chi_{l-1} LR^3 L \to \psi_{l-1} LR^2,\\ (\chi_l LR^3 L \to \psi_l LR^2, \emptyset)) \ldots)), \emptyset), \emptyset),\\ P_1(\tau) = (\chi_3 L^2 \to LR^3 L, \emptyset),\\ P_2(\tau) = (\chi_2 L^2 \to (LR \to LR^2, \emptyset), \emptyset),\\ P_3(\tau) = (\chi_2 L^2 \to (L^2 R \to (LR^2, LR^3), \emptyset), \emptyset),\\ P_4(\tau) = (\chi_3 L^2 \to (LR \to (\chi_0 LR^3 L \to L^2 R^2,\\ (\chi_1 LR^3 L \to RLR^2, \emptyset)), \emptyset), \emptyset),\\ P_5(\tau) = (\chi_4 L^2 \to \tau(LR^3 L, (I, F)LR, \tau(LR^4 L, R), R^2), \emptyset),\\ P_6(\tau) = (\chi_4 L^2 \to (LR \to (LR^2 \to \tau(LR^4 L, (I, F)^3 LR,\\ \tau(L, LR, L^2 R^2, R^3), \tau(L, LR, RLR^2, R^3),\\ L^2 R^2, RLR^2, R^3), \tau(LR^3 L, R)), \emptyset), \emptyset),\\ P_7(\tau) = (\chi_4 L^2 \to (\chi_< (LR^3 L, LR^2 L) \to\\ \tau(LR^4 L, \sigma_2 (LR^3 L, R)), \emptyset), \emptyset),\\ P_8(\tau) = (\chi_3 L^2 \to (\chi_< (LR^3 L, LR^2 L) \to\\ \tau(LR, \sigma_1 (LR^3 L, L^2 R, R^3)), \emptyset), \emptyset),\end{aligned}$$

---

[78] In some of the right-hand expressions, application of the operation Π to more than two arguments occurs. The meaning of such abbreviations has been defined at the beginning of Section 3.

$$P_9(\tau)=(\chi_3 L^2 \rightarrow (\chi_0 \tau(LR^3L^2, (I, F)LRL, R, RL)\rightarrow R, \emptyset)(I, M^2), \emptyset).$$

Thus the $\mathfrak{S}$-computability of ω_ν in $\{\psi_1, \ldots, \psi_l, M^2\}$ is established. For proving the $\mathfrak{S}$-computability of ω in $\{\psi_1, \ldots, \psi_l\}$, we can use the same construction, but with $\emptyset$ instead of $(\chi_9 LRL \rightarrow P_9(\tau), \emptyset)$ in the expression for $P(\tau)$ (note that M^2 does not occur in the expressions for $P_0(\tau), \ldots, P_8(\tau)$).

Suppose now that the element φ of $\mathcal{F}$ belongs to $PC(A, \psi_1, \ldots, \psi_l)$. Then there is some element e of A^* such that

$$\varphi(q)=\omega(\langle e, \langle\langle q\rangle\rangle\rangle)$$

for all q in M. Let $\bar{e}$ be the constant function, assigning the value e to all elements of M. The above equality can be rewritten into the form

$$\varphi=\omega(\bar{e}, T, I, F)$$

(again the extension of the operation Π to the case of more than two arguments is used, for short). From the definition of the set A^*, it easily follows that, for each element of A^*, the corresponding constant function is $\mathfrak{S}$-elementary in $\mathcal{C}_A$. In particular, so is the function $\bar{e}$, and hence φ is $\mathfrak{S}$-computable in $\mathcal{C}_A \cup \{\psi_1, \ldots, \psi_l\}$. Since $\mathfrak{S}$-computability is equivalent to $\mathfrak{M}_B$-computability in the considered case, the statement (i) of the theorem is thus proved. The statement (ii) is treated in a quite similar way. ∎

Corollary 1. Under the assumptions of Theorem 1, the following equivalences hold:

(i) $\varphi \in PC(A, \psi_1, \ldots, \psi_l)$ iff φ can be generated from elements of the set $\mathcal{C}_A \cup \{\psi_1, \ldots, \psi_l, L, R\}$ by means of composition, $\mathfrak{M}_B$-combination and $\mathfrak{M}_B$-iteration;

(ii) $\varphi \in SC(A, \psi_1, \ldots, \psi_l)$ iff φ can be generated from elements of the set $\mathcal{C}_A \cup \{\psi_1, \ldots, \psi_l, (B^*)^2, L, R\}$ by means of composition, $\mathfrak{M}_B$-combination and $\mathfrak{M}_B$-iteration.

Proof. The elements T and F can be generated from L by means of the above three operations due to the fact that $L^3[L, L](u)=0$ for all u in B^*. ∎

We carried out the proof of Theorem 1 with fixed ψ_1,

$\ldots, \psi_l$ and used the non-parameterized version of our First Recursion Theorem. The only place, where $\psi_1, \ldots, \psi_l$ take part in some construction from the proof, is the definition of the mapping P_0. We could use the same construction to define P_0 as a mapping of $\mathcal{F}^{l+1}$ into $\mathcal{F}$, including also ψ_1, $\ldots, \psi_l$ into the list of the arguments of P_0. After doing the same for the composite mapping P, the parameterized version of the First Recursion Theorem can be applied, and so we can proof that all operators prime computable with constants from A are $\mathfrak{M}_B$-computable in $\mathcal{C}_A$, and all operators search computable with constants from A are $\mathfrak{M}_B$-computable in $\mathcal{C}_A \cup \{M^2\}$. The converse statements are also true, as an analysis of the proof of Propositions I.7.1 and I.7.2 follows. Thus prime and search computability are equivalent to $\mathfrak{M}_B$-computability (and to $\mathfrak{S}_{\mathrm{m}}(\mathfrak{M}_B)$-computability) in suitable sets of elements of $\mathcal{F}_{\mathrm{m}}(B^*)$ not only in the case of functions, but also in the case of operators.

The characterizations from Corollary 1 can be useful in various proofs concerning prime and search computability, especially in the direction from such a computability to other properties. They have been used, for instance, in Ditchev [1981, 1983, 1984, 1987] and Soskov [1983, 1984, 1987].

Since prime and search computability turn out to be particular cases of computability in combinatory spaces, the general theorems concerning the last computability are applicable to prime and search computability. For example, our First Recursion Theorem from Section 4 implies a First Recursion Theorem for prime computability and one for search computability. As we mentioned in Section I.7, there is such a theorem for search computability in Moschovakis [1969], but only for the case of total single-valued $\psi_1, \ldots, \psi_l$, and one for prime computability is briefly mentioned without complete formulation. The results established until now in this book show that both theorems are valid without any restrictions on $\psi_1, \ldots, \psi_l$.

We shall note two known facts which are immediate corollaries of Theorem 1 (to be more precise, the implication from the theorem to them is immediate only in the case when one-argument functions are considered, but this restriction is not essential, as mentioned in Subsection (II) of Section I.7).

Corollary 2 (Lemma 31 of Moschovakis [1969]). Prime computability implies search computability.

Corollary 3. Search computability from A in $\psi_1, \ldots, \psi_l$ is equivalent to prime computability from A in $\psi_1, \ldots, \psi_l$, $(B^*)^2$.[79]

The last corollary enables obtaining the First Recursion Theorem for search computability as a corollary from the First Recursion Theorem for prime computability.

Remark 1. Besides the characterization of prime computability from Theorem 1, other characterizations of it as $\mathfrak{U}$-computability are also possible. Two such characterizations can be found, in essence, in the thesis Soskova [1979]. In the first of them, $\mathfrak{M}_B$ is changed by restricting L and R to $B^* \setminus B^{\circ}$. In this case one has to add to the set $\mathcal{C}_A \cup \{\psi_1, \ldots, \psi_l\}$ also the total function χ defined by the condition that $\chi(u)=0$ for all u in B, and $\chi(u)=1$ for all other u in B^*. The second characterization concerns functions in the closure of B under ordered pairs (without using the element O). It turns out that the prime computability for such functions is equivalent to $\mathfrak{U}$-computability, where $\mathfrak{U}$ is the computational structure from Example I.1.7.

(IV) **Application to Friedman-Shepherdson computability.** In this subsection, the assumptions and the denotations of the previous one remain valid, and a certain additional assumption will be made.

In Section 2, we mentioned the Soskov's characterization (Soskov [1987]) of the Friedman-Shepherdson computability.[80] Suppose $\mathcal{A}$ is a partial algebraic structure with the carrier B, $\psi_1, \ldots, \psi_l$ are elements of $\mathcal{F}_m(B^*)$ representing the primitive functions and the primitive predicates of $\mathcal{A}$ (the elements 0 and 1 of B^* used for the representation of the truth values), and φ is an element of $\mathcal{F}_m(B^*)$ representing some partial operation in B. Under this addi-

[79] As Professor Y. N. Moschovakis informed us in 1975, the validity of this statement has been noticed several years before by some of his students.

[80] I.e. computability over an abstract structure by means of recursively enumerable definitional schemes (Friedman [1971], Shepherdson [1975]). As shown by Soskov, the same notion of computability can be introduced also by using recursively enumerable determinants in the sense of Ershov [1981].

tional assumption, Soskov's characterization enables supplementing Theorem 1 by the following statement, which shows that Friedman-Shepherdson computability is again a particular case of computability in iterative combinatory spaces:

(iii) the partial operation represented by φ is Friedman-Shepherdson computable in $\mathcal{A}$ iff φ is $\mathfrak{M}_B$-computable in $\{\psi_1, \dots, \psi_l, \mathbb{N}^2\}$.

As noted in Soskov [1987], a characterization of this sort of the notions of prime, search and Friedman-Shepherdson computability makes clear that absolute prime computability in $\psi_1, \dots, \psi_l$ implies Friedman-Shepherdson computability in $\mathcal{A}$, which in its turn implies search computability in $\psi_1, \dots, \psi_l$. It is so, since $\{\psi_1, \dots, \psi_l\}$ is a subset of $\{\psi_1, \dots, \psi_l, \mathbb{N}^2\}$, and $\mathbb{N}^2$ is $\mathfrak{M}_B$-computable in $(B^*)^2$ (as seen from the equality $\mathbb{N}^2 = \rho(B^*)^2\rho$, where $\rho = (\chi_{\mathrm{N}} \to I, \emptyset)$, and χ_{N}, I are the same as in the proof of Theorem 1). Of course, immediately the corollary follows that Friedman-Shepherdson computability in $\mathcal{A}$ is equivalent to absolute prime computability in $\psi_1, \dots, \psi_l, \mathbb{N}^2$ (by Theorem 18 of Soskov [1987], this is equivalent to a particular case of Theorem 5 of the same paper). Extending this equivalence to the case of operators, we see the validity of a First Recursion Theorem also for the Friedman-Shepherdson computability.

Exercises

1. Let $\mathfrak{U} = \langle M, J, L, R, T, F, H\rangle$ be the computational structure corresponding to a given FP-system, and let $\mathcal{B}$ be some subset of $\mathcal{F}_{\mathrm{p}}(M)$ containing the functions tl, apndl, eq and $\bar{\emptyset}$ (for the denotations, cf. Section I.4). Prove the $\mathfrak{U}$-computability in $\mathcal{B}$ of the functions distl, distr and trans, determined by the equalities

$$\mathrm{distl}(\langle s, \emptyset\rangle) = \emptyset,$$

$$\mathrm{distl}(\langle s, \langle t_1, \dots, t_k\rangle\rangle) = \langle\langle s, t_1\rangle, \dots, \langle s, t_k\rangle\rangle,$$

$$\mathrm{distr}(\langle \emptyset, s\rangle) = \emptyset,$$

$$\mathrm{distr}(\langle\langle t_1, \dots, t_k\rangle, s\rangle) = \langle\langle t_1, s\rangle, \dots, \langle t_k, s\rangle\rangle,$$

$$\mathrm{trans}(\emptyset) = \emptyset, \quad \mathrm{trans}(\langle \emptyset, \dots, \emptyset\rangle) = \emptyset,$$

$$\mathrm{trans}(\langle\langle t_{11}, \dots, t_{1n}\rangle, \dots, \langle t_{m1}, \dots, t_{mn}\rangle\rangle) = \langle\langle t_{11}, \dots, t_{m1}\rangle, \dots, \langle t_{1n}, \dots, t_{mn}\rangle\rangle$$

and by the condition that distl, distr and trans are defined only for such types of objects which are indicated as arguments in the left-hand sides of the corresponding equalities.

Hint. Use the First Recursion Theorem, or try a direct construction. Results from the exercises to Section I.7 also can be useful. In the case of the function **trans**, it could be convenient to start by the construction of a function which is $\mathfrak{U}$-computable in $\mathcal{B}$ and acts on ordered pairs in the same way as **trans**.

2. To the assumptions of the previous exercise, add the assumption that $\mathbb{N} \subset M$ and the functions $\bar{0}$, $\bar{1}$ belong to $\mathcal{B}$, as well as the function $+$ which is defined only for ordered pairs of natural numbers and assigns to each such pair the sum of its components. Prove the $\mathfrak{U}$-computability in $\mathcal{B}$ of the function **length** which is defined only for the finite sequences of elements of M and assigns to each such sequence its length.

3. Give a direct construction (not using the First Recursion Theorem) of elements χ_N, $\chi_<$, σ_1 and σ_2 of $\mathrm{COMP}_{\mathfrak{S}}(\emptyset)$ with the properties needed for the proof of Theorem 1.

4. Prove that $\mathfrak{M}_B$-computability in $\{\psi_1, \ldots, \psi_l, \mathbb{N}^2\}$ is equivalent to $\mathfrak{M}_B$-computability in $\{\psi_1, \ldots, \psi_l, B^* \times \{0, 1\}\}$. Prove also the equality

$$B^* \times \{0, 1\} = ((B^*)^2 \to T, F).$$

6. Normal Form Theorems for computable elements and mappings in iterative combinatory spaces

In this section, the same assumptions will be made as in Section 4. Namely an iterative combinatory space

$$\mathfrak{S} = \langle \mathcal{F}, I, \mathcal{C}, \Pi, L, R, \Sigma, T, F \rangle$$

and a subset $\mathcal{B}$ of $\mathcal{F}$ will be supposed to be given.

We start by recalling a fact mentioned in Section 4 after the end of the proof of the First Recursion Theorem. We noted that a mapping E of $\mathcal{F}$ into $\mathcal{F}$ exists such that E is $\mathfrak{S}$-elementary in $\mathcal{B}$, and, for all θ in $\mathcal{F}$, the components $\Delta_1(\theta), \ldots, \Delta_l(\theta)$ of the least solution of the considered system of inequalities 4.(3) are expressed by means of the formulas 4.(10). The system 4.(3) in question had a special form, namely each inequality in it had some of the forms 4.(5)–4.(9). However, the above statement remains valid also in the general case, i.e. for an arbitrary system 4.(3) with $\Gamma_1, \ldots, \Gamma_l$ $\mathfrak{S}$-computable in $\mathcal{B}$. It is so, since the least solution of such a general system consists of the first l components of a system having a greater number of unknowns and containing only inequalities of the forms 4.(5)-4.(9).

Let us now apply this to the system consisting of the single inequality

$$\tau \geq \Gamma(\theta),$$

where Γ is an arbitrary mapping $\mathfrak{S}$-computable in $\mathcal{B}$. Since the least solution τ of this system is equal to $\Gamma(\theta)$, we get the following result:

Proposition 1. If Γ is a mapping of $\mathcal{F}$ into $\mathcal{F}$ $\mathfrak{S}$-computable in $\mathcal{B}$, then a mapping E of $\mathcal{F}$ into $\mathcal{F}$ exists such that E is $\mathfrak{S}$-elementary in $\mathcal{B}$, and, for all θ in $\mathcal{F}$, the equality

$$\Gamma(\theta) = L[\mathrm{E}(\theta), (L \to F, T)R](I, \bar{1}\bar{0}) \tag{1}$$

holds.

There is no problem in generalizing this proposition to mappings Γ of $\mathcal{F}^n$ into $\mathcal{F}$, for arbitrary positive integers n. The change will be only that E will be then also a mapping of $\mathcal{F}^n$ into $\mathcal{F}$. We shall not give the corresponding formulation explicitly. However, it is worth giving the formulation, so to say, for $n = 0$.

Proposition 2. If φ is an element of $\mathcal{F}$ $\mathfrak{S}$-computable in $\mathcal{B}$ then an element ε of $\mathcal{F}$ exists such that ε is $\mathfrak{S}$-elementary in $\mathcal{B}$, and the equality

$$\varphi = L[\varepsilon, (L \to F, T)R](I, \bar{1}\bar{0}) \tag{2}$$

holds.

We think no argumentation is needed for the truth of Proposition 2 after Proposition 1 is established.

The expressions in the right-hand sides of (1) and (2) can be considered as normal form representations of $\Gamma(\theta)$ and of φ, respectively. Of course, the elements $(I, \bar{1}\bar{0})$, $(L \to F, T)R$, L occurring in these expressions have their origin from the coding adopted in the proof of the First Recursion Theorem. These elements can be replaced by some other ones by changing the mentioned coding. Of course, making changes in the heavy proof in question is not a pleasant task, and it is not clear enough to what extent modifications can be made. Fortunately, their is an easier way to make such modifications, namely by application of Corollary II.6.5. Here is a result which follows from Propositions 1 and 2 in this way.

Theorem 1. Let $\langle \Xi, \xi_0, \xi_1, \xi_2 \rangle$ be a ternary join mechanism in $\mathcal{F}$ having the following properties: (a) the mapping Ξ and the elements ξ_0, ξ_1, ξ_2 are $\mathfrak{S}$-elementary in $\mathcal{B}$; (b) the elements ξ_0, ξ_1, ξ_2 are normal. Let χ and ρ be elements of $\mathcal{F}$ which satisfy the conditions (ii) and (iii) of Corollary II.6.5.[81] Then each mapping Γ of $\mathcal{F}$ into $\mathcal{F}$ which is $\mathfrak{S}$-computable in $\mathcal{B}$ can be represented in the form

(3) $$\Gamma(\theta) = \rho[\mathrm{E}(\theta), \chi]\xi_1$$

with some mapping E $\mathfrak{S}$-elementary in $\mathcal{B}$, and each element φ of $\mathrm{COMP}_{\mathfrak{S}}(\mathcal{B})$ can be represented in the form

$$\varphi = \rho[\varepsilon, \chi]\xi_1$$

with some element ε from $\mathrm{ELEM}_{\mathfrak{S}}(\mathcal{B})$.

Proof. Let Γ be a mapping of $\mathcal{F}$ into $\mathcal{F}$ which is $\mathfrak{S}$-computable in $\mathcal{B}$. By Proposition 1, the mapping Γ is representable in the form

(4) $$\Gamma(\theta) = \rho_0[\mathrm{E}_0(\theta), \chi_0]\alpha_0,$$

where E_0 is a mapping $\mathfrak{S}$-elementary in $\mathcal{B}$, and ρ_0, χ_0, α_0 belong to $\mathrm{ELEM}_{\mathfrak{S}}(\emptyset)$. Then, by Corollary II.6.5, the equality (3) holds with

$$\mathrm{E}(\theta) = \Xi(\xi_0, \xi_2\alpha_0, (\chi_0 \to \xi_2\mathrm{E}_0(\theta), \xi_0\rho_0)),$$

and this E is $\mathfrak{S}$-elementary in $\mathcal{B}$ by the assumption (a) about the given join mechanism. The case of an element φ of $\mathrm{COMP}_{\mathfrak{S}}(\mathcal{B})$ is similar.∎

Making use of Example II.6.4, we obtain the following simple looking particular case of the above theorem.

Corollary 1. Each mapping Γ of $\mathcal{F}$ into $\mathcal{F}$ which is $\mathfrak{S}$-computable in $\mathcal{B}$ can be represented in the form

$$\Gamma(\theta) = R[\mathrm{E}(\theta), L](T, (T, I))$$

with some mapping E $\mathfrak{S}$-elementary in $\mathcal{B}$, and each element φ of $\mathrm{COMP}_{\mathfrak{S}}(\mathcal{B})$ can be represented in the form

---

[81] I.e. $\rho\xi_0 = I$, and $\chi\xi_0 = F\eta_0$, $\chi\xi_1 = T\eta_1$, $\chi\xi_2 = T\eta_2$ for some normal elements η_0, η_1, η_2. We note that these conditions will be surely satisfied by

$$\rho = \Xi(I, I, I), \quad \chi = \Xi(F, T, T),$$

and these concrete ρ, χ are $\mathfrak{S}$-elementary in $\mathcal{B}$, as far as the mapping Ξ is $\mathfrak{S}$-elementary in $\mathcal{B}$. In the natural examples known to us, the mapping Ξ is always $\mathfrak{S}$-elementary in $\emptyset$, and so will be the concrete ρ, χ constructed above.

(5) $$\varphi = R[\varepsilon, L](T, (T, I))$$

with some element ε from $\mathrm{ELEM}_{\mathfrak{S}}(\mathcal{B})$.

Many other particular cases can be obtained by using other ternary join mechanisms in $\mathcal{F}$ (for example, that ones from Exercise II.6.5). Of course, all these results can be immediately generalized to mappings Γ of $\mathcal{F}^n$ into $\mathcal{F}$, for arbitrary positive integers n.

Remark 1. Some other normal form representations of computable elements and mappings of certain more special kinds can be obtained by applying the normal form theorems from Georgieva [1980] and Ivanov [1980, 1986] to the companion operative space of the given combinatory space.

A comparison is appropriate of the normal form representations obtained in this section and the Kleene normal form of the partial recursive functions. The main difference between them lies in the fact that the Kleene normal form uses μ-operation, and our normal form uses iteration instead. In order that both normal form make sense, let $\mathfrak{S}$ be the combinatory space $\mathfrak{S}_p(\mathfrak{U})$ corresponding to a standard computational structure $\mathfrak{U} = \langle \mathbb{N}, J, L, R, T, F, H \rangle$ over the natural numbers, assuming that J, L, R are primitive recursive. Let $\mathcal{B} = \{S, P\}$, where as usually $S = \lambda u.\, u+1$, $P = \lambda u.\, u \dot{-} 1$. Then all elements of $\mathcal{F}_p(\mathbb{N})$ $\mathfrak{S}$-elementary in $\mathcal{B}$ are primitive recursive, and therefore Corollary 1 implies representability of all unary partial recursive functions φ in the form (5) with primitive recursive ε. Obviously, the representation (5) is different from the Kleene representation. However, (5) is sufficient in the considered case to see that φ can be obtained from some primitive recursive functions by means of substitution and a single application of the μ-operation. It is so, since the equality

$$[\varepsilon, L](u) = \varepsilon^{\theta(u)}(u),$$

holds, where

$$\theta(u) = \mu t\, [L(\varepsilon^t(u)) = 0].$$

The question is justified, whether a normal form theorem is possible which directly comprises the Kleene Normal Form theorem as a particular case. There are examples showing that one can hardly expect a natural generalization of the Kleene theorem to the case of arbitrary iterative combinatory spaces. Namely an iterative combinatory space $\mathfrak{S}$ and a set $\mathcal{B}$ of its elements can be indicated (see Exercises 2 and 3) such that:

(i) for the elements of the space, there is a natural notion of μ-recursiveness in $\mathcal{B}$;

(ii) there are elements of $\mathrm{COMP}_{\mathfrak{S}}(\mathcal{B})$ which are not μ-recursive in $\mathcal{B}$.

On the other hand, a natural generalization of the Kleene theorem to this case must imply that all elements $\mathfrak{S}$-computable in $\mathcal{B}$ are μ-recursive in $\mathcal{B}$.

Exercises

1. Show that the statements of Propositions 1 and 2 are particular cases of the statement of Theorem 1, i.e. in any given iterative combinatory space $\langle \mathcal{F}, I, \mathcal{C}, \Pi, L, R, \Sigma, T, F\rangle$ the assumptions of Theorem 1 can be satisfied by

$$\rho = L, \quad \chi = (L \rightarrow F, T)R$$

and some appropriate ternary join mechanism $\langle \Xi, \zeta_0, \zeta_1, \zeta_2\rangle$ with

$$\zeta_1 = (I, \overline{1}\,\overline{0}).$$

2. (Cf. Ivanov [1977, 1978]) Let the combinatory space $\mathfrak{S} = \langle \mathcal{F}, I, \mathcal{C}, \Pi, L, R, \Sigma, T, F\rangle$ be the space $\mathfrak{S}_p(\mathfrak{U})$, where $\mathfrak{U} = \langle \mathbb{N}, J, L, R, T, F, H\rangle$ is a standard computational structure on the natural numbers, and the functions J, L, R are primitive recursive. Let $S = \lambda u.\, u+1$, $P = \lambda u.\, u \dot{-} 1$. Consider the mappings Γ_1 and Γ_2 of $\mathcal{F}$ into $\mathcal{F}$ defined as follows:

$$\Gamma_1(\theta) = R[(PL, \theta R), L],$$
$$\Gamma_2(\theta) = R[(L, SR), \theta](I, F).$$

Prove the equalities

$$\Gamma_1(\theta)(J(t, u)) \simeq \theta^t(u),$$
$$\Gamma_2(\theta)(u) \simeq \mu t[\theta(u, t) = 0],$$

where θ ranges over $\mathcal{F}_p(\mathbb{N})$, and t, u range over $\mathbb{N}$. Prove that, for all $\psi_1, \ldots, \psi_l$ and φ in $\mathcal{F}_p(\mathbb{N})$, the following statements hold:

(i) the function φ is primitive recursive in $\psi_1, \ldots, \psi_l$ iff φ can be generated from $I, F, S, L, R, \psi_1, \ldots, \psi_l$ by means of composition, Π and Γ_1;

(ii) the function φ is μ-recursive in $\psi_1, \ldots, \psi_l$ iff φ can be generated from $I, F, S, L, R, \psi_1, \ldots, \psi_l$ by means of composition, Π, Γ_1 and Γ_2.

3. Let $\mathfrak{S}$ be the combinatory space from Exercises 2.9-2.18, and let the functions J, L, R be primitive recursive.

Consider the mappings Γ_1 and Γ_2 of $\mathcal{F}$ into $\mathcal{F}$ defined in the same way as in the previous exercise, but with $I_{\mathbb{N}}^{\sim}$, $L^{\sim}$, $R^{\sim}$, $F^{\sim}$, $S^{\sim}$, $P^{\sim}$ instead of I, L, R, F, S, P, respectively. Prove the existence of an element of $\mathrm{COMP}_{\varnothing}(\{S^{\sim}, P^{\sim}, (\mathbb{N}^2)^{\sim}\})$ which cannot be generated from $I_{\mathbb{N}}^{\sim}$, $F^{\sim}$, $S^{\sim}$, $L^{\sim}$, $R^{\sim}$, $(\mathbb{N}^2)^{\sim}$ by means of composition, Π, Γ_1 and Γ_2.

Hint. Prove that, whenever an element $\langle f, A\rangle$ can be generated from $I_{\mathbb{N}}^{\sim}$, $F^{\sim}$, $S^{\sim}$, $L^{\sim}$, $R^{\sim}$, $(\mathbb{N}^2)^{\sim}$ by means of composition, Π, Γ_1 and Γ_2, then A is in the arithmetical hierarchy (cf. Rogers [1967, § 14.1] for the definition). Then use Exercise 2.13 and the existence of Π_1^1-sets which are not in the arithmetical hierarchy (cf. Rogers [1967, § 16.1]).

4. Prove the strengthening of Theorem 1 which arises after replacing the words "with some mapping E $\varnothing$-elementary in $\mathcal{B}$" by the text "with some mapping E of the form

$$E(\theta) = (\pi \to \sigma, \varkappa(\theta\nu, I))$$

with π, σ, $\varkappa$, ν $\varnothing$-elementary in $\mathcal{B}$".

Hint. Use Exercise 4.1 to represent Γ in the form (4) with E_0 having the form

$$E_0(\theta) = (\pi_0 \to \sigma_0, (\theta L, R^{m+2})),$$

where m is some natural number, and π_0, σ_0 are $\varnothing$-elementary in $\varnothing$. Show that, for each θ in $\mathcal{F}$, the element $E_0(\theta)$ is the component τ_4 of the least solution $\langle\tau_1, \tau_2, \tau_3, \tau_4\rangle$ of the system

$$\tau_1 \ge \tau_2 \alpha_0,$$
$$\tau_2 \ge (\chi_0 \to \tau_3, \rho_0),$$
$$\tau_3 \ge (\pi_0 \to \tau_2 \sigma_0, \tau_4),$$
$$\tau_4 \ge \tau_2 (\theta L, R^{m+2}).$$

To apply Theorem II.6.3 to this system, use the join mechanism $\langle\Xi', \zeta_0, \zeta_1, \zeta_2\bar{0}, \zeta_2\bar{1}, \zeta_2 R_*^2\rangle$, where

$$\Xi'(\tau_0, \tau_1, \tau_2, \tau_3, \tau_4) = \Xi(\tau_0, \tau_1, \Pi_*(\tau_2, \tau_3, \tau_4)).$$

At the application of the theorem, use an appropriate element ε of the form

$$(\pi \to \sigma, \zeta_2\bar{0}(\theta L\rho', R^{m+2}\rho')).$$

Taking this $\mathcal{E}$ as $E(\theta)$, apply Exercise II.1.16 to represent $E(\theta)$ in the needed form.

7. Universal computable elements in iterative combinatory spaces

Again an iterative combinatory space

$$\mathfrak{S} = \langle \mathcal{F}, I, \mathcal{C}, \Pi, L, R, \Sigma, T, F \rangle$$

and a subset $\mathcal{B}$ of $\mathcal{F}$ will be supposed to be given. For the sake of simplicity, we shall suppose that T and F belong to $\mathcal{C}$. Our exposition will be close in its spirit to the content of a manuscript of L. Ivanov written in 1977.

We shall first give a series of definitions.

Definition 1. For each subset $\mathcal{G}$ of $\mathcal{C}$, the *combination closure of* $\mathcal{G}$ is the least subset of $\mathcal{F}$ containing $\mathcal{G}$ and closed under the operation Π.

Obviously, the combination closure of any subset of $\mathcal{C}$ is again a subset of $\mathcal{C}$.

Definition 2. Let ω be an element of $\mathcal{F}$, and $\mathcal{A}$ be a subset of $\mathcal{C}$. An element θ of $\mathcal{F}$ will be called *canonically $\mathcal{A}$-expressible through* ω if $\theta = \omega(z, I)$ for some element z of the combination closure of $\mathcal{A} \cup \{T, F\}$.[82] The element ω will be called *$\mathcal{A}$-universal for* a given subset $\mathcal{G}$ of $\mathcal{F}$ if all elements of $\mathcal{G}$ are canonically $\mathcal{A}$-expressible through ω.

Definition 3. An element of $\mathcal{F}$ will be called *absolutely normal* if this element belongs to the combination closure of the set $\{I, T, F\}$.

Obviously, all absolutely normal elements of $\mathcal{F}$ are normal and $\mathfrak{S}$-elementary in $\emptyset$.

Definition 4. An element ω of $\mathcal{F}$ will be called *completely universal for* a given subset $\mathcal{G}$ of $\mathcal{F}$ if ω is $\emptyset$-universal for $\mathcal{G} \cup \{L, R\}$, and absolutely normal elements

---

[82] Note that the condition $\theta = \omega(z, I)$ is equivalent to the condition that $\theta x = \omega(z, x)$ for all x in $\mathcal{C}$. The element z of $\mathcal{C}$ can be regarded as a code for the element θ of $\mathcal{F}$. In the book Skordev [1980] only some special elements of the combination closure of $\{T, F\}$ play the role of such codes, namely the elements of the form $\bar{n}T$, used there also for the representation of the natural numbers (cf. Exercises 8-10 after this section).

$\eta_0, \eta_1, \eta_2, \eta_3$ exist such that the following equalities hold for all a, x, y in $\mathcal{C}$:

$$\omega(\eta_0 a, I) = a,$$
$$\omega(\eta_1(x, y), I) = \omega(x, I)\omega(y, I),$$
$$\omega(\eta_2(x, y), I) = (\omega(x, I), \omega(y, I)),$$
$$\omega(\eta_3(x, y), I) = [\omega(x, I), \omega(y, I)].$$

The existence of a completely universal element is not obvious even for the empty subset of $\mathcal{F}$. Postponing the existence problem, we shall prove first a proposition showing the usefulness of the completely universal elements in case they exist.

Proposition 1. Let ω be completely universal for the subset $\mathcal{B}$ of $\mathcal{F}$. Then, for every subset $\mathcal{A}$ of $\mathcal{C}$, the element ω is $\mathcal{A}$-universal for the set $\mathrm{COMP}_{\mathfrak{S}}(\mathcal{B}\cup\mathcal{A})$.

Proof. Let $\mathcal{A}$ be an arbitrary subset of $\mathcal{C}$, and $\mathcal{A}'$ be the combination closure of $\mathcal{A}\cup\{T, F\}$. For any fixed z in $\mathcal{A}'$, the set $\{\eta\in\mathcal{F}: \eta z\in\mathcal{A}'\}$ is closed under Π and contains $\{I, T, F\}$. Hence $\eta z\in\mathcal{A}'$ for any z in $\mathcal{A}'$ and any absolutely normal element η of $\mathcal{F}$. Having this property at our disposal, we can use induction along the generation of the elements of $\mathrm{COMP}_{\mathfrak{S}}(\mathcal{B}\cup\mathcal{A})$ for proving that any such element is canonically $\mathcal{A}$-representable through ω. The only step in this proof, which needs explicit mentioning, is the verification that T and F are canonically $\mathcal{A}$-representable through ω - the first equality in Definition 4 is used at this step. ■

The main result in this section reads as follows.

Theorem 1 (Existence of computable completely universal elements). Let the subset $\mathcal{B}$ of $\mathcal{F}$ be finite. Then there is an element ω of $\mathrm{COMP}_{\mathfrak{S}}(\mathcal{B})$ such that ω is completely universal for $\mathcal{B}$.

Proof. We set

$$\eta_i = \bar{i}, \quad i = 0, 1, 2, 3,$$

which implies $R^{i+1}\eta_i = I$. We shall construct mappings Γ_1, Γ_2, Γ_3 which are $\mathfrak{S}$-elementary in $\emptyset$ and satisfy the following conditions for all x, y in $\mathcal{C}$ and all τ in $\mathcal{F}$:

$$\Gamma_1(\tau)(\eta_1(x, y), I) = \tau(x, I)\tau(y, I);$$
$$\Gamma_2(\tau)(\eta_2(x, y), I) = (\tau(x, I), \tau(y, I));$$
$$\Gamma_3(\tau)(\eta_3(x, y), I) = [\tau(x, I), \tau(y, I)].$$

If $\zeta = (\eta_i(x, y), I)$ then $I = R\zeta$, $(x, y) = R^{i+1}L\zeta$, hence

$$x = LR^{i+1}L\zeta, \quad y = R^{i+2}L\zeta.$$

Since

$$\tau(x, I)\tau(y, I) = \tau(x, \tau(y, I))$$

for all x, y in $\mathcal{C}$, it is clear that we can define Γ_1 in the following way:

$$\Gamma_1(\tau) = \tau(LR^2L, \tau(R^3L, R))$$

(the right distributivity of the normal elements with respect to Π is used). Even simpler we see that an appropriate definition of Γ_2 can be the following one:

$$\Gamma_2(\tau) = (\tau(LR^3L, R), \tau(R^4L, R)).$$

More problems arise in connection with the definition of Γ_3, since there is no right distributivity of the normal elements with respect to iteration. In this case we shall use the fact that

$$[\tau(x, I), \tau(y, I)] =$$
$$[\tau(L^2, R)((x, y), I), \tau(RL, R)((x, y), I)] =$$
$$R[(L, \tau(L^2, R)), \tau(RL, R)]((x, y), I),$$

by Corollary II.3.1. Hence Γ_3 can be defined by means of the equality

$$\Gamma_3(\tau) = R[(L, \tau(L^2, R)), \tau(RL, R)](R^4L, R).$$

After having the mappings Γ_1, Γ_2, Γ_3, we can write the equalities from Definition 4 in the form

(1) $\qquad \omega(\eta_0 a, I) = RL(\eta_0 a, I);$

(2) $\qquad \omega(\eta_i(x, y), I) = \Gamma_i(\omega)(\eta_i(x, y), I), \quad i = 1, 2, 3.$

We shall look for an element ω, which satisfies these equalities (for all a, x, y in $\mathcal{C}$) and, in addition, the equalities

(3) $\qquad \omega(\overline{3+k}T, I) = \psi_k R(\overline{3+k}T, I), \quad k = 1, \ldots, m,$

where $\psi_1, \ldots, \psi_m$ are the elements of $\mathcal{B}\cup\{L, R\}$.

Taking into account the equalities

$$LR^j\,\bar{i} = F\,\overline{i-j-1}, \quad j = 0, \ldots, j-1,$$
$$LR^i\,\bar{i} = T,$$

we see that the equalities (1), (2) and (3) will be surely satisfied if the following equality holds:

$$\omega = (L \to RL, (LR \to \Gamma_1(\omega), (LR^2 \to \Gamma_2(\omega), (LR^3 \to \Gamma_3(\omega),$$
$$(LR^4 \to \psi_1 R, \ldots, (LR^{m+2} \to \psi_{m-1}R, \psi_m R)\ldots)))))))).$$

By the First Recursion Theorem, the above equality is satisfied by some ω in $\mathrm{COMP}_{\otimes}(\mathcal{B})$, and the fact of satisfying the equality guarantees that ω is completely universal for $\mathcal{B}$. ■

Remark 1. If ω is such as in the above theorem then clearly $\mathrm{COMP}_{\otimes}(\mathcal{B})$ is equal to the set of the elements of $\mathcal{F}$ which are canonically $\emptyset$-representable through ω.

For the computable completely universal elements an analogue holds of the $s-m-n$ Theorem from the ordinary recursive function theory. Without aiming at thorough formal analogy, we shall formulate this analogue as follows.

Theorem 2 ($s-m-n$ Theorem for computable completely universal elements). Let $\omega \in \mathrm{COMP}_{\otimes}(\mathcal{B})$, and let ω be completely universal for $\mathcal{B}$. Then an element σ of $\mathrm{ELEM}_{\otimes}(\emptyset)$ exists such that $\sigma(z, x)$ belongs to the combination closure of $\{z, x, T, F\}$, and

$$\omega(z, (x, y)) = \omega(\sigma(z, x), y)$$

for all x, y, z in $\mathcal{C}$.

Proof. By Proposition 1, the element ω is $\emptyset$-universal for $\mathrm{COMP}_{\otimes}(\mathcal{B})$. Therefore the elements I and ω are canonically $\emptyset$-representable through ω, i.e. $I = \omega(z_0, I)$, $\omega = \omega(z_1, I)$ for some elements z_0, z_1 of the combination closure of $\{T, F\}$. We choose such z_0, z_1, as well as elements η_0, η_1, η_2 with the properties from Definition 4. Then we have

$$\begin{aligned}\omega(z, (x, y)) = \omega(z, (x, I))y = \\ \omega(z, (\omega(\eta_0 x, I), \omega(z_0, I)))y = \\ \omega(\omega(\eta_0 z, I), \omega(\eta_2(\eta_0 x, z_0), I)y = \\ \omega(z_1, I)\omega(\eta_2(\eta_0 z, \eta_2(\eta_0 x, z_0)), I)y = \\ \omega(\eta_1(z_1, \eta_2(\eta_0 z, \eta_2(\eta_0 x, z_0))), I)y = \\ \omega(\eta_1(z_1, \eta_2(\eta_0 z, \eta_2(\eta_0 x, z_0))), y).\end{aligned}$$

Thus it is sufficient to choose σ so that

$$\sigma(z, x) = \eta_1(z_1, \eta_2(\eta_0 z, \eta_2(\eta_0 x, z_0)))$$

for all z, x in $\mathcal{C}$, and this can be achieved by setting

$$\sigma = \eta_1(z_1, \eta_2(\eta_0 L, \eta_2(\eta_0 R, z_0))).$$ ■

Another form of the $s-m-n$ Theorem is given in the exercises.

One of the most impressive applications of the $s-m-n$

Theorem in the ordinary recursive function theory is the proof of the Second Recursion Theorem. The above $s-m-n$ Theorem makes possible a similar application.

Theorem 3 (Parameterized Second Recursion Theorem for computable completely universal elements). Let $\omega \in \mathrm{COMP}_{\otimes}(\mathcal{B})$, and let ω be completely universal for $\mathcal{B}$. Then an element ρ of $\mathrm{ELEM}_{\otimes}(\emptyset)$ exists such that ρz belongs to the combination closure of $\{z, T, F\}$, and

$$\omega(\rho z, y) = \omega(\omega(z, \rho z), y)$$

for all y, z in $\mathcal{C}$.

Proof. Let σ be an element such as in Theorem 2. We find an element z_2 of the combination closure of $\{T, F\}$ such that, for all x, y, z in $\mathcal{C}$, the equality

$$\omega(z_2, (z, (x, y))) = \omega(\omega(z, \sigma(x, x)), y)$$

holds. Such a z_2 exists by Proposition 1 and by the fact that, for all x, y, z in $\mathcal{C}$, the equality

$$\omega(\omega(z, \sigma(x, x)), y) = \omega(\omega(L, \sigma(LR, LR)), R^2)(z, (x, y))$$

holds, and $\omega(\omega(L, \sigma(LR, LR)), R^2) \in \mathrm{COMP}_{\otimes}(\mathcal{B})$. By the properties of σ, we have the equalities

$$\omega(z_2, (z, (x, y))) = \omega(\sigma(z_2, z), (x, y)) = \omega(\sigma(\sigma(z_2, z), x), y).$$

Therefore

$$\omega(\sigma(\sigma(z_2, z), x), y) = \omega(\omega(z, \sigma(x, x)), y)$$

for all x, y, z in $\mathcal{C}$, and, in particular, this will be true for arbitrary y, z in $\mathcal{C}$ and $x = \sigma(z_2, z)$. Thus it is sufficient to choose ρ so that

$$\rho z = \sigma(\sigma(z_2, z), \sigma(z_2, z)),$$

i.e. to set

$$\rho = \sigma(\sigma(z_2, I), \sigma(z_2, I)). \blacksquare$$

Some other forms of the Second Recursion Theorem are given in the exercises.

Exercises

1. (Second form of the $s-m-n$ Theorem for computable completely universal elements) Let ω be an element of $\mathrm{COMP}_{\otimes}(\mathcal{B})$ completely universal for $\mathcal{B}$. Let $\mathcal{A}$ be a subset of $\mathcal{F}$, and φ belong to $\mathrm{COMP}_{\otimes}(\mathcal{B} \cup \mathcal{A})$. Then an element ψ of

$\mathrm{ELEM}_{\otimes}(\mathcal{A})$ exists such that ψx belongs to the combination closure of $\mathcal{A} \cup \{x, T, F\}$, and

$$\varphi(x, y) = \omega(\psi x, y)$$

for all x, y in $\mathcal{C}$.

2. (Non-parameterized Second Recursion Theorem for computable completely universal elements). Let ω be an element of $\mathrm{COMP}_{\otimes}(\mathcal{B})$ completely universal for $\mathcal{B}$. Let $\mathcal{A}$ be a subset of $\mathcal{F}$, and θ belong to $\mathrm{COMP}_{\otimes}(\mathcal{B} \cup \mathcal{A})$. Then there is an element e of the combination closure of $\mathcal{A} \cup \{T, F\}$ such that, for all y in $\mathcal{C}$, the equality

$$\omega(e, y) = \omega(\theta e, y)$$

holds.

3. (Second form of the non-parameterized Second Recursion Theorem) Let ω be an element of $\mathrm{COMP}_{\otimes}(\mathcal{B})$ completely universal for $\mathcal{B}$. Let $\mathcal{A}$ be a subset of $\mathcal{F}$, and φ belong to $\mathrm{COMP}_{\otimes}(\mathcal{B} \cup \mathcal{A})$. Then there is an element e of the combination closure of $\mathcal{A} \cup \{T, F\}$ such that, for all y in $\mathcal{C}$, the equality

$$\omega(e, y) = \varphi(e, y)$$

holds.

4. (Second form of the Parameterized Second Recursion Theorem). Let ω be an element of $\mathrm{COMP}_{\otimes}(\mathcal{B})$ completely universal for $\mathcal{B}$. Let $\mathcal{A}$ be a subset of $\mathcal{F}$, and φ belong to $\mathrm{COMP}_{\otimes}(\mathcal{B} \cup \mathcal{A})$. Then an element ψ of $\mathrm{ELEM}_{\otimes}(\mathcal{A})$ exists such that ψx belongs to the combination closure of $\mathcal{A} \cup \{x, T, F\}$, and

$$\omega(\psi x, y)) = \varphi(\psi x, (x, y))$$

for all x, y in $\mathcal{C}$.

5. (Effectiveness of the computable mappings) Let ω be an element of $\mathrm{COMP}_{\otimes}(\mathcal{B})$ completely universal for $\mathcal{B}$. Let $\mathcal{A}$ be a subset of $\mathcal{F}$, and Γ be a mapping of $\mathcal{F}$ into $\mathcal{F}$ which is $\otimes$-computable in $\mathcal{B} \cup \mathcal{A}$. Prove the existence of an element γ of $\mathrm{ELEM}_{\otimes}(\mathcal{B} \cup \mathcal{A})$ such that, for any z in $\mathcal{C}$, γz belongs to the combination closure of $\mathcal{A} \cup \{x, T, F\}$, and the equality

$$\Gamma(\omega(z, I)) = \omega(\gamma z, I)$$

holds. Generalize the result to mappings of $\mathcal{F}^{l}$ into $\mathcal{F}$, where l is an arbitrary positive integer.

Hint. Use Exercise 2.19 and the $s-m-n$ Theorem. In the

case of $l > 1$, start with representing $(z_1, I), \dots, (z_l, I)$ in the form $\pi_1(z, I), \dots, \pi_l(z, I)$, where $z = (z_1, \dots, z_l)$, and $\pi_1, \dots, \pi_l$ are $\mathfrak{S}$-elementary in $\emptyset$.

6. Let the set $\mathcal{B}$ be finite. Prove the existence of an element ω of $\mathrm{COMP}_{\mathfrak{S}}(\mathcal{B})$ such that the set $\mathrm{ELEM}_{\mathfrak{S}}(\mathcal{B})$ is equal to the set of the elements which are canonically $\emptyset$-expressible through ω.

7. Suppose the set $\mathcal{B}$ is finite, all elements of $\mathrm{ELEM}_{\mathfrak{S}}(\mathcal{B})$ are normal, and there is an element τ of $\mathrm{ELEM}_{\mathfrak{S}}(\mathcal{B})$ such that $\tau x \neq x$ for all x in $\mathcal{C}$. Prove the existence of a normal element of $\mathrm{COMP}_{\mathfrak{S}}(\mathcal{B})$ which is not $\mathfrak{S}$-elementary in $\mathcal{B}$.

8. Let the set $\mathcal{B}$ be finite. Prove the existence of an element ω of $\mathrm{COMP}_{\mathfrak{S}}(\mathcal{B})$ with the following properties:

(i) $\mathrm{COMP}_{\mathfrak{S}}(\mathcal{B}) = \{\omega_n : n \in \mathbb{N}\}$, where $\omega_n = \omega(\bar{n}T, I)$;

(ii) there are two-argument primitive recursive functions h_1, h_2, h_3 such that, for all i, j in $\mathbb{N}$,

$$\omega_{h_1(i,j)} = \omega_i \omega_j, \quad \omega_{h_2(i,j)} = (\omega_i, \omega_j), \quad \omega_{h_3(i,j)} = [\omega_i, \omega_j].$$

9. Let ω be an element of $\mathrm{COMP}_{\mathfrak{S}}(\mathcal{B})$ with the properties from the previous exercise. Prove the existence of a two-argument primitive recursive function f with the following property: for each one-argument recursive function g, there is a natural number k such that

$$\omega_{g(n)} = \omega_{f(k,n)}$$

for all n in $\mathbb{N}$.

Hint. Use Corollary 3.2 to show that, for each one-argument recursive function g, there is a natural number k such that

$$\omega_{g(n)} = \omega(\omega_k \bar{n}T, I)$$

for all n in $\mathbb{N}$. Prove the existence of a primitive recursive function f_0 such that $\bar{n}T = \omega_{f_0(n)}$ for all n in $\mathbb{N}$.

10. Let ω be an element of $\mathrm{COMP}_{\mathfrak{S}}(\mathcal{B})$ with the properties from Exercise 8. Prove the following Second Recursion Theorem: for any element φ of $\mathrm{COMP}_{\mathfrak{S}}(\mathcal{B})$, there is a natural number m such that

$$\omega(\bar{m}T, y) = \varphi(\bar{m}T, y)$$

for all y in $\mathcal{C}$ (compare with Exercise 3).

Hint. Represent $\varphi(\bar{n}T, I)$ in the form $\omega_{h(n)}$, where h is some primitive recursive function. Then apply the statement of the previous exercise to the function

$$g(n)=h(f(n, n)),$$

where f is the same as there.

8. A notion of search computability in iterative combinatory spaces

In this section (including the exercises), an iterative combinatory space

$$\mathfrak{S}=\langle\mathcal{F}, I, \mathcal{C}, \Pi, L, R, \Sigma, T, F\rangle,$$

a subset $\mathcal{B}$ of $\mathcal{F}$ and an element U of $\mathcal{F}$ are supposed to be given, and the following assumptions concerning U are made:

(i) for all x in $\mathcal{C}$, the inequalities

$$U \geq x, \quad x \geq xU$$

hold;

(ii) for all φ in $\mathcal{F}$ and all x in $\mathcal{C}$, the inequality

$$(\varphi x, I)U \geq (\varphi x, U)$$

holds;[83]

(iii) $U \geq L$, $U \geq R$.

Example 1. Let $\mathfrak{S}=\mathfrak{S}_m(\mathfrak{U})$, where $\mathfrak{U}=\langle M, J, L, R, T, F, H\rangle$ is a computational structure. Then the above assumptions are satisfied if we set $U=M^2$. The same is true also in the case when $\mathfrak{S}=\mathfrak{S}_m(\mathfrak{U}, E)$, where $\mathfrak{U}$ is as above, and E is some set having no common elements with the set M.

Example 2. If $\mathfrak{S}=\mathfrak{S}_p(\mathfrak{U})$, where $\mathfrak{U}$ is the same as in the previous example, then no element U in $\mathcal{F}$ exists satisfying the above assumptions.

Intuitively, the element U must be considered as describing unrestricted search in the data domain, i.e. choice with unbounded non-determinism.

Remark 1. The element U is not necessarily the greatest element of $\mathcal{F}$. This can be seen from the second part of

---

[83] This assumptions is surely satisfied in the case when the combinatory space $\mathfrak{S}$ is symmetric.

Example 1. From the proposition which will be proved below, it follows that U is always the least upper bound of the set $\mathcal{C}$, hence U is unique (if it exists at all).

Proposition 1. Let σ and γ be elements of $\mathcal{F}$ such that $\sigma \geq \gamma x$ for all x in $\mathcal{C}$. Then $\sigma x \geq \gamma$ for all x in $\mathcal{C}$, and $\sigma \geq \gamma U$.[84]

Proof. Let x be an arbitrary element of $\mathcal{C}$. Then, for all y in $\mathcal{C}$, we have

$$\sigma x y = \sigma x \geq \gamma y x = \gamma y,$$

and from here the inequality $\sigma x \geq \gamma$ follows. Now we use this inequality in the following way: for arbitrary x in $\mathcal{C}$, we have

$$\sigma x = \sigma x x \geq \sigma x U x \geq \gamma U x,$$

hence $\sigma \geq \gamma U$. ∎

Definition 1. An element of $\mathcal{F}$ will be called *search* $\mathfrak{S}$*-computable in* $\mathcal{B}$ if this element is $\mathfrak{S}$-computable in $\mathcal{B} \cup \{U\}$.

Remark 2. If $\mathfrak{S} = \mathfrak{S}_m(\mathfrak{M}_B)$, where $\mathfrak{M}_B$ is the Moschovakis structure based on B (the computational structure from Example I.1.2), and $\mathcal{B} = \mathcal{C}_A \cup \{\psi_1, \ldots, \psi_l\}$, then, by Theorem 5.1, the elements of $\mathcal{F}$ search $\mathfrak{S}$-computable in $\mathcal{B}$ are exactly the unary functions in SC(A, $\psi_1, \ldots, \psi_l$).[85] This can be regarded as a justification for the terminology introduced by the above definition. Of course, the general theory of $\mathfrak{S}$-computability can be applied to the introduced notion, but nevertheless additional work is needed in order that the study of this notion becomes motivated enough. We shall show that an interesting general statement concerning search $\mathfrak{S}$-computability can be proved. To do this, we shall first study the properties of the element U in more detail.

Proposition 2. For all x in $\mathcal{C}$ and all φ in $\mathcal{F}$, the equalities

---

[84] The assumptions concerning U are not needed for the truth of the first statement in the conclusion.

[85] Another example deserving attention can be obtained by taking $\mathfrak{S} = \mathfrak{S}_m(\mathfrak{U})$, $\mathcal{B} = \{\lambda u.\, u+1, \lambda u.\, u \dot{-} 1\}$, where $\mathfrak{U}$ is a standard computational structure over the natural numbers. In this case, as we know from Section I.6, the elements of $\mathcal{F}$ search $\mathfrak{S}$-computable in $\mathcal{B}$ are exactly the recursively enumerable binary relations.

$$xU = x, \quad Ux = U, \quad (\varphi x, I)U = (\varphi x, U)$$

hold.

Proof. For all x in $\mathcal{C}$, we have

$$xU \geq xx = x,$$

and since we have also the converse inequality, the equality $xU = x$ follows. The proof of the second equality is more complicated. By application of Proposition 1 to $\delta = U$, $\gamma = I$, we see that $Ux \geq I$ for all x in $\mathcal{C}$. Therefore (using also $x = xU$), we have

$$Ux = Uxy = UxUy \geq IUy = Uy$$

for all x, y in $\mathcal{C}$. Hence $Ux = Uy$ for all x, y in $\mathcal{C}$. Let x be an arbitrary element of $\mathcal{C}$. Then

$$Uxy = Ux = Uy$$

for all y in $\mathcal{C}$, and consequently $Ux = U$. For the proof of the third equality, suppose some φ from $\mathcal{F}$ and some x from $\mathcal{C}$ are given. Then, for all y in $\mathcal{C}$, we have

$$(\varphi x, U) \geq (\varphi x, y) = (\varphi x, I)y,$$

and therefore, by Proposition 1, the inequality

$$(\varphi x, U) \geq (\varphi x, I)U$$

holds. Since we have also the converse inequality, the proof is completed. ∎

Proposition 3. The equalities $L(I, U) = R(U, I) = I$ hold.

Proof. For all x in $\mathcal{C}$, we have

$$L(I, U)x = L(x, Ux) = L(x, I)Ux = xU = x,$$

$$R(U, I)x = R(Ux, x) = R(I, x)Ux = xU = x. \blacksquare$$

Proposition 4. Let γ, δ, φ be elements of $\mathcal{F}$, and let the inequality $\delta \geq \gamma(\varphi, x)$ hold for all x in $\mathcal{C}$. Then also the inequality $\delta \geq \gamma(\varphi, U)$ holds.[86]

Proof. Let y be an arbitrary element of $\mathcal{C}$. Then, for all x in $\mathcal{C}$, we have

$$\delta y \geq \gamma(\varphi, x)y = \gamma(\varphi y, I)x.$$

From here, making use of Propositions 1 and 2, we conclude that

$$\delta y \geq \gamma(\varphi y, I)U = \gamma(\varphi y, U) = \gamma(\varphi, U)y.$$

---

[86] The case of $\varphi = I$ of this proposition is due to L. Ivanov.

Since y is arbitrary, this completes the proof. ∎

Proposition 5. For all ψ in $\mathcal{F}$, the equality

$$(I, U)\psi(I, U) = (\psi(L, LR), R^2)(I, U)$$

holds.

Proof. For all y, z in $\mathcal{C}$, we have

$$(I, y)\psi(I, z) = (\psi(I, z), y) =$$
$$(\psi(L, LR), R^2)(I, (z, y)) \le (\psi(L, LR), R^2)(I, U).$$

From here, by Proposition 4, it follows that, for all y in $\mathcal{C}$, we have

$$(\psi(L, LR), R^2)(I, U) \ge (I, y)\psi(I, U) = (\psi(I, U), y).$$

Applying Proposition 4 once more, we get

$$(\psi(L, LR), R^2)(I, U) \ge (\psi(I, U), U).$$

On the other hand, for all x in $\mathcal{C}$, we have

$$(\psi(L, LR), R^2)(I, x) = (\psi(I, Lx), Rx) \le (\psi(I, Ux), Ux) =$$
$$(\psi(I, U), U),$$

hence, again by Proposition 4, the inequality

$$(\psi(L, LR), R^2)(I, U) \le (\psi(I, U), U).$$

Thus we proved the equality

$$(\psi(L, LR), R^2)(I, U) = (\psi(I, U), U),$$

and it remains only to note that, due to the second equality in Proposition 2, also the equality

$$(I, U)\psi(I, U) = (\psi(I, U), U)$$

holds. ∎

Proposition 6. For all ψ and χ in $\mathcal{F}$, the following equality holds:

$$[\psi(I, U), \chi] = L[\psi(L, LR), R^2), \chi L](I, U).$$

Proof. Let

$$\iota_1 = [\psi(I, U), \chi], \quad \iota_2 = [(\psi(L, LR), R^2), \chi L],$$
$$\sigma_1 = \psi(I, U), \quad \sigma_2 = (\psi(L, LR), R^2).$$

Then

$$L\iota_2(I, U) = (\chi L \to L\iota_2\sigma_2, L)(I, U),$$

and, making use of Exercise II.1.20 and Propositions 2, 3, 5, we get

$$L\iota_2(I, U) = (\chi \to L\iota_2\sigma_2(I, U), L(I, U)) =$$
$$(\chi \to L\iota_2(I, U)\sigma_1, I).$$

This equality implies the inequality

$$L\iota_2(I, U) \geq \iota_1 .$$

To prove the converse equality, it is sufficient (by Proposition 4) to prove that, for all x in $\mathcal{C}$, the inequality

$$\iota_1 \geq L\iota_2(I, x)$$

holds. This is equivalent to proving the inequality

$$\iota_1 L \underset{\mathcal{A}}{\geq} L\iota_2 ,$$

where $\mathcal{A}$ is the set of all elements of the form (y, x) with $y, x \in \mathcal{C}$. The set $\mathcal{A}$ is invariant with respect to σ_2, by Exercises II.1.39. On the other hand, for all y, x in $\mathcal{C}$, we have

$$(\chi L \to \iota_1 L\sigma_2 , L)(y, x) = (\chi \to \iota_1 L(\psi(I, Lx), Rx), I)y \leq$$
$$(\chi \to \iota_1 L(\psi(I, Ux), Ux), I)y = (\chi \to \iota_1 L(I, U)\sigma_1 , I)y =$$
$$(\chi \to \iota_1 \sigma_1 , I)y = \iota_1 L(y, x),$$

and this completes the proof.■

Now the main result about search $\mathfrak{S}$-computability will be formulated.

Theorem 1 (canonical representation of search $\mathfrak{S}$-computable elements). Each element of $\mathcal{F}$ search $\mathfrak{S}$-computable in $\mathcal{B}$ can be represented in the form $\varphi(I, U)$, where φ is some element of $\mathcal{F}$ $\mathfrak{S}$-computable in $\mathcal{B}$.

Proof. During the proof, we shall call *canonically representable* the elements of $\mathcal{F}$ having the form $\varphi(I, U)$, where $\varphi \in \mathrm{COMP}_{\mathfrak{S}}(\mathcal{B})$. The proof will contain several lemmas.

Lemma 1. All elements of $\mathrm{COMP}_{\mathfrak{S}}(\mathcal{B})$ are canonically representable.

Proof. If θ belongs to $\mathrm{COMP}_{\mathfrak{S}}(\mathcal{B})$ then θ is canonically representable, by the equality

$$\theta = \theta L(I, U),$$

which follows from Proposition 3.■

Lemma 2. The element U is canonically representable.

Proof. Using the equality

$$U = R(I, U),$$

which follows from Corollary II.1.2.■

Lemma 3. The multiplication preserves the canonical representability.

Proof. If $\theta = \theta_1\theta_2$, where $\theta_i = \varphi_i(I, U)$, $i = 1, 2$, and φ_1, φ_2 are $\mathfrak{S}$-computable in $\mathcal{B}$, then, by Proposition 5, the equality

$$\theta = \varphi_1(\varphi_2(L, LR), R^2)(I, U)$$

holds, and clearly $\varphi_1(\varphi_2(L, LR), R^2)$ is also $\mathfrak{S}$-computable in $\mathcal{B}$. ∎

Lemma 4. If θ is canonically representable, and ρ belongs to $\mathrm{COMP}_{\mathfrak{S}}(\mathcal{B})$, then (ρ, θ) is also canonically representable.

Proof. If $\theta = \varphi(I, U)$ then

$$(\rho, \theta) = (\rho L, \varphi)(I, U),$$

by Exercise II.1.15 and Proposition 2.

Lemma 5. The operation Π preserves the canonical representability.

Proof. Application of Lemmas 1, 3, 4 and of the equality

$$(\theta_1, \theta_2) = (R, \theta_2 L)(I, \theta_1),$$

which follows from Exercise II.1.14 and Corollary II.1.2. ∎

Lemma 6. If θ is canonically representable then the element $[\theta, L]$ is also canonically representable.

Proof. If $\theta = \varphi(I, U)$, and $\varphi \in \mathrm{COMP}_{\mathfrak{S}}(\mathcal{B})$, then, by Proposition 6, the equality

$$\theta = L[\varphi(L, LR), R^2), L^2](I, U)$$

holds, and clearly $L[\varphi(L, LR), R^2), L^2] \in \mathrm{COMP}_{\mathfrak{S}}(\mathcal{B})$. ∎

Lemma 7. The iteration preserves the canonical representability.

Proof. Application of Lemmas 1, 3, 5 and of the equality

$$[\theta_1, \theta_2] = R[(\theta_2, I)\theta_1 R, L](\theta_2, I),$$

known from Corollary II.5.2. ∎

Having Lemmas 1, 2, 3, 5 and 7 at our disposal, we prove by induction that all elements of $\mathcal{F}$ search $\mathfrak{S}$-computable in $\mathcal{B}$ are canonically representable. ∎

In Section I.6, we represented an arbitrary recursively enumerable binary relation θ in the form

$$\theta = \Delta(\theta)\Pi(I, \mathbb{N}^2),$$

where $\Delta(\theta)$ is some partial recursive function, and $I = I_{\mathbb{N}}$. Of course, the possibility of such a representation is a particular case of Theorem 1, but this cannot be regarded as an application of the theorem, since the above representation has been already used for proving the $\mathfrak{S}$-computability of θ in the set $\{\lambda u.\, u+1, \lambda u.\, u\dot{-}1, \mathbb{N}^2\}$. A real application can be done in the case mentioned in Remark 2, and the result obtained in this way easily implies the following statement:

Corollary 1. Let A be a subset of B^*, $\psi_1, \ldots, \psi_l$ be partial multiple-valued functions in B^*, and θ be an n-argument function from $SC(A, \psi_1, \ldots, \psi_l)$. Then there is an $n+1$-argument function φ in $PC(A, \psi_1, \ldots, \psi_l)$ such that, for all $q_1, \ldots, q_n$ in B^*, the equality

$$\theta(q_1, \ldots, q_n) = \bigcup\{\varphi(q_1, \ldots, q_n, r) : r \in B^*\}$$

holds.

No doubt, a direct proof of the above result must be surely possible. Such a direct proof can be based on the idea of replacing finitely many arbitrary choices in B^* by a single choice of an appropriate more complicated element of B^*. However, we note that the Normal Form Theorem in Moschovakis [1969] (Theorem 1 of the paper) gives directly the result of Corollary 1 only under the extra assumption that $\psi_1, \ldots, \psi_l$ are single-valued and total.

Having in mind Exercise 5.4, one could try to give an abstract treatment of the Friedman-Shepherdson computability by studying $\mathfrak{S}$-computability in $\mathcal{B}\cup\{(U \to T, F)\}$. Additional assumptions will be probably needed for the success of such an attempt.

A certain drawback of the considerations in this section is that iterative combinatory spaces having an element U with the properties (i)-(iii) are encountered not too often. In connection with this, we mention the paper Ivanov [1981], since another generalization of search computability is studied there, which covers also some iterative combinatory spaces without such an element U.

Exercises

1. Prove the equality $U^2 = U$.

Hint. Use the first two equalities in Proposition 2.

2. Let $\gamma, \delta, \varphi, \psi$ be elements of $\mathcal{F}$, and let the inequality $\delta \geq \gamma(\varphi x, \psi)$ hold for all x in $\mathcal{C}$. Prove that also the inequality $\delta \geq \gamma(\varphi U, \psi)$ holds.

3. Let $\mathcal{H}$ be the set of all elements θ of $\mathcal{F}$ which satisfy the inequality $\theta \leq U$, and let $\mathcal{B} \subseteq \mathcal{H}$, $(I \to I, I) \in \mathcal{H}$. Prove that all elements of $\mathcal{F}$ search $\mathfrak{S}$-computable in $\mathcal{B}$ belong to $\mathcal{H}$.

4. Prove that each element θ of the set $\mathrm{ELEM}_{\otimes}(\mathcal{B}\cup\{U\})$ can be represented in the form $\theta=\varphi(I, U)$, where φ is some element of the set $\mathrm{ELEM}_{\otimes}(\mathcal{B})$.

9. On the formalization of the proof of the First Recursion Theorem

In Section II.8, some formal systems have been introduced aiming at a strength sufficient for the formalization of the theory presented in this book. It would be very tiring to carry out a systematic investigation of all proofs for making clear their formalizability. Therefore we shall concentrate our attention on the heaviest of the proofs, namely the proof of the First Recursion Theorem, presented in Section 4. We shall discuss the problem of formulating and proving the statement of the theorem by the means of the formal system A′ from Section II.8.

First of all, we note that the First Recursion Theorem is formulated in Section 4 in a way not directly translatable in the language of A′. One of the difficulties lies in the using the notion of mapping, say, of $\mathcal{F}^{l}$ into $\mathcal{F}$, since the system A′ has no variables for such mappings. Another difficulty lies in using the notion of a subset of $\mathcal{F}$, namely a subset $\mathcal{B}$ of $\mathcal{F}$ is supposed to be given. We must show first a way for obviating these difficulties. We shall propose now a metamathematical statement which can be regarded as a refinement of the mathematical result in question.

Instead of considering mappings, we shall consider functional expressions containing variables for elements of $\mathcal{F}$. Such variables will be used both for the unknowns and for the parameters. As to the set $\mathcal{B}$, Propositions 2.5 and 2.6 enable reduction of the general case to the case of $\mathcal{B}=0$, and in this case the second difficulty disappears. In the light of this, we think the following statement is acceptable as a metamathematical counterpart of the First Recursion Theorem from Section 4.

Theorem 1. There is an algorithm which transforms each non-empty finite sequence $\mathcal{U}_1, \ldots, \mathcal{U}_l$ of functional expressions into a sequence of functional expressions $\mathcal{V}_1, \ldots, \mathcal{V}_l$ containing none of the variables $f_1, \ldots, f_l$ and such that the following two formulas are deducible in the system A:

(1) $\quad f_1=\mathcal{V}_1 \,\&\cdots\&\, f_l=\mathcal{V}_l \Rightarrow f_1=\mathcal{U}_1 \,\&\cdots\&\, f_l=\mathcal{U}_l,$

(2) $\quad f_1\geq\mathcal{U}_1 \,\&\cdots\&\, f_l\geq\mathcal{U}_l \Rightarrow f_1\geq\mathcal{V}_1 \,\&\cdots\&\, f_l\geq\mathcal{V}_l.$

Proof. By Theorem II.8.1, it is sufficient to show that the above formulas are deducible in the system A'. Therefore only deducibility in A' will be considered throughout this proof. The way of proceeding will be by showing that the proof of the First Recursion Theorem from Section 4 can be formalized in A'. We shall stress on the places where the formalization encounters difficulties. To be close to that proof, we shall restrict ourselves to the case when there is at the most one variable different from $f_1, \dots, f_l$ which may occur in some $\mathcal{U}_r$, and we shall assume that this is the variable f_{l+1} (it will appear in the places where θ occurs in the original proof, and $f_1, \dots, f_l$ will appear in the places where $\tau_1, \dots, \tau_l$ occur, respectively).

It is possible to give a formalized counterpart of the reduction of the general case in the First Recursion Theorem to the special case when each one of the inequalities has some of the forms 4.(5)-4.(9). This will be a constructively described transformation of finite sequences of functional expressions into other such sequences, and the description must be a part of the description of the algorithm. Moreover, for each concrete system of functional expressions, there is a deducible in A' formal counterpart of the statement describing the interdependence between the least solution of the initially given system of inequalities and of the new one obtained from it. We leave the corresponding details to the reader.

From now on, we assume that the mentioned reduction is carried out, and we have a concrete non-empty finite sequence $\mathcal{U}_1, \dots, \mathcal{U}_l$ of functional expressions each of them either being some of the expressions Λ, L, R, T, F, f_{l+1} or having some of the forms $f_j f_i$, (f_i, f_j), $(f_i \supset f_j, f_k)$ with i, j, k from the set $\{1, \dots, l\}$. Then an explicit construction can be given of the expressions $\mathcal{V}_1, \dots, \mathcal{V}_l$ by simply rewriting in the language of A' the expressions for $E_0(\theta)$, ..., $E_{2l+1}(\theta)$, then the expression for $E(\theta)$ formed from them, and, at last, the expressions for $\Delta_1(\theta), \dots, \Delta_l(\theta)$ (we recall that functional expressions representing the particular natural numbers have been introduced in Section 3, namely an arbitrary natural number n is represented by the functional expression $(F,)^n(T,)$, denoted by n^*). The more difficult thing is to show the deducibility in A' of the corresponding formulas (1) and (2). This will be done by formalization of the corresponding part of the proof from Section 4.

Some portions of the proof can be carried out in A′ without essential troubles. For example the definition of the notion of a coding element can be easily formulated in the language of A′. The definition of the set $\mathcal{P}_c$ is the first place, where the formalization is not obvious, since it is not clear how to express in the language of A′ the property of an element of $\mathcal{F}$ to be the product of finitely many coding elements.

In an informal presentation of the proof, the definition of $\mathcal{P}_c$ can be given in the form of an recursive definition, namely: $y \in \mathcal{P}_c$ iff $y = \overline{0}c$ or $y = \eta z$ for some coding element η and some element z of $\mathcal{P}_c$. One possible way to transform this in an explicit definition (again in the non-formal language) is to define $\mathcal{P}_c$ as the least subset $\mathcal{A}$ of $\mathcal{C}$ with the properties that $\overline{0}c \in \mathcal{A}$, and whenever η is some coding element, and z is an element of $\mathcal{A}$, then ηz belongs to $\mathcal{A}$ too. Of course, before giving such a definition, one first proves that such a least subset exists, and after giving the definition, one shows that the equivalence from the formulated recursive definition is actually true for the explicitly defined $\mathcal{P}_c$. All this can be carried out in the system A′, due to the existence of variables for subsets of $\mathcal{C}$, to the presence of the comprehension scheme II.6.(19), and to a certain monotonicity of the condition in the right-hand side of the recursive definition in question. We shall describe below a general method of using such recursive definitions within the system A′.

For arbitrary natural numbers j and k, let $s_j \subseteq s_k$ be an abbreviation for the formula $\forall c_0 (c_0 \in s_j \Rightarrow c_0 \in s_k)$. Then the following lemma holds.

Lemma 1. Let $\Phi(c_0, s_0)$ be a formula of the system A′ such that there are no free occurrences of s_1 in this formula, and s_1 is free for s_0 in the formula. Let the following formula be deducible in the system A′:

(3) $$s_0 \subseteq s_1 \Rightarrow (\Phi(c_0, s_0) \Rightarrow \Phi(c_0, s_1))$$

(where $\Phi(c_0, s_1)$ means $\Phi(c_0, s_0)(s_1/s_0)$). Then the following formula is also deducible in A′:

(4) $$\exists s_0 (\forall c_0 (c_0 \in s_0 \Leftrightarrow \Phi(c_0, s_0)) \;\&\; \forall s_1 (\forall c_0 (\Phi(c_0, s_1) \Rightarrow c_0 \in s_1) \Rightarrow s_0 \subseteq s_1)).$$

Proof. We start with an application of the comprehension scheme II.6.(19) giving the formula

(5) $\exists s_0 \forall c_0 (c_0 \in s_0 \Leftrightarrow$
$$\forall s_1 (\forall c_0 (\Phi(c_0, s_1) \Rightarrow c_0 \in s_1) \Rightarrow c_0 \in s_1))$$

(this formula states the existence of the intersection of all subsets of $\mathscr{C}$ which, taken as values of s_1, satisfy the formula $\forall s_1 (\forall c_0 (\Phi(c_0, s_1) \Rightarrow c_0 \in s_1))$. The rest of the proof is a verification that any set s_0 with the property stated in the formula (5) has also the property stated in the formula (4). Of course, the only problem is to show that the property of s_0 from (5) implies

$$\forall c_0 (c_0 \in s_0 \Leftrightarrow \Phi(c_0, s_0)).$$

Namely here the assumed deducibility of (3) is used. Making use of it, one shows that

$$\forall c_0 (\Phi(c_0, s_0) \Rightarrow c_0 \in s_0)$$

is implied by the property of s_0 stated in (5). To show that the same property implies also

$$\forall c_0 (c_0 \in s_0 \Rightarrow \Phi(c_0, s_0)),$$

one uses the above fact, as well as the deducibility of the formula

$$\exists s_1 \forall c_0 (c_0 \in s_1 \Leftrightarrow \Phi(c_0, s_0))$$

(this formula is obtainable by one more application of the comprehension scheme II.6.(19)). The deducibility of (3) is used again in this last part of the proof.∎

Clearly the way to consider $\mathscr{P}_c$ within the system A′ on the base of the above lemma is to apply the lemma to a formula $\Phi(c_0, s_0)$ expressing the condition that the value of c_0 is equal to $\bar{0}c$ or to ηz, where η is some coding element, and z belongs to the value of s_0. Such a formula is, for example,

(6) $c_0 = 0^* c_1 \vee \exists f_0 \exists c_2 (c_0 = f_0 c_2 \;\&\; \Psi(f_0) \;\&\; c_2 \in s_0)$,

where $\Psi(f_0)$ is a formula expressing the statement that the value of f_0 is a coding element (f_0 being the only free variable of $\Psi(f_0)$), and the variable c_1 is intended to have the value c. It is obvious that, at the above choice of the formula $\Phi(c_0, s_0)$, the corresponding formula (3) is deducible in the system A′.

The considerations made until now enable the formalizability in A′ of the reasonings about $\mathscr{P}_c$ which use the following two properties of this set: (a) the equivalence, formulated as a recursive definition of $\mathcal{S}_c$, and (b) the fact that $\mathcal{S}_c$ is contained in each subset $\mathcal{A}$ of $\mathcal{C}$ such that $c \in \mathcal{A}$, and whenever η is some coding element, and z is an element of $\mathcal{A}$, then ηz belongs to $\mathcal{A}$ too. Further we shall consider only a non-formalized version of the proof of the First Recursion Theorem, and we shall use the set $\mathcal{S}_c$ in this version, but with the restriction to base all our non-tautological reasonings about $\mathcal{S}_c$ only on the proper- (a) and (b) of this set.

Having the set $\mathcal{S}_c$ at our disposal, we can immediately define the set $\mathcal{T}_c$, ant its properties can be reduced to the properties of $\mathcal{S}_c$.

The next place with non-obvious formalization is the definition of proportionality. We shall reduce also this definition to the recursive definition of a certain subset of $\mathcal{C}$. Given two elements z_1, z_2 of $\mathcal{C}$, we can replace the definition of the property of y_1, y_2 to be proportional to z_1, z_2 by the definition of the set of all elements of $\mathcal{C}$ having the form (y_1, y_2), where y_1, y_2 are proportional to z_1, z_2. Denoting this set by $\mathcal{P}(z_1, z_2)$, we can introduce it non-formally by the following recursive definition: $x \in \mathcal{P}(z_1, z_2)$ iff $x = (z_1, z_2)$ or there are a coding element η and elements y_1, y_2 of $\mathcal{C}$ such that $x = (\eta y_1, \eta y_2)$ and $(y_1, y_2) \in \mathcal{P}(z_1, z_2)$. Lemma 1 enables the formalization of such reasonings about $\mathcal{P}(z_1, z_2)$ which are based on (c) the above equivalence and (d) the fact that $\mathcal{P}(z_1, z_2)$ is contained in each subset $\mathcal{A}$ of $\mathcal{C}$ such that $(z_1, z_2) \in \mathcal{A}$, and whenever $(y_1, y_2) \in \mathcal{A}$ and η is some coding element, then $(\eta y_1, \eta y_2) \in \mathcal{A}$ too. Namely it is appropriate to apply Lemma 1 with the following formula in the role of $\Phi(c_0, s_0)$:

(7) $$c_0 = (c_1, c_2) \vee \exists f_0 \exists c_3 \exists c_4 (c_0 = (f_0 c_3, f_0 c_4) \ \& \ \Psi(f_0) \ \& \ (c_3, c_4) \in s_0),$$

where $\Psi(f_0)$ is the same as in the case of the previous recursive definition, and c_1, c_2 are intended to have values z_1 and z_2, respectively).

In the sequel, the proportionality of y_1, y_2 to z_1, z_2

will be replaced by the condition that $(y_1, y_2) \in \mathcal{P}(z_1, z_2)$, and our non-tautological reasonings about $\mathcal{P}(z_1, z_2)$ will be based on its properties (c) and (d).

For example, Lemma 4.1 will be formulated so:

Let z_1, z_2 be given elements of $\mathcal{C}$, λ_1, λ_2 be given elements of $\mathcal{F}$. Whenever x, y_1, y_2 are elements of $\mathcal{C}$, and $(y_1, y_2) \in \mathcal{P}(z_1, z_2)$, let the inequality

$$\lambda_1(x, y_1) \geq \lambda_2(x, y_2)$$

hold. Then, for any choice of the coding element η, the inequality

$$\lambda_1 \varepsilon(x, \eta y_1) \geq \lambda_2 \varepsilon(x, \eta y_2)$$

holds under the same conditions on x, y_1, y_2.

Similar changes must be done in its proof. The first place in the proof, where properties of the set $\mathcal{P}(z_1, z_2)$ must be used, is in the investigation of the case when the considered inequality has the form $\tau_r \geq \tau_j \tau_i$. In this case one has to make the conclusion $(\bar{i}\bar{j}y_1, \bar{i}\bar{j}y_2) \in \mathcal{P}(z_1, z_2)$ from the assumption $(y_1, y_2) \in \mathcal{P}(z_1, z_2)$, and obviously this can be done on the basis of the property (c). The situation is similar in all other places of the proof of this lemma, where properties of $\mathcal{P}(z_1, z_2)$ must be used.

The formulation of Lemma 4.2 needs no modification. In a place of the proof, the fact that $y \in \mathcal{G}_c$ is used to conclude that $y = \bar{0}c$ or $y = \eta y_0$, where η is some coding element, and y_0 is again an element of $\mathcal{G}_c$. Of course, this can be done on the base of the property (a) of $\mathcal{G}_c$. However, the last sentence in the proof of the lemma needs a more detailed argumentation now. In that sentence the case is considered of $y = \eta y_0$, where η and y_0 are as above. It is claimed that the inequality

$$\lambda_1 \varepsilon(x, y) \geq \lambda_2 \varepsilon(x, y)$$

can be obtained by application of Lemma 1 to $z_1 = z_2 = \bar{0}c$. To make such an application in the new form of presentation, we have to verify that, whenever x, y_1, y_2 are elements of $\mathcal{C}$, and (y_1, y_2) belongs to $\mathcal{P}(\bar{0}c, \bar{0}c)$, then the inequality

$$\lambda_1(x, y_1) \geq \lambda_2(x, y_2)$$

holds, the inequality $\lambda_1 \underset{\mathcal{T}_c}{\geq} \lambda_2$ being assumed. To make the verification, it is sufficient to be able to conclude $y_1 = y_2 \in \mathcal{G}_c$ from the fact that y_1, y_2 are elements of $\mathcal{C}$, and (y_1, y_2) belongs to $\mathcal{P}(\bar{0}c, \bar{0}c)$. The possibility of such a conclusion needs an argumentation now. But this is not the only problem. The application of Lemma 1 to the considered case would give the conclusion that

$$\lambda_1 \varepsilon(x, \eta y_0) \geq \lambda_2 \varepsilon(x, \eta y_0)$$

in case we know that $(y_0, y_0) \in \mathcal{P}(\bar{0}c, \bar{0}c)$, and instead of this condition we have $y_0 \in \mathcal{G}_c$. All these problems can be solved by proving the following three statements:

I. Let z be some element of $\mathcal{C}$. Whenever y_1, y_2 are elements of $\mathcal{C}$, and (y_1, y_2) belongs to $\mathcal{P}(z, z)$, then $y_1 = y_2$.

II. Let z be some element of $\mathcal{C}$. Whenever y_1, y_2 are elements of $\mathcal{C}$, and (y_1, y_2) belongs to $\mathcal{P}(\bar{0}c, z)$, then $y_1 \in \mathcal{G}_c$.

III. Whenever $y \in \mathcal{G}_c$, then $(y, y) \in \mathcal{P}(\bar{0}c, \bar{0}c)$.

The statements I and II follow in an obvious way from the property (d) of the sets $\mathcal{P}(z_1, z_2)$ (in the proof of II, also the property (a) of $\mathcal{G}_c$ is used). The statement III follows from the property (b) of $\mathcal{G}_c$ (the property (c) of the set $\mathcal{P}(z_1, z_2)$ with $z_1 = z_2 = z$ is also used in the proof).

The statement and the proof of Lemma 4.3 can be formalized without any difficulty. As to the Lemma 4.4, there are more problems connected with its proof, and we shall consider them a little later. Of course, the formulation of the lemma at the present approach will be the following one:

(#) For each z_0 in $\mathcal{C}$, there is an element γ of $\mathcal{F}$ such that, whenever y, z are elements of $\mathcal{C}$, and (y, z) belongs to $\mathcal{P}(\bar{0}c, z_0)$, then the equality $\gamma y = z$ holds.

For the time being, we shall show how to present the rest of the proof of the First Recursion Theorem in the needed form, possibly using also the above statement.

The new formulation of Lemma 4.5 reads as follows.

Whenever x, y, z_0, z belong to $\mathcal{C}$, and (y, z) belongs to $\mathcal{P}(\bar{0}c, z_0)$, then

$$\iota(x, z) \geq \iota(I, z_0) L \iota(x, y).$$

For the proof of this statement one more statement is needed in addition to the new version (#) of Lemma 4.4 which still waits to be proved. This other statement is the following one.

IV. Let z_0 be an arbitrary element of $\mathcal{E}$. Whenever $y \in \mathcal{G}_c$, then there is some z in $\mathcal{E}$ such that (y, z) belongs to $\mathcal{P}(\bar{0}c, z_0)$.

The proof of this statement can be easily carried out on the base of the property (b) of $\mathcal{G}_c$ (using also the property (c) for the set $\mathcal{P}(\bar{0}c, z_0)$).

Making use of the statement IV and the formulated above version (#) of Lemma 4.4, we can carry out the proof of the new version of Lemma 4.5 without necessity of other changes in the proof from Section 4, except replacing the condition that y, z are proportional to $\bar{0}c, z_0$ by the condition that $(y, z) \in \mathcal{P}(0\bar{c}, z_0)$.

The formulation of Lemma 4.6 remains without modification, and the modification in the proof is obvious. No changes are needed in the formulations and in the proofs of Lemmas 4.7–4.10. No changes are needed also in the concluding part of the proof.

So the only remaining obstacle is the difficulty in using the proof of Lemma 4.4 for the purpose of proving the version (#) of this lemma. We are going now to explain how to overcome the obstacle in question.

Looking at the proof of Lemma 4.4, we see a strong presence in that proof of the idea of conversion of a finite sequence $\eta_1, \ldots, \eta_p$ of coding elements into the sequence $\eta_p, \ldots, \eta_1$. This idea appears in the form of transformation of products $\eta_1 \ldots \eta_p \bar{0} z$ into products $\eta_p \ldots \eta_1 x$. Unfortunately, the language of the system A′ does not give a simple means to express the statement that two given elements of $\mathcal{E}$ can be represented as such two products, with some given z, x and one and the same finite sequence $\eta_1, \ldots, \eta_p$ of coding elements. To obviate this difficulty, we shall consider, for any given x, y in $\mathcal{E}$, a subset $Q(x, y)$ of $\mathcal{E}$ with the following non-formal definition: this subset consists of the elements of the form

$$(z, \ \eta_p \ldots \eta_1 y),$$

where $\eta_1, \ldots, \eta_p$ are coding elements, and z is an element of $\mathcal{C}$ such that the equality

$$x=\eta_1 \dots \eta_p z$$

holds. So to say, $Q(x, y)$ consists of all elements of $\mathcal{C}$ which can be obtained from (x, y) by consecutive moving of coding elements from the left of x to the left of y. The same set can be introduced, again non-formally, also by the following recursive definition: $w \in Q(x, y)$ iff $w=(x, y)$ or there are a coding element η and elements x', y' of $\mathcal{C}$ such that $w=(x', \eta y')$ and $(\eta x', y') \in Q(x, y)$. Lemma 1 enables the formalizability of this recursive definition. Therefore reasoning about the sets $Q(x, y)$ is acceptable from the point of view of formalization in A' if such reasoning is based on (e) the above equivalence and/or on (f) the fact that $Q(x, y)$ is contained in each subset $\mathcal{A}$ of $\mathcal{C}$ with the property that $(x, y) \in \mathcal{A}$, and whenever η is a coding element, and $(\eta x', y') \in \mathcal{A}$ for some elements x', y' of $\mathcal{C}$, then $(x', \eta y') \in \mathcal{A}$ too.

Some properties of the sets $Q(x, y)$ will be formulated now as statements V, VI, VII, and these statements will be proved in a way which can be formalized in the system A'.

V. Whenever x, y, x', y' are elements of $\mathcal{C}$, and (x', y') belongs to $Q(x, y)$, then $Q(x', y') \subseteq Q(x, y)$.

Proof. Application of the property (e) of $Q(x, y)$ and the property (f) of $Q(x', y')$. ∎

VI. For each z in $\mathcal{C}$ and for each y in the set $\mathcal{P}_c$, there is an element x of $\mathcal{C}$ such that $(z, y) \in Q(x, \bar{0}c)$.

Proof. Let $\mathcal{A}$ be the set of all elements y of $\mathcal{C}$ such that for each z in $\mathcal{C}$ the condition $(z, y) \in Q(x, \bar{0}c)$ is satisfied for some x in $\mathcal{C}$. The element $\bar{0}c$ belongs to $\mathcal{A}$, since $(z, \bar{0}c) \in Q(z, \bar{0}c)$ for each z in $\mathcal{C}$, by the property (e). Suppose $y \in \mathcal{A}$, and let η be an arbitrary coding element. We shall show that $\eta y \in \mathcal{A}$ too. To do this, we take an arbitrary element z of $\mathcal{C}$ and choose x in $\mathcal{C}$ such that $(\eta z, y) \in Q(x, \bar{0}c)$. Then $(z, \eta y) \in Q(x, \bar{0}c)$, again by the property (e). So we established that $\eta y \in \mathcal{A}$. From the proved properties of $\mathcal{A}$, making use of the property (b) of $\mathcal{P}_c$, we conclude that $\mathcal{P}_c \subseteq \mathcal{A}$. ∎

VII. Whenever x, y, z, x_0, z_0 are elements of $\mathcal{C}$ satisfying the conditions

$$(x, y) \in Q(x_0, \bar{0}c), \quad (y, z) \in \mathcal{P}(\bar{0}c, z_0),$$

then

$$(z_0, x_0) \in Q(z, x).$$

Proof. Let x_0, z_0 be some fixed elements of $\mathcal{C}$. We shall denote by $\mathcal{A}$ be the set of all elements (y, z), where y, z

belong to $\mathcal{C}$, and, for all x in $\mathcal{C}$, the implication

(8) $$(x, y) \in Q(x_0, \bar{0}c) \Longrightarrow (z_0, x_0) \in Q(z, x)$$

holds. The proof will be carried out by proving that $\mathcal{A}$ contains the set $\mathcal{P}(\bar{0}c, z_0)$, and this will be done by using the property (d).

Let us first show that $(\bar{0}c, z_0) \in \mathcal{A}$. Suppose x is an element of $\mathcal{C}$ satisfying the condition $(x, \bar{0}c) \in Q(x_0, \bar{0}c)$. Then, by the property (e), the equality

$$(x, \bar{0}c) = (x_0, \bar{0}c)$$

holds or there are some coding element η and some element y of $\mathcal{C}$ such that $\bar{0}c = \eta y$. The second case is obviously impossible and therefore $x = x_0$. Hence the condition $(z_0, x_0) \in Q(z_0, x)$ is satisfied. Thus the implication (8) holds when $\bar{0}c$ and z_0 are substituted for y and z, respectively.

Now suppose that (y, z) is some element of $\mathcal{A}$, and η is some coding element. We shall show that $(\eta y, \eta z)$ also belongs to $\mathcal{A}$. For that purpose, suppose that x is an element of $\mathcal{C}$ satisfying the condition

$$(x, \eta y) \in Q(x_0, \bar{0}c).$$

Then, by the property (e), $\eta y = \bar{0}c$ or there are elements x', y' of $\mathcal{C}$ and coding element η' such that

$$(x, \eta y) = (x', \eta' y'), \quad (\eta' x', y') \in Q(x_0, \bar{0}c).$$

The first case is impossible, and the equality in the second case implies (as easily seen) the equalities

$$x = x', \quad \eta = \eta', \quad y = y'.$$

Therefore

$$(\eta x, y) \in Q(x_0, \bar{0}c).$$

From here, making use of the assumption that $(y, z) \in \mathcal{A}$, we conclude that

$$(z_0, x_0) \in Q(z, \eta x).$$

By the property (e),

$$(z, \eta x) \in Q(\eta z, x),$$

and this, together with the statement V, implies the inclusion

$$Q(z, \eta x) \subseteq Q(\eta z, x).$$

Therefore

$$(z_0, x_0) \in Q(\eta z, x),$$

and we see that the implication (8) holds when ηy, ηz are substituted for y, z, respectively. ■

We are now ready to give a proof of (#) in the needed style.

Proof of (#). As in the proof of Lemma 4.4, we construct an element ρ of $\mathcal{F}$ such that, for all x, z in $\mathcal{C}$ and all coding elements η, the equalities

$$\rho(x, \bar{0}z) = x, \quad \rho(x, \eta z) = \rho(\eta x, z)$$

hold. From these properties of ρ, we make the following conclusion: (g) whenever $x_0 \in \mathcal{C}$ and $w \in Q(x_0, \bar{0}c)$, then the equality $\rho w = x_0$ holds. To make this conclusion, we take a fixed element x_0 of $\mathcal{C}$ and denote by $\mathcal{A}$ the set of all elements w of $\mathcal{C}$ satisfying the condition $\rho w = x_0$. From the properties of ρ, it follows that $(x_0, \bar{0}c) \in \mathcal{A}$, and, whenever η is a coding element, x, z are elements of $\mathcal{C}$, and $(\eta x, z) \in \mathcal{A}$, then $(x, \eta z) \in \mathcal{A}$ too. By the property (f), this implies the inclusion $Q(x_0, \bar{0}c) \subseteq \mathcal{A}$.

Suppose now an arbitrary element z_0 of $\mathcal{C}$ is given. As in the proof of Lemma 4.4, we set

$$\gamma = \rho(z_0, \rho(\bar{0}c, I)).$$

Let y, z be elements of $\mathcal{C}$ such that

$$(y, z) \in \mathcal{P}(\bar{0}c, z_0). \tag{9}$$

We have to show that $\gamma y = z$, i.e. to prove the equality $\rho(z_0, \rho(\bar{0}c, y)) = z$.

By the statement II, $y \in \mathcal{P}_c$. Hence, by the statement VI, there is an element x_0 of $\mathcal{C}$ such that

$$(\bar{0}c, y) \in Q(x_0, \bar{0}c). \tag{10}$$

From here, by the property (g) of ρ, the equality

$$\rho(\bar{0}c, y) = x_0$$

follows, and it remains to prove that $\rho(z_0, x_0) = z$.

We note that (9) and (10) enable an application of the statement VII, and the conclusion from its application is

$$(z_0, x_0) \in Q(z, \bar{0}c).$$

Using once more the property (g), we get the needed equality. ■

So we kept our promise to prove the statement (#) in a way which can be formalized in the system A'. Thus we showed the formalizability in A' of the proof of the First Recursion Theorem from Section 4. ■

By proving Theorem 1, we demonstrated that the formal systems from Section II.8 are sufficient for the formalization of quite complicated proofs from the theory of iterative combinatory spaces. The observant readers will probably notice that we used the comprehension scheme II.6.(19) to a very small extent. Its main application was to make possible some recursively defined subsets of $\mathcal{C}$ to be used in a formalized way, and in fact the recursive definitions in question had a quite special form. Therefore it is natural to try to weaken the formal systems, and in particular the mentioned scheme, without losing the validity of a theorem of the sort of Theorem 1. A possible weakening of the used formal systems is indicated in the exercises.

Exercises

1. Show that the formulas (6) and (7) in the proof of Theorem 1 can be replaced by some formulas having bound variables only of the type c_i. Describe a method for a similar modification of other formulas having connection with the notion of a coding element.

2. Show that the use of the sets $Q(x, y)$ in the proof of Theorem 1 can be replaced by the use of a set $\mathcal{R}$ which is connected with them in the following way: $\mathcal{R}$ consists of all elements $(\omega, (x, y))$, such that x, y belong to $\mathcal{C}$, and ω belongs to $Q(x, y)$. For the set $\mathcal{R}$, give a recursive definition which is in the scope of Lemma 1, and trace out the changes in the proof of the statement (#) due to using $\mathcal{R}$ instead of the sets $Q(x, y)$.

3. Show that Theorem 1 remains valid if we make the following changes in the system A: (i) on the formula Φ in the comprehension scheme II.6.(19), we impose the restriction that bound variables only of the type c_i can occur in Φ; (ii) we add as a new axiom the formula (4) for the case when $\Phi(c_0, s_0)$ is a translation in A of the formula

$$c_0 \in s_2 \vee \exists c_1 ((c_0, c_1) \in s_3 \ \& \ c_1 \in s_0).$$

APPENDIX

A SURVEY OF EXAMPLES OF COMBINATORY SPACES

1. Introductory remarks

In the preceding chapter, a notion of computability has been introduced for the case of iterative combinatory spaces and some general theorems have been proved for this notion. In this way, a generalization of a certain part of the ordinary theory of computability has been obtained. Of course, an important thing for a generalization is the variety of examples covered by it. From the first two chapters of the book, it is clear that our generalization covers various examples, and, in particular, some ones connected with notions of principal interest. However, it does not become clear how large is the class of all possible examples and whether there are such ones which are essentially different from the mentioned so far. We cannot give an exhaustive answer to the first of these questions, but we shall show that the diversity of the examples of iterative combinatory spaces is considerably greater from what is shown by the examples presented up to now.

We shall start our review in the next section with a short recapitulation of the examples mentioned in the preceding text of the book. We hope this will help the reader to have a better orientation, all the more that some of these examples have been given not in the main text, but only in some exercises, and, in addition, some of the examples could be accompanied with more detailed intuitive explanations. In the next several sections some classes of other examples will be presented. As it is clear from what has been said above, we shall be interested mainly in iterative combinatory spaces. However, also certain examples will be given of combinatory spaces which are of some interest without being necessarily iterative.

2. A recapitulation of the examples presented so far

Almost all examples of iterative combinatory spaces considered yet are based on computational structures. The definition of the notion of computational structure has been introduced in Section I.1. Intuitively, a computational structure is an infinite set supplied with a pairing mechanism and with a mechanism for coding truth and falsity. A computational structure

$$\mathfrak{A} = \langle M, J, L, R, T, F, H \rangle$$

will be supposed to be given in the rest of this section, except for the last paragraph (the functions J, L, R form the pairing mechanism, and the functions T, F together with the predicate H form the mechanism for coding truth and falsity).

One and the same computational structure can be used as a base for the construction of different combinatory spaces. Each of them is characterized by the choice of a partially ordered semigroup $\mathcal{F}$, and the elements of $\mathcal{F}$ can be intuitively regarded as representing the behaviour of devices which transform elements of M into elements of M. The nature of the devices can be different in different examples, and the consideration of non-deterministic devices is a fruitful source for the construction of combinatory spaces. Computational procedures are considered a particular case of devices.

The semigroup multiplication in $\mathcal{F}$ corresponds to *sequential composition* (piping) of devices.[87] The semigroup $\mathcal{F}$ must have an identity I with the intuitive interpretation that I represents a device carrying out the identical transformation of the elements of M into themselves. The functions L, R, T, F must be elements of $\mathcal{F}$ in the simplest cases, or, in the more complicated ones, there must be elements of $\mathcal{F}$ corresponding to L, R, T, F in some natural way. The function J is used for the definition of the combination operation Π in $\mathcal{F}$. The operation Π corresponds to combining two devices so that both of them must be consecu-

[87] The device obtained by sequential composition of two devices proceeds as follows: the first of the given devices must be applied to the input data, and if its work terminates successfully, then the corresponding output data is used as input data for the second of the given devices. The output of the second device is considered as the output of the sequential composition of both.

tively applied to the given input data, and then the function J must be applied to the obtained pair of results; this kind of combining of the devices will be called *J-combination*.[88] The predicate H is used for the definition of the branching operation Σ in $\mathcal{F}$. The operation Σ corresponds to *if H-combination* of three devices, i.e. to combining them in such a way that the result of the application of the first one to the input data determines which of the other two devices to be applied to the same data.[89] In the case of an iterative combinatory space, an operation called iteration is determined implicitly through the already listed ones, and it corresponds to an operation of **while** *H-combination* of two devices.[90]

---

[88] More precisely, the device obtained by such a kind of combination proceeds as follows. The first of the given devices must be applied to the input data, and if its work terminates successfully producing some output data s, then the second of the given devices must be applied to the same input data as the first one. If the work of the second device also terminates successfully and produces some output data t, then the object $J(s, t)$ is considered as the output data of the compound device.

[89] A more precise description of the action of such a combination of three devices reads as follows. The first of the given devices must be applied to the input data. The work of the composed device may successfully terminate only in case the work of the first device terminates successfully and produces an output data r belonging to the domain of the predicate H. In such a case, if the value of $H(r)$ is **true** then the second of the given devices must be applied to the initial input data, and the result of its work is considered as the result of the compound device, otherwise the third of the given devices must be used in the same way instead of the second one.

[90] The **while** H-combination of two devices proceeds as follows. The processing of the input data starts with an application of the second device to it, and a successful termination of the complete process is possible only in the case when this application produces an output r belonging to the domain of the predicate H. In this case, if the value of $H(r)$ is **false** then the complete process terminates and the initial input data is considered as the output data, otherwise the first device is applied to the initial input data, and if this application produces some output v then v is taken as new initial input data, and everything is repeated from the beginning.

The semigroup $\mathcal{F}$ must contain a subset $\mathcal{C}$ which consists of the constant mappings of M into itself in the simplest cases, or, in the more complicated ones, of some elements representing these mappings in some natural sense.

The simplest example of a combinatory space corresponds to the study of deterministic devices by means of the extensional description of their input-output behaviour. In order that the combinatory space is an iterative one, we must allow some devices producing no output for some (or even for all) input data. Mathematically, this case is characterized by choosing $\mathcal{F}$ to be the partially ordered semigroup $\mathcal{F}_p(M)$ of all partial mappings of M into M (cf. Section I.2 for the definition of $\mathcal{F}_p(M)$). The corresponding combinatory space is denoted by $\mathfrak{S}_p(\mathfrak{U})$ (cf. Example II.1.2). The description of the above listed operations in this combinatory space can be found in Section I.2, and a characterization of the computable elements and of the computable mappings is given in Section I.3 for the case when $\mathfrak{U}$ is a standard computational structure over the natural numbers, and the functions $\lambda u.u+1$ and $\lambda u.u\dot{-}1$ are among the elements of $\mathcal{F}$ taken as primitive ones (we recall the fact that the $\mathfrak{U}$-computability defined in any of the cases considered in Chapter I is equivalent to the computability in the corresponding combinatory space).

A more complicated example corresponds to the study of non-deterministic devices again by means of the extensional description of their input-output behaviour (the deterministic devices regarded as a particular case of the non-deterministic ones). In this case one chooses $\mathcal{F}$ to be the partially ordered semigroup $\mathcal{F}_m(M)$ of all binary relations in M (cf. Section I.5 for the definition of $\mathcal{F}_m(M)$). The corresponding combinatory space is denoted by $\mathfrak{S}_m(\mathfrak{U})$ (cf. Example II.1.1). The description of the operations in this combinatory space can be found in Section I.5. In Section I.6, a characterization is given of the computable elements in this combinatory space for the case when $\mathfrak{U}$ is a standard computational structure over the natural numbers, and the functions $\lambda u.u+1$ and $\lambda u.u\dot{-}1$ together with the relation $\mathbb{N}^2$ are among the elements of $\mathcal{F}$ taken as primitive ones (the semigroup $\mathcal{F}_p(M)$ is considered as a subsemigroup of $\mathcal{F}_m(M)$ according to identification of the functions with their graphs). A characterization of the computable elements is given also for the case when $\mathfrak{U}$ is the Moschovakis structure $\mathfrak{M}_B$ over a given set B (cf. Example I.1.2 for the definition of $\mathfrak{M}_B$, and Section I.7 and Subsection (III) of Section III.5 for the mentioned characterization).

The next kind of examples correspond to the study of the data processing devices by means of an enriched extensional description of their input-output behaviour including also the error messages arising in some cases of unsuccessful termination. A set E is supposed to be given such that $E \cap M = \emptyset$ (the elements of E represent the possible error messages). The semigroups $\mathcal{F}_{\mathrm{p}}(M)$ and $\mathcal{F}_{\mathrm{m}}(M)$ are extended to semigroups $\mathcal{F}_{\mathrm{p}}(M, E)$ and $\mathcal{F}_{\mathrm{m}}(M, E)$, respectively (cf. Section I.8), and the corresponding combinatory spaces are denoted by $\mathfrak{S}_{\mathrm{p}}(\mathfrak{U}, E)$ and $\mathfrak{S}_{\mathrm{m}}(\mathfrak{U}, E)$ (cf. Examples II.1.4 and II.1.3). A description of the operations in these combinatory spaces can be found in Section I.8. For the case when $\mathfrak{U}$ is a standard computational over the natural numbers, characterizations of the computable elements for different choices of the primitive elements of $\mathcal{F}$ can be found in Theorem I.8.1 and Exercises I.8.5–I.8.7.

In the case when E consists of a single element, then the combinatory space $\mathfrak{S}_{\mathrm{m}}(\mathfrak{U}, E)$ is isomorphic to another one whose semigroup consists of all ordered pairs $\langle f, A\rangle$ with $f \in \mathcal{F}_{\mathrm{m}}(M)$ and $A \subseteq M$ (cf. Exercise I.8.5). This combinatory space is due to S. Nikolova and the ordered pairs $\langle f, A\rangle$ mentioned above have the following intuitive interpretation as descriptions of devices: f is the usual extensional description of the input-output behaviour of the device, and A consists of those input data which are safe with respect to rise of failures (i.e. no termination with an error message is possible when starting with them).

In the combinatory space from the above paragraph, the partial ordering in its semigroup is defined by means of the following equivalence:

$$\langle f, A\rangle \geq \langle g, B\rangle \iff f \geq g \ \& \ A \subseteq B.$$

A change only in the definition of the partial ordering, turns this combinatory space in a quite different one. The new partial ordering is defined by means of a more complicated equivalence, namely

$$\langle f, A\rangle \geq \langle g, B\rangle \iff f \geq g \ \& \ A \supseteq B \ \& \\ \forall u \in B \, \forall v \, (\langle u, v\rangle \in f \implies \langle u, v\rangle \in g)$$

(cf. Exercise II.4.13). The change in the partial ordering leads to a quite different operation of iteration (see Exercises II.4.17 and II.4.18). The characterization of the iteration in the combinatory space obtained in this way supports the following intuitive interpretation of the ordered pairs $\langle f, A\rangle$ as an appropriate one for this case: f is again the usual extensional description of the input-output behaviour of the device, but A consists now of those input data for which the work of the device necessarily terminates (i.e. the work terminates in all possible variants of proceeding starting with the input data in question). In the case when $\mathfrak{A}$ is a standard computational structure over the natural numbers, and a combinatory space of the above type is considered, the elements of the semigroup are characterized which are computable with respect to some naturally chosen primitive elements (cf. Exercises III.2.9–III.2.18).

In Exercise II.4.19 a modification of this kind of combinatory spaces is noted. Namely a change is made in the definition of the semigroup of the space by including the additional requirement $A \subseteq \operatorname{dom} f$ imposed on the pairs $\langle f, A\rangle$.[91] Such pairs can be used for the same kind of description as above in the case of devices with no possibility of unsuccessful termination (they can be used also for a description of arbitrary devices, but with a slightly different intuitive interpretation of A: the elements of A must be those input data for which the work of the device necessarily terminates successfully).

In Exercise II.4.11 a fuzzy analogue is given of the combinatory space $\mathfrak{S}_m(\mathfrak{A})$. In the combinatory space constructed there, the semigroup consists of all $\mathbb{L}$-fuzzy binary relations in M, where $\mathbb{L}$ is a lattice satisfying some not very restrictive assumptions. An explicit characterization of the iteration in such a combinatory space is given in Exercise II.4.16. In Exercise III.2.8, the case is considered when $\mathfrak{A}$ is a standard computational structure over the natural numbers, and the lattice $\mathbb{L}$ is a finite linearly ordered set. In this case, a characterization is given of the $\mathbb{L}$-fuzzy binary relations which are computable with respect to some naturally chosen primitive ones.

From the preceding chapters also some possibilities are seen how to construct new combinatory spaces starting from already constructed ones. Remark II.1.6 indicates a way for modifying the branching operation, and it can be easily seen that such a modification will produce an iterative combinatory space if the given combinatory space is iterative. A bit later a possibility to modify the elements L and R in some cases is mentioned. Some exercises also indicate ways for the construction of new combinatory spaces. For example, Exercise II.1.40 introduces the power-space $\mathfrak{S}^K$, where $\mathfrak{S}$

---

[91] This is actually the case studied in the earlier publications of the author.

is an arbitrary combinatory space, and K is an arbitrary non-empty set (this construction will be generalized in Section 10). Exercise II.3.9 shows that $\mathfrak{S}^K$ is iterative, whenever $\mathfrak{S}$ is iterative. Exercises II.4.10 and II.4.21 illustrate the possibility to consider subspaces of some combinatory spaces.

Exercise II.4.22 indicates some examples of iterative combinatory spaces, which are constructed by using, so to say, generalized computational structures. Namely M, J, T and F are used, which are as in an ordinary computational structure, but L and R could be not single-valued on the elements not belonging to $\operatorname{rng} J$, and also the mechanism for the interpretation of the elements of M as truth and falsity could be ambiguous for the elements not belonging to $\operatorname{rng} T \cup \operatorname{rng} F$. We shall discuss again some situations of a similar sort in Section 5.

3. Further examples of combinatory spaces consisting of fuzzy binary relations

In this section, a part of the assumptions and the notations from Exercise II.4.11 will be adopted, namely the following ones. We suppose that a set M and a lattice $\mathbb{L}$ are given. We assume that $\mathbb{L}$ has a greatest element $\mathbb{1}$ and a least element $\mathbf{0}$, where $\mathbb{1} \neq \mathbf{0}$, and the range of each mapping μ of M into $\mathbb{L}$ has a least upper bound in $\mathbb{L}$ with the property that

$$l \wedge \sup \operatorname{rng} \mu = \sup \{ l \wedge \mu(u) : u \in M \}$$

for all l in $\mathbb{L}$. We shall denote by $\mathcal{F}$ the set of all $\mathbb{L}$-fuzzy binary relations in M, i.e. all mappings of M^2 into $\mathbb{L}$. The set $\mathcal{F}$ will be considered with the composition operation defined by means of the equality

$$\varphi \psi = \lambda u w . \sup \{ \psi(u, v) \wedge \varphi(v, w) : v \in M \}$$

and with the partial ordering defined by means of the equivalence

$$\varphi \geq \psi \Longleftrightarrow \forall u v (\varphi(u, v) \geq \psi(u, v)).$$

For each subset f of M^2, we shall denote by $f^{\sim}$ the element of $\mathcal{F}$ defined by

$$f^{\sim}(u, v) = \begin{cases} \mathbb{1} & \text{if } \langle u, v \rangle \in f, \\ \mathbf{0} & \text{if } \langle u, v \rangle \notin f. \end{cases}$$

We adopt the following definition: a $\mathbb{L}$-*fuzzy partial ordering on* M is any element ε of $\mathcal{F}$ which satisfies the conditions

$$\varepsilon(u, u) = \mathbb{1}, \quad \varepsilon(u, v) \wedge \varepsilon(v, w) \leq \varepsilon(u, w),$$
$$\varepsilon(u, v) = \mathbb{1} \ \& \ \varepsilon(v, u) = \mathbb{1} \implies u = v$$

for all u, v, w in M. Obviously, if f is an ordinary partial ordering in M then $f^{\sim}$ is a $\mathbb{L}$-fuzzy partial ordering in M. In particular, so is $I_M^{\sim}$.

Remark 1. The condition that $\varepsilon(u, v) \wedge \varepsilon(v, w) \leq \varepsilon(u, w)$ for all u, v, w in M is obviously equivalent to the inequality $\varepsilon^2 \leq \varepsilon$.

If ε is a $\mathbb{L}$-fuzzy partial ordering in M then we shall denote by $\mathcal{F}_\varepsilon$ the set of all θ in $\mathcal{F}$ which satisfy the conditions

$$\varepsilon\theta = \theta\varepsilon = \theta.$$

Clearly, these conditions are satisfied for all θ in $\mathcal{F}$ if $\varepsilon = I_M^{\sim}$. In the case of an arbitrary $\mathbb{L}$-fuzzy partial ordering ε in M, the above equalities are equivalent to the inequalities

$$\varepsilon\theta \leq \theta, \quad \theta\varepsilon \leq \theta,$$

expressible as the conditions that

$$\theta(u, v) \wedge \varepsilon(v, w) \leq \theta(u, w), \quad \varepsilon(u, v) \wedge \theta(v, w) \leq \theta(u, w)$$

for all u, v, w in M. Consequently, $\varepsilon \in \mathcal{F}_\varepsilon$ (i.e. $\varepsilon^2 = \varepsilon$), whenever ε is a $\mathbb{L}$-fuzzy partial ordering in M. The equality $\varepsilon^2 = \varepsilon$ implies that $\varepsilon f^{\sim} \varepsilon \in \mathcal{F}_\varepsilon$ for each subset f of M^2.

For each $\mathbb{L}$-fuzzy partial ordering ε in M, the subset $\mathcal{F}_\varepsilon$ of $\mathcal{F}$ is obviously closed under composition, and ε is an identity of the subsemigroup $\mathcal{F}_\varepsilon$ of $\mathcal{F}$. Of course, $\mathcal{F}_\varepsilon = \mathcal{F}$ in the case when $\varepsilon = I_M^{\sim}$. We shall consider $\mathcal{F}_\varepsilon$ as a partially ordered semigroup, using the partial ordering induced by the partially ordering in $\mathcal{F}$. A subset $\mathcal{C}_\varepsilon$ of this semigroup will be defined as follows. For each s in M, we set

$$\bar{s} = M \times \{s\},$$

i.e. $\bar{s}$ is the constant function assigning the value s to all elements of M. The equality $\bar{s}^{\sim} \varepsilon = \bar{s}^{\sim}$ is easily verified, and therefore $\varepsilon \bar{s}^{\sim} \in \mathcal{F}_\varepsilon$. We note that

$$(\varepsilon \bar{s}^{\sim})(u, w) = \varepsilon(s, w)$$

for all u, w in M. The set of all elements $\varepsilon \bar{s}^{\sim}$, where $s \in M$, will be denoted by $\mathcal{C}_\varepsilon$ (in the case when $\varepsilon = I_M^{\sim}$ this set coincides with the set $\mathcal{C}$ from Exercise II.4.11).

We shall generalize the construction from Exercise II.4.11 in the following way. The set M will be supposed to be the first component of a computational structure

$$\mathfrak{A} = \langle M, J, L, R, T, F, H \rangle.$$

In addition, an $\mathbb{L}$-fuzzy partial ordering ε in M will be supposed to be given such that the following conditions are satisfied for all s, t, s', t', u, v in M:

(i) $\varepsilon(J(s, t), J(s', t')) = \varepsilon(s, s') \wedge \varepsilon(t, t')$;

(ii) if f is some of the functions L, R then

$$u \in \operatorname{dom} f \;\&\; v \in \operatorname{dom} f \Longrightarrow \varepsilon(u, v) \le \varepsilon(f(u), f(v));$$

(iii) if f is some of the functions T, F then

$$\varepsilon(u, v) \le \varepsilon(f(u), f(v));$$

(iv) $u \in \operatorname{dom} H \;\&\; v \in \operatorname{dom} H \;\&\; H(u) \ne H(v) \Longrightarrow \varepsilon(u, v) = \mathbf{0}$.

Example 1. For an arbitrary choice of the computational structure $\mathfrak{A}$, the above conditions are obviously satisfied if $\varepsilon = I_M^{\sim}$.

Example 2. Suppose M is an infinite set, and J is an injection of M^2 into M. Let ε be an $\mathbb{L}$-fuzzy partial ordering in M such that $\mathbf{0} \in \operatorname{rng} \varepsilon$ and the condition (i) is satisfied for all s, t, s', t' in M. In particular, ε could be the image $\ge^{\sim}$ of a partial ordering $\ge$ in M such that

(1) $$J(s, t) \ge J(s', t') \Longleftrightarrow s \ge s' \;\&\; t \ge t'$$

for all s, t, s', t' in M (besides the trivial case when $\varepsilon = I_M^{\sim}$, such is the case also when M is the partially ordered semigroup of an operative space, and J is the operation Π_* in it). We shall now define L, R, T, F, H so that $\langle M, J, L, R, T, F, H \rangle$ will be a computational structure, and the conditions (ii)-(iv) will be also satisfied. We define the functions L and R by the conditions that

$$\operatorname{dom} L = \operatorname{dom} R = \operatorname{rng} J,$$

and

$$L(J(s, t)) = s, \quad R(J(s, t)) = t$$

for all s, t in M. For the definition of T, F, H, we construct elements a and b such that

$$\varepsilon(a, b) = \varepsilon(b, a) = \mathbf{0}.$$

Namely we set $a = J(a_0, b_0)$, $b = J(b_0, a_0)$, where a_0, b_0 are elements of M satisfying the condition that $\varepsilon(a_0, b_0) = \mathbf{0}$. Then we set

$$T = \bar{a}, \quad F = \bar{b}, \quad \operatorname{dom} H = \{a, b\}, \quad H(a) = \text{true}, \quad H(b) = \text{false}.$$

Example 3. Let $\mathfrak{A} = \langle M, J, L, R, T, F, H\rangle$ be the Moschovakis structure based on an arbitrary set B, and ε_0 be an arbitrary $\mathbb{L}$-fuzzy partial ordering in B. Then there is a unique $\mathbb{L}$-fuzzy partial ordering ε in M satisfying the conditions (i)-(iv) such that ε is an extension of ε_0, and $\varepsilon(u, v) = \mathbf{0}$ whenever u and v are elements of M not belonging to one and the same of the three sets B, $\{O\}$ and $M \setminus B^{\circ}$ (where $B^{\circ} = B \cup \{O\}$).

Proof of the statement of Example 3. The imposed conditions on ε determine uniquely its values on the pairs not in $(M \setminus B^{\circ})^2$, and the condition (i) requires that

$$\varepsilon(\langle s, t\rangle, \langle s', t'\rangle) = \varepsilon(s, s') \wedge \varepsilon(t, t') \tag{2}$$

for all s, t, s', t' in M. Therefore an easy induction shows the uniqueness of ε if such an ε exists at all.

To prove the existence, we define ε by recursion along the construction of the elements of M from elements of B°. Namely we define first ε on the pairs not in $(M \setminus B^{\circ})^2$ by setting $\varepsilon(u, v) = \varepsilon_0(u, v)$ in the case when both u and v belong to B, $\varepsilon(O, O) = \mathbb{1}$, and $\varepsilon(u, v) = \mathbf{0}$ when u and v are elements of M not belonging to one and the same of the sets B, $\{O\}$ and $M \setminus B^{\circ}$. Then we use the equality (2) to extend the definition of ε to all pairs in $(M \setminus B^{\circ})^2$.

The $\mathbb{L}$-fuzzy relation ε defined in this way satisfies the condition (i), and it is an extension of ε_0 assigning the value $\mathbf{0}$ to the pairs indicated in the statement of the example. It remains to prove that ε is a $\mathbb{L}$-fuzzy partial order in M, and the conditions (ii)-(iv) are also satisfied.

By induction on u we show that $\varepsilon(u, u) = \mathbb{1}$ for all u in M. An induction on v shows the validity of the inequality $\varepsilon(u, v) \wedge \varepsilon(v, w) \leq \varepsilon(u, w)$ for all u, v, w in M. The implication $\varepsilon(u, v) = \mathbb{1}$ & $\varepsilon(v, u) = \mathbb{1} \Longrightarrow u = v$ also can be proved by induction (for example, on u). Of course, all these inductions make use of the corresponding properties of ε_0.

To verify the fact that condition (ii) is satisfied, we consider separately the case when u and v are both in B, the case when they are both in $\{O\}$, the case when they are both in $M \setminus B^{\circ}$, and the case when they are not in one and the same of these three sets. The validity of the condition (iii) follows immediately from the fact that T and F are constant functions, and the validity of (iv) follows directly from the definition of $\mathcal{E}$. ∎

Now we shall define an operation $\Pi_{\mathcal{E}}$ from $\mathcal{F}_{\mathcal{E}}^{2}$ to $\mathcal{F}$ by means of the equality

$$\Pi_{\mathcal{E}}(\varphi, \psi)(u, v) = \sup\{\varphi(u, s) \wedge \psi(u, t) \wedge \mathcal{E}(J(s, t), v) : s \in M,\ t \in M\}$$

(in the case when $\mathcal{E} = I_{M}^{\sim}$, this operation coincides with the operation Π from Exercise II.4.11). We shall prove now that $\operatorname{rng}\Pi_{\mathcal{E}} \subseteq \mathcal{F}_{\mathcal{E}}$. Let φ, ψ be arbitrary elements of $\mathcal{F}_{\mathcal{E}}$, and r, u, v, w be arbitrary elements of M. Then

$$\begin{aligned}
&\Pi_{\mathcal{E}}(\varphi, \psi)(u, v) \wedge \mathcal{E}(v, w) = \\
&\sup\{\varphi(u, s) \wedge \psi(u, t) \wedge \mathcal{E}(J(s, t), v) \wedge \mathcal{E}(v, w) : s \in M,\ t \in M\} \le \\
&\qquad \sup\{\varphi(u, s) \wedge \psi(u, t) \wedge \mathcal{E}(J(s, t), w) : s \in M,\ t \in M\} = \\
&\qquad\qquad \Pi_{\mathcal{E}}(\varphi, \psi)(u, w),
\end{aligned}$$

$$\begin{aligned}
&\mathcal{E}(r, u) \wedge \Pi_{\mathcal{E}}(\varphi, \psi)(u, v) = \\
&\sup\{\mathcal{E}(r, u) \wedge \varphi(u, s) \wedge \psi(u, t) \wedge \mathcal{E}(J(s, t), v) : s \in M,\ t \in M\} = \\
&\sup\{(\mathcal{E}(r, u) \wedge \varphi(u, s)) \wedge (\mathcal{E}(r, u) \wedge \psi(u, t)) \wedge \mathcal{E}(J(s, t), v) : s \in M,\ t \in M\} \le \\
&\qquad \sup\{\varphi(r, s) \wedge \psi(r, t) \wedge \mathcal{E}(J(s, t), v) : s \in M,\ t \in M\} = \\
&\qquad\qquad \Pi_{\mathcal{E}}(\varphi, \psi)(r, v).
\end{aligned}$$

Thus $\Pi_{\mathcal{E}}$ is a binary operation in $\mathcal{F}_{\mathcal{E}}$.

Let an operation Σ from $\mathcal{F}_{\mathcal{E}}^{3}$ into $\mathcal{F}$ be defined by means of the same expression as in Exercise II.4.11, namely

$$\Sigma(\chi, \varphi, \psi)(u, v) = ((H\chi)(u, \mathrm{true}) \wedge \varphi(u, v)) \vee ((H\chi)(u, \mathrm{false}) \wedge \psi(u, v)),$$

where

$$(H\chi)(u, p) = \sup\{\chi(u, s) : s \in H^{-1}(p)\}.$$ [92]

Let χ, φ, ψ be arbitrary elements of $\mathcal{F}_{\mathcal{E}}$. We shall show that $\Sigma(\chi, \varphi, \psi)$ belongs to $\mathcal{F}_{\mathcal{E}}$ too. For that purpose, we first note that, for all r, u in M, we have

$$\begin{aligned}
&\mathcal{E}(r, u) \wedge (H\chi)(u, p) = \sup\{\mathcal{E}(r, u) \wedge \chi(u, s) : s \in H^{-1}(p)\} \le \\
&\qquad \sup\{\chi(r, s) : s \in H^{-1}(p)\} = (H\chi)(r, p)
\end{aligned}$$

---

[92] Note that the defining equality of Σ can be written in the following simple form in the conditions of Example 2:

$$\Sigma(\chi, \varphi, \psi)(u, v) = (\chi(u, a) \wedge \varphi(u, v)) \vee (\chi(u, b) \wedge \psi(u, v)).$$

($p \in \{\text{true}, \text{false}\}$). Now it is easy to prove that, for all r, u, v, w in M, the inequalities

$$\Sigma(\chi, \varphi, \psi)(u, v) \wedge \mathcal{E}(v, w) \le \Sigma(\chi, \varphi, \psi)(u, w),$$
$$\mathcal{E}(r, u) \wedge \Sigma(\chi, \varphi, \psi)(u, v) \le \Sigma(\chi, \varphi, \psi)(r, v)$$

hold.

The facts concerning the generalization promised above are formulated in the following proposition.

Proposition 1. The 9-tuple

$$\langle \mathcal{F}_{\mathcal{E}}, \mathcal{E}, \mathcal{C}_{\mathcal{E}}, \Pi_{\mathcal{E}}, \mathcal{E}L^{\sim}\mathcal{E}, \mathcal{E}R^{\sim}\mathcal{E}, \Sigma, \mathcal{E}T^{\sim}, \mathcal{E}F^{\sim} \rangle$$

is a symmetric and iterative combinatory space. The iteration in this combinatory space can be characterized in the same way as in Exercise II.4.16, namely

$$[\sigma, \chi](u, w) = \sup\{\rho_m(u, w) : m \in \mathbb{N}\},$$

where

$$\rho_m(u, w) = \sup\{ \bigwedge_{j=0}^{m-1} ((H\chi)(v_j, \text{true}) \wedge \sigma(v_j, v_{j+1})) \wedge (H\chi)(v_m, \text{false}) : v_0, v_1, \ldots, v_m \in M, \ v_0 = u, \ v_m = w\}.$$

We leave the proof to the reader, restricting ourselves only to giving the following brief instructions:

1. Verify that

$$(\theta \mathcal{E} \bar{s}^{\sim})(u, v) = \theta(s, v)$$

for all θ in $\mathcal{F}_{\mathcal{E}}$ and all s, u, v in M.

2. Making use of condition (ii), verify that

$$(\mathcal{E}L^{\sim}\mathcal{E})(J(s, t), v) = \mathcal{E}(s, v), \quad (\mathcal{E}R^{\sim}\mathcal{E})(J(s, t), v) = \mathcal{E}(t, v)$$

for all s, t, v in M.

3. Making use of condition (i), verify that, for all s, t in M, the equality

$$\Pi_{\mathcal{E}}(\mathcal{E}\bar{s}^{\sim}, \mathcal{E}\bar{t}^{\sim}) = \mathcal{E}\bar{u}^{\sim}$$

holds, where $u = J(s, t)$.

4. Verify that

$$(\mathcal{E}T^{\sim})(u, s) = \mathcal{E}(T(u), s), \quad (\mathcal{E}F^{\sim})(u, s) = \mathcal{E}(F(u), s)$$

for all u, s in M.

5. Making use of condition (iv), note that

$$\mathcal{E}(T(u), s) = 0$$

for all s in $H^{-1}(\text{false})$, and

$$\mathcal{E}(F(u), s) = 0$$

for all s in $H^{-1}(\text{true})$.

6. Verify that, for all θ in $\mathcal{F}_{\mathcal{E}}$ and all u in M,

$$\sup\{\theta(u, v) \wedge (H\mathcal{E})(v, p) : v \in M\} = (H\theta)(u, p)$$

($p \in \{\text{true}, \text{false}\}$).

7. To prove that the combinatory space is iterative and to obtain the expression for the iteration, use the Level Omega Iteration Lemma (Proposition II.4.4) and Proposition II.4.6.

Remark 2. In the combinatory space from Proposition 1, the elements $\mathcal{E}L^{\sim}\mathcal{E}$ and $\mathcal{E}R^{\sim}\mathcal{E}$ of $\mathcal{F}_{\mathcal{E}}$ can be replaced, respectively, by the $\mathbb{L}$-fuzzy relations L' and R' in M which are defined as follows:

$$L'(u, v) = \sup\{\mathcal{E}(u, J(v, w)) : w \in M\},$$

$$R'(u, v) = \sup\{\mathcal{E}(u, J(w, v)) : w \in M\}.$$

These $\mathbb{L}$-fuzzy relations also belong to $\mathcal{F}_{\mathcal{E}}$, and the statement of Proposition 1 remains valid after doing the mentioned replacement. We leave the verification of this to the reader, and we restrict ourselves only to the following hint: prove that

$$L'(J(s, t), v) = \mathcal{E}(s, v), \quad R'(J(s, t), v) = \mathcal{E}(t, v)$$

for all s, t, v in M (compare with instruction 2 for the proof of Proposition 1).

It is probably worthwhile to reformulate the result from Proposition 1 for the case when $\mathbb{L}$ has only two elements and hence $\mathcal{E}$ is the image $\geq^{\sim}$ of some ordinary partial ordering $\geq$ in M. We shall give the reformulation in the terms of ordinary relations and of such an ordinary partial ordering.

The conditions (i)–(iv) in such a situation require that for all s, t, s', t' in M the equivalence (1) holds, the functions L, R, T, F are monotonically increasing (L and R in their domains), and the inequality $u \geq v$ is impossible when

$$u \in \operatorname{dom} H \ \&\ v \in \operatorname{dom} H \ \&\ H(u) \neq H(v)$$

The partially ordered semigroup $\mathcal{F}$ is actually $\mathcal{F}_{\mathrm{m}}(M)$ in this situation, and its subsemigroup $\mathcal{F}_{\mathcal{E}}$, which will be denoted by $\mathcal{F}_{\geq}$ now, consists of the elements θ of $\mathcal{F}_{\mathrm{m}}(M)$ satisfying the condition that

$$\langle u, v\rangle \in \theta \ \&\ v \geq w \implies \langle u, w\rangle \in \theta,$$

$$u \geq v \ \&\ \langle v, w\rangle \in \theta \implies \langle u, w\rangle \in \theta$$

for all u, v, w in M. Of course, this subsemigroup will be considered with the partial ordering by inclusion, which is the partial ordering induced from $\mathcal{F}_{\mathrm{m}}(M)$. If s is an element of M then the $\mathbb{L}$-fuzzy relation $\varepsilon\bar{s}^{\sim}$ is the image of the relation

$$\bar{s}_{\geq}=\{\langle u, v\rangle\in M^2 :\ s\geq v\}.$$

The last relation is an element of $\mathcal{F}_{\geq}$, and the set of all such relations will be denoted by $\mathcal{C}_{\geq}$. Instead of $\varepsilon L^{\sim}\varepsilon$, $\varepsilon R^{\sim}\varepsilon$, εT and εF, the relations

$$L_{\geq}=\{\langle u, v\rangle\in M^2 :\ \exists u'\exists v'(u\geq u' \ \&\ \langle u', v'\rangle\in L \ \&\ v'\geq v)\},$$
$$R_{\geq}=\{\langle u, v\rangle\in M^2 :\ \exists u'\exists v'(u\geq u' \ \&\ \langle u', v'\rangle\in R \ \&\ v'\geq v)\},$$
$$T_{\geq}=\{\langle u, v\rangle\in M^2 :\ T(u)\geq v\},$$
$$F_{\geq}=\{\langle u, v\rangle\in M^2 :\ F(u)\geq v\}$$

will be considered. The binary operation $\Pi_{\geq}$ corresponding to Π_{ε} is defined as follows:

$$\Pi_{\geq}(\varphi, \psi)=\{(u, v)\in M^2 :\ \exists s\exists t(\langle u, s\rangle\in\varphi \ \&\ \langle u, t\rangle\in\psi \ \&\ J(s, t)\geq v)\}.$$

The operation Σ in $\mathcal{F}_{\geq}$ is the restriction of the operation Σ from $\mathcal{F}_{\mathrm{m}}(M)$. In this denotations, the particular case of Proposition 1 corresponding to the considered situation yields the following result (which, of course, can be proved also in a direct way):

Corollary 1. The 9-tuple

$$\langle\mathcal{F}_{\geq}, \geq, \mathcal{C}_{\geq}, \Pi_{\geq}, L_{\geq}, R_{\geq}, \Sigma, T_{\geq}, F_{\geq}\rangle$$

is a symmetric and iterative combinatory space. The iteration in this combinatory space can be characterized in the same way as in the combinatory space $\mathcal{F}_{\mathrm{m}}(M)$.

Remark 3. In the above situation, it is obvious that, for all s, t in M, the inclusion $\bar{s}_{\geq}\supseteq\bar{t}_{\geq}$ holds iff the inequality $s\geq t$ holds in M. Therefore the set $\mathcal{C}_{\geq}$, considered as a partially ordered set by using the partial ordering induced from $\mathcal{F}_{\geq}$, turns out to be isomorphic to the partially ordered set M. This, together with Example 2 or Example 3, gives an affirmative answer to a question of D. Vakarelov, namely the question whether a combinatory space $\langle\mathcal{F}, I, \mathcal{C}, \Pi, L, R, \Sigma, T, F\rangle$ is possible such that the partial ordering in $\mathcal{C}$ induced from $\mathcal{F}$ is different from the identity relation in $\mathcal{C}$.

Remark 4. The above remark, together with Example 2 and the elementary properties of combinatory spaces, shows which are, up to isomorphism, the possible partial orderings induced in the sets $\mathcal{C}$ from the partially ordered sets $\mathcal{F}$ in combinatory spaces $\langle \mathcal{F}, I, \mathcal{C}, \Pi, L, R, \Sigma, T, F\rangle$. The characterization (given in Skordev [1976c]) is the following one: exactly those partial orderings can be obtained in this way at which the Cartesian square of the partially ordered set is isomorphic to some subset of this set (the Cartesian square considered with the natural partial ordering in it which corresponds to the partial ordering in the given set).

4. Probabilistic examples of iterative combinatory spaces

(I) **On data processing devices with probabilistic nondeterminism.** A computational structure

$$\mathfrak{U} = \langle M, J, L, R, T, F, H\rangle$$

will be supposed to be given throughout the whole section. We shall construct some combinatory spaces whose elements could be used as descriptions of probabilistic data processing devices transforming elements of M into elements of M. The devices in question are thought about as nondeterministic ones whose output depends on the input in a probabilistic manner. To be a little more precise, we must say that each concrete application of the device to the input data should produce at the most one corresponding output, but different concrete applications to the same input data could lead to different results (including possibly productive termination of a concrete application and no output data produced by another one); in general, the result of the application could be not predictable, but certain statements about the result must have a definite probability. Such non-deterministic data processing devices and the computability of random functions by them have been studied, for example, in the papers Santos [1969, 1971].

(II) **The case of discrete probability distributions.** The simplest case of probabilistic non-determinism of data processing within the given set M is that one when, given any input data from M, there is a corresponding discrete distribution function indicating the probabilities of all elements of M to be the output data, as well as the probability that no output data will be produced (of course, in the case of an uncountable set M most of its elements must have a zero probability to be the output data). The information about the distribution functions corresponding to all possible input data can be collected in a function θ from M^2 to the interval $[0, 1]$ of the real line $\mathbb{R}$ so that, for any fixed u in M, the function $\lambda v.\, \theta(u, v)$ assigns to all elements of M their probabilities to be the output data when the device is applied with input data u. Clearly, for all u in M, the inequality

$$\sum_v \theta(u, v) \le 1 \tag{1}$$

must be satisfied,[93] the value of the left-hand side expression being the probability of a productive termination of the work at the input data u, and the difference between 1 and this value being the probability that no output data will be produced by this work.[94]

We shall now construct a combinatory space corresponding to such a kind of mathematical description of the behaviour of some non-deterministic devices. To do this, we have to make clear what mathematical operations on the functions θ correspond to those three ways of combining devices which have been used until now as an intuitive background for the definition of composition, combination and branching (see, for example, the beginning of Section 2).

We shall denote by $\mathcal{F}$ the set of all functions θ from M^2 to $[0, 1]$ which satisfy the inequality (1) for all u in M. This set will be considered as a partially ordered one by supplying it with the partial ordering defined as follows: $\varphi \ge \psi$ iff $\varphi(u, v) \ge \psi(u, v)$ for all u, v in M. For any two elements φ and ψ of $\mathcal{F}$, we define a real-valued function $\varphi\psi$ by means of the equality

$$\varphi\psi = \lambda u w . \sum_v \psi(u, v)\varphi(v, w).$$

This is a matrix product known from the theory of Markov processes, at least for the case when M is countable. The function $\varphi\psi$ is easily seen to belong again to $\mathcal{F}$. If φ

---

[93] The convention is adopted that, for each each non-negative real-valued function τ on M, the symbol $\sum_v \tau(v)$ denotes the least upper bound of all finite sums

$$\tau(v_1) + \ldots + \tau(v_n),$$

where $v_1, \ldots, v_n$ are distinct elements of M.

[94] We note that, in the case of a countable set M, the non-negative real-valued functions θ on M^2 satisfying, for all u in M, the equality

$$\sum_v \theta(u, v) = 1$$

are called *stochastic matrices* and the numbers $\theta(u, v)$ are named *transition probabilities*.

and ψ describe the behaviour of some devices then, under a certain independence assumption, the function $\varphi\psi$ describes the behaviour of their sequential composition, where the device described by ψ is applied first, and its output data is taken as the input data for the device described by φ. The multiplication in $\mathcal{F}$, defined in this way, is obviously monotonically increasing, and it turns out to be associative. So $\mathcal{F}$ becomes a partially ordered semigroup.

Using similar intuitive considerations and the mentioned independence assumption, we see that the following definitions of a binary operation Π and a ternary operation Σ are in concordance with the other two ways of combining devices:

$$\Pi(\varphi, \psi)(u, v) = \begin{cases} \varphi(u, L(v))\psi(u, R(v)) & \text{if } v \in \operatorname{rng} J, \\ 0 & \text{if } v \notin \operatorname{rng} J, \end{cases}$$

$$\Sigma(\chi, \varphi, \psi)(u, v) = ((H\chi)(u, \text{true})\varphi(u, v)) + ((H\chi)(u, \text{false})\psi(u, v)),$$

where

$$(H\chi)(u, p) = \sum_s \chi(u, s) \| s \in H^{-1}(p) \|,$$

$\|A\|$ denoting 1 if A is true and 0 otherwise. The operations Π and Σ can be shown to transform elements of $\mathcal{F}$ again into elements of $\mathcal{F}$ (for the proof of the statement concerning Π, it is useful to verify that

$$\sum_v \Pi(\varphi, \psi)(u, v) = \sum_s \varphi(u, s) \sum_t \psi(u, t)$$

for all φ, ψ in $\mathcal{F}$ and all u in M).

If f is a partial function from M to M, then we shall represent f by the elements $f^{\sim}$ of $\mathcal{F}$ defined as follows:

$$f^{\sim}(u, v) = \| \langle u, v \rangle \in f \|.$$

Let $\mathcal{C}$ consist of the elements of $\mathcal{F}$ representing in this sense the constant total mappings of M into M. Then the following proposition holds:

Proposition 1. The 9-tuple

$$\mathfrak{S} = \langle \mathcal{F}, I_M^{\sim}, \mathcal{C}, \Pi, L^{\sim}, R^{\sim}, \Sigma, T^{\sim}, F^{\sim} \rangle$$

is a symmetric and iterative combinatory space. For arbitrary σ, χ in $\mathcal{F}$ and arbitrary u, w in M the equality

$$[\sigma, \chi](u, w) = \sum_{m=0}^{\infty} \rho_m(u, w)$$

holds, where

$$\rho_m(u, w) = \sum_{v_0}\sum_{v_1}\cdots\sum_{v_m} \| v_0 = u \ \& \ v_m = w \| \times$$

$$(H\chi)(v_m, \text{false}) \prod_{j=0}^{m-1} ((H\chi)(v_j, \text{true})\sigma(v_j, v_{j+1})).$$

The verification of the conditions from the definition of the notion of a symmetric combinatory space will be left to the reader (an analogy with the proof of Proposition 3.1 can be instructive). When properties of the operation Π are considered, it is convenient to use the following equality:

$$\Pi(\varphi, \psi)(u, v) = \sum_s\sum_t \varphi(u, s)\psi(u, t)I_M^{\sim}(J(s, t), v).$$

The fact that $\mathfrak{S}$ is iterative can be established by using the Level Omega Iteration Lemma (Proposition II.4.4), and the expression for the iteration can be obtained by supplementing an application of Proposition II.4.6.

In the case when $\mathfrak{U}$ is a standard computational structure over the natural numbers all sums with summation variables ranging over M can be written as ordinary infinite series. For example, the condition (1) can be written as follows:

$$\sum_{v=0}^{\infty} \theta(u, v) \le 1.$$

In this case, a characterization will be given for the elements of $\mathcal{F}$ computable in the set

$$\mathcal{B} = \{S^{\sim}, P^{\sim}, \alpha\},$$

where $S = \lambda u.\, u+1$, $P = \lambda u.\, u \dot{-} 1$, and

$$\alpha(u, v) = \begin{cases} \frac{1}{2} & \text{if } v \le 1, \\ 0 & \text{if } v > 1 \end{cases}$$

(the element α of $\mathcal{F}$ characterizes the "data processing by tossing a coin").[95] The characterization reads as follows.

Theorem 1. Let the computational structure $\mathfrak{U}$ be a standard computational structure over the natural numbers, and let $\mathfrak{S}$ be the iterative combinatory space from Proposition 1. Then, for each element θ of $\mathcal{F}$, the following two

---

[95] We note that in Tabakov [1977] the set of the elements of $\mathcal{F}$ is studied which are in a certain sense primitive recursive with respect to α (this element is denoted there by A).

conditions are equivalent:

(a) θ is $\mathfrak{S}$-computable in $\mathcal{B}$;

(b) the set of all quadruples $\langle k, l, u, v\rangle$ of natural numbers satisfying the inequality

$$\frac{k}{l+1} < \theta(u, v)$$

is recursively enumerable.

Proof. The implication from (a) to (b) is proved by a more or less straightforward induction along the construction of θ. Suppose now θ satisfies the condition (b). We shall give an outline of the proof that condition (a) will be also satisfied.

We first note that, for each one-argument partial recursive function f, the corresponding element $f^{\sim}$ of $\mathcal{F}$ is $\mathfrak{S}$-computable in $\{S^{\sim}, P^{\sim}\}$ (hence also in $\mathcal{B}$). This follows from Theorem I.3.1 and the fact that the mapping $\lambda f.\, f^{\sim}$ is a homomorphism with respect to the operations composition, combination and iteration in the iterative combinatory spaces $\mathfrak{S}_p(\mathfrak{U})$ and $\mathfrak{S}$.

The next step is to prove the $\mathfrak{S}$-computability in $\mathcal{B}$ of the element β of $\mathcal{F}$ defined in the following way:

$$\beta(u, v) = \begin{cases} 2^{-u} & \text{if } v < 2^u, \\ 0 & \text{if } v \geq 2^u \end{cases}$$

(in Tabakov [1977] this element is denoted by **eqm**). The intuitive idea of the proof is based on the fact that a binary representation of an arbitrary number which is less than 2^u can be obtained starting from the empty string by appending u times a 0 or 1 digit on the right, and if the choice of this digit is realized by tossing a coin then, for each number less than 2^u, the probability that a representation of this number will be obtained is equal to 2^{-u} (the empty string is considered as a representation of the number 0). We hope the reader will be able to transform this intuitive idea into an actual proof (we note only that the mentioned non-deterministic appending of a 0 or 1 digit on the right can be described by the element

$$(\lambda t.\, 2L(t) + R(t))^{\sim}\Pi(I_{\mathbb{N}}^{\sim}, \alpha)$$

of $\mathcal{F}$).

The following lemma will be needed further, and the idea of its proof is similar to an idea used in Tabakov [1977].

Lemma 1. Let h be a one-argument recursive function, and G be such a two-argument recursive function that, for all natural numbers u, the equality

$$\sum_{v=0}^{\infty} G(u, v) = 2^{h(u)}$$

holds.[96] Then the element $\lambda uv.\, G(u, v)2^{-h(u)}$ of $\mathcal{F}$ is $\mathfrak{S}$-computable in $\mathcal{B}$.

Proof. We construct a two-argument recursive function H such that

$$H(u, r) = \max\{i: \sum_{v<i} G(u, v) \leq r\}$$

whenever $r < 2^{h(u)}$. It is easy to see that, for any fixed u and i, the equation $H(u, r) = i$ has exactly $G(u, i)$ solutions r satisfying the inequality $r < 2^{h(u)}$. Using this fact, one can verify the equality

$$\lambda uv.\, G(u, v)2^{-h(u)} = (\lambda t.\, H(L(t), R(t)))^{\sim}\Pi(I_{\mathbb{N}}^{\sim}, \beta h^{\sim}).$$ ∎

Now, leaving the details to the reader, we note that the following conclusion can be drawn from the assumption that θ satisfies the condition (b):

There are three-argument recursive functions X and Y such that

$$\theta(u, v) = \lim_{n\to\infty} \frac{X(n, u, v)}{Y(n, u, v)+1} \tag{2}$$

for all u, v in $\mathbb{N}$, and

$$\frac{X(n, u, v)}{Y(n, u, v)+1} \leq \frac{X(n+1, u, v)}{Y(n+1, u, v)+1}$$

for all n, u, v in $\mathbb{N}$.

Making use of the representation (2), we shall obtain an appropriate representation of $\theta(u, v)$ as the sum of an infinite series.

Lemma 2. There are two-argument recursive functions A, D, an one-argument recursive function B and a three-argument recursive function C such that

---

[96] I.e., for all u in $\mathbb{N}$, there is a natural number n such that

$$\sum_{v=0}^{n} G(u, v) = 2^{h(u)},$$

and $G(u, v) = 0$ for all v greater than n.

$$0 < A(m, u) \leq 2^{B(m)}, \quad \sum_{v=0}^{m} C(m, u, v) = 2^{D(m,u)}$$

$$C(m, u, v) = 0, \quad v = m+1, m+2, m+3, \ldots$$

for m, u in $\mathbb{N}$, and

$$\theta(u, v) = \sum_{m=0}^{\infty} \frac{A(0,u)}{2^{B(0)}} \frac{A(1,u)}{2^{B(1)}} \cdots \frac{A(m-1,u)}{2^{B(m-1)}} \left(1 - \frac{A(m,u)}{2^{B(m)}}\right) \frac{C(m,u,v)}{2^{D(m,u)}}$$

for all u, v in $\mathbb{N}$.

Proof. Let $\mathbb{D}$ be the set of all rational numbers of the form $\frac{m}{2^n}$, where m and n are natural numbers. We shall note three statements concerning the construction of numbers from $\mathbb{D}$ having certain connections with given rational numbers. In fact we shall need the translations of these statements in the terms of existence of certain recursive functions, but, for the sake of brevity, formulations will be given using the words "one can effectively find".

The first statement asserts the well-known density of the set $\mathbb{D}$ in the set of the non-negative rational numbers, namely:

1. For any two rational numbers a and b satisfying the inequalities $0 \leq a < b$, one can effectively find a number d from $\mathbb{D}$ satisfying the inequalities $a < d < b$.

To have an explicit example of a translation of the kind mentioned above, we shall reformulate this statement as follows: there are four-argument recursive functions M and N such that

$$\frac{i}{j+1} < \frac{M(i,j,k,l)}{2^{N(i,j,k,l)}} < \frac{k}{l+1}$$

whenever i, j, k, l are natural numbers satisfying the inequalities

$$\frac{i}{j+1} < \frac{k}{l+1}.$$

The next statement of such a nature is the following one:

2. Let $a_1, \ldots, a_s$ and $b_1, \ldots, b_s$ be rational numbers satisfying the inequalities

$$0 \leq a_i < b_i, \quad i = 1, \ldots, s,$$

and let d be a number from $\mathbb{D}$ satisfying the inequalities

$$a_1 + \ldots + a_s < d < b_1 + \ldots + b_s.$$

Then one can effectively find numbers $d_1, \ldots, d_s$ from $\mathbb{D}$ satisfying the conditions

$$a_i < d_i < b_i, \quad i=1, \ldots, s,$$
$$d_1 + \ldots + d_s = d.$$

The proof of this statement is by induction, and it makes use of Statement 1. The translation of Statement 2 in the terms of existence of certain recursive functions will be omitted (as well as the translation of the next one).

3. Let $a_1, \ldots, a_s$ be non-negative rational numbers satisfying the inequality

$$a_1 + \ldots + a_s \leq 1,$$

and e be a positive rational number. Then one can effectively find numbers c and $d_1, \ldots, d_s$ from $\mathbb{D}$ satisfying the conditions

$$c < 1, \quad d_1 + \ldots + d_s = 1,$$
$$a_i - e \leq c d_i \leq a_i, \quad i=1, \ldots, s.$$

For proving this statement, we first consider the case when $a_1 + \ldots + a_s = 0$. In this case, we set $c=0$, $d_1 = 1$, $d_2 = \ldots = d_s = 0$. Otherwise, using statement 1, we choose a number c from $\mathbb{D}$ satisfying the inequalities

$$\max\{a_1 - e, 0\} + \ldots + \max\{a_s - e, 0\} < c < a_1 + \ldots + a_s,$$

and then, making use of statement 2, we choose numbers $d_1, \ldots, d_s$ from $\mathbb{D}$ such that

$$\frac{\max\{a_i - e, 0\}}{c} \leq d_i \leq \frac{a_i}{c}, \quad i=1, \ldots, s,$$
$$d_1 + \ldots + d_s = 1.$$

The numbers $c, d_1, \ldots, d_s$ constructed in this way satisfy the formulated conditions.

Now we set

$$\theta_n(u, v) = \frac{X(n, u, v)}{Y(n, u, v)+1},$$

and we address ourselves to the construction of functions A, B, C, D with the needed properties. This will be done in the form of a construction of the functions

$$U(n, u) = \frac{A(n, u)}{2^{B(n)}}, \quad V(n, u, v) = \frac{C(n, u, v)}{2^{D(n, u)}}$$

whose values will belong to $\mathbb{D}$ for all m, u, v in $\mathbb{N}$. These functions must be effectively computable, and they must satisfy the following conditions:

(3) $$0 < U(m, u) \leq 1, \quad \sum_{v=0}^{m} V(m, u, v) = 1,$$

(4) $$V(m, u, v) = 0, \quad v = m+1, m+2, m+3, \ldots$$

(5) $$\theta(u, v) = \sum_{m=0}^{\infty} W(m, u, v),$$

where

$$W(m, u, v) = U(0, u) \ldots U(m-1, u)(1 - U(m, u))V(m, u, v).$$

We shall ensure the validity of the equality (5) by constructing the functions U and V so that the inequalities

(6) $$\theta_n(u, v) - \frac{1}{2^n} \leq \sum_{m<n} W(m, u, v) \leq \theta_n(u, v)$$

will hold whenever $v < n$.

The functional values $U(n, u)$, $V(n, u, v)$ will be defined for any fixed value of u by recursion on n. If $n = 0$ then the requirement concerning the inequalities (6) is trivially satisfied. Suppose now that, for all values of m which are less than some given natural number n, values from the set $\mathbb{D}$ are effectively assigned to the expressions of the form $U(m, u)$ and $V(m, u, v)$ in such a way that for all v which are less than n the inequalities (6) hold, and for all m which are less than n the conditions (3) and (4) are satisfied. Then the expressions $W(m, u, v)$ with $m < n$ will also make sense. We set

$$a_v = \frac{\theta_{n+1}(u, v) - \sum_{m<n} W(m, u, v)}{U(0, u) \ldots U(n-1, u)}, \quad v = 0, 1, \ldots, n.$$

The numbers $a_0, a_1, \ldots, a_n$ are rational. They are non-negative, as it follows from the inequalities

$$\theta_{n+1}(u, v) \geq \theta_n(u, v) \geq 0$$

and the validity of (4) and (6) for $m < n$ and $v < n$, respectively. By (1)-(4), we have the inequalities

$$\sum_{v=0}^{n} \theta_{n+1}(u, v) \leq \sum_{v=0}^{n} \theta(u, v) \leq 1$$

and the equalities

$$\sum_{v=0}^{n} \sum_{m<n} W(m, u, v) = \sum_{m<n} \sum_{v=0}^{n} W(m, u, v) =$$

$$\sum_{m<n} U(0, u) \dots U(m-1, u)(1 - U(m, u)) =$$

$$\sum_{m<n} (U(0, u) \dots U(m-1, u) - U(0, u) \dots U(m, u)) =$$

$$1 - U(0, u) \dots U(n-1, u).$$

Hence

$$a_0 + a_1 + \dots + a_n \leq 1,$$

and therefore, by Statement 3, one can effectively find numbers c and $d_0, d_1, \dots, d_n$ from the set $\mathbb{D}$ such that

$$c < 1, \quad d_0 + \dots + d_n = 1,$$

$$a_v - \frac{1}{2^{n+1}} \leq c d_v \leq a_v, \quad v = 0, \dots, n.$$

We set

$$U(n, u) = 1 - c,$$

$$V(n, u, v) = d_v, \quad v = 0, \dots, n,$$

$$V(n, u, v) = 0, \quad v = n+1, n+2, n+3, \dots$$

Then the inequalities

$$\theta_{n+1}(u, v) - \frac{1}{2^{n+1}} \leq \sum_{m<n+1} W(m, u, v) \leq \theta_{n+1}(u, v)$$

will hold whenever $v < n+1$, and for all m which are less than $n+1$ the conditions (3) and (4) will be satisfied. ∎

Having now Lemmas 1 and 2 at our disposal, we shall represent the element θ of $\mathcal{F}$ in a form showing its $\mathfrak{S}$-computability in $\mathcal{B}$. For that purpose, making use of the functions A, B, C, D from Lemma 2, we define one-argument recursive functions h, h' and two-argument recursive functions G, G' in the following way:

$$h(u) = B((u)_0), \quad h'(u) = D((u)_0, (u)_1),$$

$$G(u, 0) = 2^{h(u)} - A((u)_0, (u)_1), \quad G(u, 1) = A((u)_0, (u)_1),$$

$$G(u, v) = 0, \quad v = 2, 3, 4, \dots,$$

$$G'(u, v) = C((u)_0, (u)_1, v).$$

Then we set

$$\varphi = \lambda u v.\, G(u, v) 2^{-h(u)}, \quad \varphi' = \lambda u v.\, G'(u, v) 2^{-h'(u)}.$$

Then φ and φ' are elements of $\mathcal{F}$, and, by Lemma 1, they are $\mathfrak{S}$-computable in $\mathcal{B}$. The $\mathfrak{S}$-computability of θ in $\mathcal{B}$

(and hence the validity of the theorem) will be established by proving the equality

$$\theta=\varphi'[f^{\sim},\varphi]g^{\sim}, \tag{7}$$

where $f=\lambda u.\,2u$, $g=\lambda u.\,3^u$.

For the proof of (7), we set $\iota=[f^{\sim},\varphi]$. An easy calculation, using the formula for the iteration from Proposition 1, shows that

$$\iota(u,2^m u)=\varphi(2^m u,0)\prod_{j=0}^{m-1}\varphi(2^j u,1)$$

for all natural numbers u, m, and $\iota(u,w)=0$ in the case when u, w are such natural numbers that $w\neq 2^m u$ for all m in $\mathbb{N}$. Therefore

$$(\varphi'\iota)(u,v)=\sum_{m=0}^{\infty}\iota(u,2^m u)\varphi'(2^m u,v)=$$

$$\sum_{m=0}^{\infty}\varphi(2^m u,0)\varphi'(2^m u,v)\prod_{j=0}^{m-1}\varphi(2^j u,1).$$

Then

$$(\varphi'\iota g^{\sim})(u,v)=(\varphi'\iota)(3^u,v)=$$

$$\sum_{m=0}^{\infty}\varphi(2^m.3^u,0)\varphi'(2^m.3^u,v)\prod_{j=0}^{m-1}\varphi(2^j.3^u,1)=$$

$$\sum_{m=0}^{\infty}\left(1-\frac{A(m,u)}{2^{B(m)}}\right)\frac{C(m,u,v)}{2^{D(m,u)}}\prod_{j=0}^{m-1}\frac{A(j,u)}{2^{B(j)}}=\theta(u,v). \blacksquare$$

Theorem 1 shows that the made choice of the set $\mathcal{B}$ leads to a natural notion of computability for random functions in $\mathbb{N}$. There are some reasons to assume that this is the most general notion of effective computability for such functions. At any rate, the class of the computable random functions will be not enlarged if we add to $\mathcal{B}$ some other elements θ of $\mathcal{F}$ satisfying the condition (b) (for example, an element θ describing "data processing by means of a dice"). It would be interesting to compare the computability notion studied in Theorem 1 with other notions from the literature, e.g. with some notions introduced by Santos in his papers mentioned in Subsection (I).

The semigroup $\mathcal{F}$ of the combinatory space from Proposition 1 is a subsemigroup of a larger one which will be denoted by $\mathcal{F}_\infty$. The elements of $\mathcal{F}_\infty$ are arbitrary mappings of M^2 into the closed interval $[0,\infty]$ (the value ∞ included). The partial ordering, the multiplication and the extensions Π_∞ and Σ_∞ of the operations Π and Σ to $\mathcal{F}_\infty$ are defined by means of the same expressions, with only one detail needing special care, namely the meaning of the products $0.\infty$ and $\infty.0$ which now may occur at the evaluation of the expressions. It turns out to be appropriate for our purposes to assign the value 0 to these products. After adopting this convention, we have the following result.

Proposition 2. The 9-tuple

$$\mathfrak{S}_\infty = \langle \mathcal{F}_\infty, I_M^\sim, \mathcal{C}, \Pi_\infty, L^\sim, R^\sim, \Sigma_\infty, T^\sim, F^\sim \rangle$$

is a symmetric and iterative combinatory space, and the iteration in this space can be expressed in the same way as in Proposition 1.

The proof of this proposition is almost the same as the proof of Proposition 1, and the modifications are mainly in the direction of simplification (since now no problems arise about the convergency of the sums with summation variables ranging over M, and nothing like the condition (1) has to be verified).

It is reasonable to look for some intuitive interpretation of the elements of $\mathcal{F}_\infty$ and of the operations on them. There is a simple interpretation for those elements of $\mathcal{F}_\infty$ whose ranges are contained in $\mathbb{N}\cup\{\infty\}$. Namely one could consider non-deterministic devices without probabilistic features, and, when given such a device, one could describe it by the element θ of $\mathcal{F}_\infty$ such that, for all u, v in M, $\theta(u, v)$ is equal to the number of the possible computation paths leading from u to v. More arbitrary elements of $\mathcal{F}_\infty$ can be used in a similar way for the description of non-deterministic devices which employ the two kind of choices: the completely free and the probabilistic ones. In this case an element of $\mathcal{F}_\infty$ could describe the device by assigning to each pair u, v of elements of M the expected number of the possible computation paths leading from u to v (the probability that a path is possible being calculated by multiplication taking into account only the probabilistic choices occurring along the path).

(III) The case of probability distributions characterized by measures on a σ-field of sets. In this subsection, the additional assumption will be made that a σ-field $\mathcal{B}$ on M is given, i.e. a set $\mathcal{B}$ of subsets of M such that $M\in\mathcal{B}$, and $\mathcal{B}$ is closed under difference and under finite and countable unions. The following coordination between $\mathcal{B}$ and the computational structure $\mathfrak{A}$ will be also assumed:

(i) for any set E belonging to $\mathcal{B}$, its pre-images

$L^{-1}(E)$, $R^{-1}(E)$, $T^{-1}(E)$, $F^{-1}(E)$ belong to $\mathfrak{B}$, and its pre-image $J^{-1}(E)$ belongs to every σ-field on M^2 containing among its elements all Cartesian products $E' \times E''$, where E' and E'' belong to $\mathfrak{B}$;

(ii) the sets $H^{-1}(\text{true})$ and $H^{-1}(\text{false})$ belong to $\mathfrak{B}$.

Example 1. Let $\mathfrak{A}$ be the Moschovakis computational structure based on an arbitrary set B, and let $\mathfrak{B}_0$ be an arbitrary σ-field on B. Let $\mathfrak{B}$ be the least one among the σ-fields $\mathfrak{H}$ on M which have the following properties: (a) $\mathfrak{B}_0 \subseteq \mathfrak{H}$; (b) $\{O\} \in \mathfrak{H}$; (c) whenever E' and E'' belong to $\mathfrak{H}$, then $E' \times E''$ also belongs to $\mathfrak{H}$. We claim that the conditions (i) and (ii) are satisfied in this case. The validity of (ii) is clear from the equalities

$$H^{-1}(\text{true}) = M \setminus B^\circ, \quad H^{-1}(\text{false}) = B^\circ, \quad B^\circ = B \cup \{O\}.$$

To verify (i) we note first that, for any E in $\mathfrak{B}$, the pre-image $L^{-1}(E)$ is either the set $E \times M$ or the union of this set with one or both of the sets $\{O\}$, B, and a similar statement concerning $R^{-1}(E)$ holds. We note also that each of the pre-images $T^{-1}(E)$, $F^{-1}(E)$ is either M or $\emptyset$. Thus the part of (i) concerning $L^{-1}(E)$, $R^{-1}(E)$, $T^{-1}(E)$, $F^{-1}(E)$ is satisfied. For the verification of the part concerning $J^{-1}(E)$, we note the equality

$$J^{-1}(E) = E \setminus B^\circ.$$

Thus it is sufficient to show that, whenever $\mathfrak{B}'$ is a σ-field on M^2 containing among its elements all Cartesian products $E' \times E''$, where E' and E'' belong to $\mathfrak{B}$, then $E \setminus B^\circ \in \mathfrak{B}'$ for all E in $\mathfrak{B}$. This can be shown by verifying that $\{E \in \mathfrak{B} : E \setminus B^\circ \in \mathfrak{B}'\}$ is one of the σ-fields $\mathfrak{H}$ on M having the properties (a)-(c).

Intuitively, we shall consider now non-deterministic devices such that, for any u in M and any E in $\mathfrak{B}$, there is a definite probability that the application of the device with input data u will produce an output data belonging to E. It is natural to require, for a fixed device and a fixed input data u, the dependence of this probability on the choice of E to be represented by a measure on $\mathfrak{B}$ with values not greater than 1.[97] For a fixed device, the infor-

---

[97] A *measure on* $\mathfrak{B}$ is a σ-additive non-negative real-valued function on $\mathfrak{B}$. It is habitual to admit ∞ as value of a measure, but we shall consider only measures with fi-

mation about the measures on $\mathfrak{B}$ corresponding to all possible input data can be collected in a function θ from $M\times\mathfrak{B}$ to the interval $[0, 1]$, such that, for any u in M, the function $\lambda E.\,\theta(u, E)$ will be the corresponding measure. Besides the condition on the function θ corresponding to this, one more condition will be imposed, due to technical reasons, in the definition which will be given below.

We shall denote by $\mathcal{F}$ the set of all functions θ from $M\times\mathfrak{B}$ to the interval $[0, 1]$ which satisfy the following conditions:

(a) for any fixed u in M, the function $\lambda E.\,\theta(u, E)$ (denoted further by $\theta(u, \cdot)$) is a measure on $\mathfrak{B}$;

(b) for any fixed E in $\mathfrak{B}$, the function $\lambda u.\,\theta(u, E)$ (denoted further by $\theta(\cdot, E)$) is Borel measurable relative to $\mathfrak{B}$.[98]

Example 2. Let $\mathfrak{U}$, $\mathfrak{B}_0$, $\mathfrak{B}$ be such as in Example 1. Let θ_0 be a mapping of $B\times\mathfrak{B}_0$ into the interval $[0, 1]$ such that $\lambda E.\,\theta_0(u, E)$ is a measure on $\mathfrak{B}_0$ for any fixed u in B, and $\lambda u.\,\theta_0(u, E)$ is Borel measurable relative to $\mathfrak{B}_0$ for any fixed E in $\mathfrak{B}_0$. It can be shown that $E\cap B\in\mathfrak{B}_0$ for any E in $\mathfrak{B}$ (by noticing that $\{E\in\mathfrak{B}: E\cap B\in\mathfrak{B}_0\}$ is one of the σ-fields $\mathfrak{K}$ on M having the properties (a)-(c)). Using this fact, we define a mapping θ of $M\times\mathfrak{B}$ into the interval $[0, 1]$ in the following way:

$$\theta(u, E) = \begin{cases} \theta_0(u, E\cap B) & \text{if } u\in B, \\ 0 & \text{if } u\notin B. \end{cases}$$

It is easy to verify that $\theta\in\mathcal{F}$ and θ is an extension of θ_0.

We shall restrict our intuitive considerations to such devices which can be described (in the already explained sense) by functions belonging to $\mathcal{F}$.

If f is a partial function in M then we define a mapping $f^{\sim}$ of $M\times\mathfrak{B}$ into the set $\{0, 1\}$ by setting

$$f^{\sim}(u, E) = \| u\in f^{-1}(E) \|$$

(the meaning of the denotations of the form $\|A\|$ has been

---

nite values.

[98] A real-valued function Z defined on M is called *Borel measurable relative to* $\mathfrak{B}$ iff, for any choice of the real number c, the set $\{u\in M: Z(u) > c\}$ belongs to $\mathfrak{B}$.

introduced in the previous section). For $\theta = f^{\sim}$, condition (a) is obviously satisfied. As to condition (b), it is equivalent to the condition that $f^{-1}(E) \in \mathscr{B}$ for every choice of E in $\mathscr{B}$. The partial functions in M which have this property are called *measurable relative to* $\mathscr{B}$. Thus we get the following result:

For any partial function f in M, $f^{\sim} \in \mathscr{F}$ iff f is measurable relative to $\mathscr{B}$.

By the assumption (i), L, R, T, F are measurable relative to $\mathscr{B}$. Obvious other examples of functions measurable relative to $\mathscr{B}$ are I_M and the constant total mappings of M into M.

Remark 1. In general, the mapping $\lambda f.\, f^{\sim}$ of the set of the measurable elements of $\mathscr{F}_p(M)$ into $\mathscr{F}$ is not necessarily an injection. This mapping is an injection iff, for any two different elements s and t of M, there is a set E from $\mathscr{B}$ such that $s \in E$, $t \notin E$.

The set $\mathscr{F}$ will be considered by the natural partial ordering defined as follows: $\varphi \geq \psi$ iff $\varphi(u, E) \geq \psi(u, E)$ for all u in M and all E in $\mathscr{B}$.

To define for the elements of $\mathscr{F}$ a multiplication corresponding to the sequential composition of devices, we need an operation of integration. The information needed for the formulation of the definition is the following one:

If Z is a bounded real-valued function defined on M and Borel measurable relative to $\mathscr{B}$, and μ is a measure on $\mathscr{B}$, then a real number $\int_M Z\,d\mu$ is defined called *the integral on M of Z with respect to μ*; the same number is denoted also by $\int_M Z(v)\,\mu(dv)$, and, of course, other variables can be used instead of v (for the definition of the integral and for its properties which are used in the sequel, cf., for example, Loève [1977]). Instead of $\int_M$ we shall write simply $\int$.

Here is the definition of the product $\varphi\psi$ of two elements φ and ψ of $\mathscr{F}$: this is a mapping of $M \times \mathscr{B}$ into $\mathbb{R}$ determined by means of the equality

$$(\varphi\psi)(u, E) = \int \varphi(v, E)\,\psi(u, dv).$$

The basic facts concerning this product are formulated in the following proposition.

Proposition 3. The set $\mathscr{F}$ is closed under the multiplication defined above. This set considered with the introduced partial ordering and multiplication is a partially ordered semigroup, and $I_M^{\sim}$ is an identity of the semigroup.

Proof. To show that $\mathcal{F}$ is closed under multiplication, suppose that φ and ψ are some elements $\mathcal{F}$. From the inequalities

$$0 \le \varphi(v, E) \le 1,$$

using the elementary properties of the integral, we get the inequalities

$$0 \le (\varphi\psi)(u, E) \le \psi(u, M).$$

These inequalities show that all values of $\varphi\psi$ belong to the interval $[0, 1]$. Making use of the σ-additivity of all functions of the form $\varphi(v, \cdot)$ and of the monotone convergence theorem, we easily see the σ-additivity of the function $(\varphi\psi)(u, \cdot)$ for any fixed u. Suppose now some fixed set E from $\mathcal{B}$ is given. We shall show the Borel measurability of the function $(\varphi\psi)(\cdot, E)$. To do this, it is sufficient to show that, for each Borel measurable bounded non-negative function Z on M, the function

$$\lambda u. \int Z(v)\psi(u, dv) \tag{8}$$

is also Borel measurable. If Z is a function of the form $\lambda v. \| v \in E_0 \|$, where E_0 is some fixed set from $\mathcal{B}$, then the corresponding function (8) is $\psi(\cdot, E_0)$, and hence this function is Borel measurable. The case of an arbitrary Borel measurable bounded non-negative function Z can be reduced to the above case by representing Z as the limit of a monotonically increasing sequence of linear combinations of functions of the above form. Thus we proved that $\varphi\psi \in \mathcal{F}$.

The multiplication in $\mathcal{F}$ is obviously monotonically increasing. To show its associativity, we suppose that elements φ, ψ and χ of $\mathcal{F}$ are given. Let us denote $\psi\chi$ by θ. Then

$$\theta(w, E) = \int \psi(u, E)\chi(w, du)$$

for all $w \in M$, $E \in \mathcal{B}$, and we have to prove the equality $(\varphi\psi)\chi = \varphi\theta$. In other words, we have to prove that, for all $w \in M$, $E \in \mathcal{B}$, the following equality holds:

$$\int \Big(\int \varphi(v, E)\psi(u, dv) \Big) \chi(w, du) = \int \varphi(v, E)\theta(w, dv).$$

To prove it, it is sufficient to show that, for each Borel measurable bounded non-negative function Z on M, the equality

$$\int \Big(\int Z(v)\psi(u, dv) \Big) \chi(w, du) = \int Z(v)\theta(w, dv) \tag{9}$$

holds for all w in M. If Z is a function of the form

$\lambda v.\ \|v\in E_0\|$, where E_0 is some fixed set from $\mathcal{B}$, then the above equality is true, since it is equivalent to the equality

$$\int \psi(u, E_0)\chi(w, du) = \theta(w, E_0).$$

The general case can be reduced to this special one in the same way as in the first part of the proof.

The equalities $I_M^{\sim}\theta = \theta I_M^{\sim} = \theta$ are also true, since they mean that

$$\int \|v\in E\|\,\theta(u, dv) = \int \theta(v, E)\,\|u\in dv\| = \theta(u, E)$$

for all $u\in M$, $E\in\mathcal{B}$. ∎

For any φ and ψ in $\mathcal{F}$, a real-valued function $\Pi(\varphi, \psi)$ will be defined on $M\times\mathcal{B}$, and this function will be shown to belong again to $\mathcal{F}$. In its definition, the σ-field $\mathcal{B}^2$ will be used, i.e. the least σ-field on M^2 containing among its elements all Cartesian products $E'\times E''$, where E' and E'' belong to $\mathcal{B}$. We shall make use also of the notion of product of two measures on $\mathcal{B}$. Namely, we shall use the fact that, to any two measures μ' and μ'' on $\mathcal{B}$, there is a unique measure μ on $\mathcal{B}^2$ (called *the product of* μ' *and* μ'' and denoted by $\mu'\times\mu''$) such that

$$\mu(E'\times E'') = \mu'(E')\,\mu''(E'')$$

for all E', E'' in $\mathcal{B}$ (for this fact and for some properties of $\mu'\times\mu''$ needed further, cf. e. g. Loève [1977]). Here is the definition of the function $\Pi(\varphi, \psi)$:

$$\Pi(\varphi, \psi)(u, E) = (\varphi(u, \cdot)\times\psi(u, \cdot))(J^{-1}(E))$$

for all $u\in M$, $E\in\mathcal{B}$. It follows from this definition that

$$\Pi(\varphi, \psi)(u, E) \le (\varphi(u, \cdot)\times\psi(u, \cdot))(M^2) = \varphi(u, M)\times\psi(u, M) \le 1.$$

A straightforward verification shows that $\Pi(\varphi, \psi)(u, \cdot)$ is a measure on $\mathcal{B}$ for any fixed u in M. For the proof of the Borel measurability of the functions $\Pi(\varphi, \psi)(\cdot, E)$, where $E\in\mathcal{B}$, we denote by $\mathcal{K}$ the set of all K from $\mathcal{B}^2$ such that the function

$$\lambda u.\ (\varphi(u, \cdot)\times\psi(u, \cdot))(K)$$

is Borel measurable relative to $\mathcal{B}$. If $K = E'\times E''$, where E' and E'' belong to $\mathcal{B}$, then $K\in\mathcal{K}$, since the above function will be the product of Borel measurable functions $\varphi(\cdot, E')$ and $\psi(\cdot, E'')$ in this case. On the other hand, $\mathcal{K}$ is closed under unions of monotonically increasing sequences and under intersections of monotonically decreasing ones. Therefore (cf., for example, Loève [1977, 1.6]) $\mathcal{K} = \mathcal{B}^2$, and, in particular, $J^{-1}(E)\in\mathcal{H}$ for all E in $\mathcal{B}$. Thus all functions $\Pi(\varphi, \psi)(\cdot, E)$, where $E\in\mathcal{B}$, are Borel measurable, and hence $\Pi(\varphi, \psi)\in\mathcal{F}$.

For any χ, φ, ψ in $\mathcal{F}$, a real-valued function $\Sigma(\chi, \varphi, \psi)$ will be defined by means of the equality

$$\Sigma(\chi, \varphi, \psi)(u, E)=(H\chi)(u, \mathrm{true})\varphi(u, E)+ (H\chi)(u, \mathrm{false})\psi(u, E),$$

where

$$(H\chi)(u, p)=\chi(u, H^{-1}(p)).$$

We have

$$\Sigma(\chi, \varphi, \psi)(u, E)\le (H\chi)(u, \mathrm{true})+(H\chi)(u, \mathrm{false})= \chi(u, H^{-1}(\mathrm{true})\cup H^{-1}(\mathrm{false}))\le 1,$$

and the verification of the conditions (a) and (b) for the function $\theta=\Sigma(\chi, \varphi, \psi)$ is straightforward. Hence this function belongs to $\mathcal{F}$ again.

For any s in M, let $\bar{s}$ be the constant mapping of M into M with value s, and let $\mathcal{C}$ be the set of all functions $\bar{s}^{\sim}$, where $s\in M$. Then the following proposition holds.

Proposition 4. The 9-tuple

$$\mathfrak{S}=\langle\mathcal{F}, I_M^{\sim}, \mathcal{C}, \Pi, L^{\sim}, R^{\sim}, \Sigma, T^{\sim}, F^{\sim}\rangle$$

is a symmetric and iterative combinatory space. For arbitrary σ, χ in $\mathcal{F}$ and arbitrary $u\in M$, $E\in\mathcal{B}$ the equality

$$[\sigma, \chi](u, E)=\sum_{m=0}^{\infty}\rho_m(u, E)$$

holds, where ρ_m is defined by means of the equality

$$\rho_m(v_0, E)=\int(\int(\dots\int(\int\|v_m\in E\|\times (H\chi)(v_m, \mathrm{false})(\prod_{j=0}^{m-1}(H\chi)(v_j, \mathrm{true}))\times$$
$$\sigma(v_{m-1}, dv_m))\sigma(v_{m-2}, dv_{m-1})\dots)\sigma(v_1, dv_2))\sigma(v_0, dv_1).$$

The proof of this proposition will be not given in details. A certain part of the things which have to be done are verifications using well-known properties of the integrals on M. We shall note only some moments from the proof, and we hope the readers who are familiar with measures and integration on abstract sets will be able to work out the whole proof.

1. One verifies that

$$(\theta\bar{s}^{\sim})(u, E) = \theta(s, E)$$

for all θ in $\mathcal{F}$, all s, u in M and all E in $\mathcal{E}$.

2. For arbitrary s, t in M, one verifies the equality

$$\Pi(\bar{s}^{\sim}, \bar{t}^{\sim}) = \bar{u}^{\sim},$$

where $u = J(s, t)$.

3. The validity of the equality

$$\Pi(I_M^{\sim}, \psi\bar{s}^{\sim})\theta = \Pi(\theta, \psi\bar{s}^{\sim})$$

is established by means of the following calculations, which make use of a representation of the product of two measures in the form of an integral and of changing of the order of integration on the basis of Fubini's theorem:

$$(\Pi(I_M^{\sim}, \psi\bar{s}^{\sim})\theta)(u, E) = \int \Pi(I_M^{\sim}, \psi\bar{s}^{\sim})(v, E)\theta(u, dv) =$$

$$\int (I_M^{\sim}(v, \cdot) \times (\psi\bar{s}^{\sim})(v, \cdot))(J^{-1}(E))\theta(u, dv) =$$

$$\int (I_M^{\sim}(v, \cdot) \times \psi(s, \cdot))(J^{-1}(E))\theta(u, dv) =$$

$$\int (\int I_M^{\sim}(v, \{p : \langle p, q\rangle \in J^{-1}(E)\})\psi(s, dq))\theta(u, dv) =$$

$$\int (\int I_M^{\sim}(v, \{p : \langle p, q\rangle \in J^{-1}(E)\})\theta(u, dv))\psi(s, dq) =$$

$$\int \theta(u, \{p : \langle p, q\rangle \in J^{-1}(E)\}))\psi(s, dq) =$$

$$(\theta(u, \cdot) \times \psi(s, \cdot))(J^{-1}(E))) =$$

$$(\theta(u, \cdot) \times (\psi\bar{s}^{\sim})(u, \cdot))(J^{-1}(E))) = \Pi(\theta, \psi\bar{s}^{\sim})(u, E).$$

The validity of the equality

$$\Pi(\varphi\bar{s}^{\sim}, I_M^{\sim})\theta = \Pi(\varphi\bar{s}^{\sim}, \theta)$$

can be established in a similar way.

4. The verification of the equalities

$$\theta\Sigma(\chi, \varphi, \psi) = \Sigma(\chi, \theta\varphi, \theta\psi),$$

$$\Sigma(I_M^{\sim}, \varphi\bar{s}^{\sim}, \psi\bar{s}^{\sim})\theta = \Sigma(\theta, \varphi\bar{s}^{\sim}, \psi\bar{s}^{\sim})$$

makes use of the linear properties of the integral.

5. To prove that the combinatory space $\mathfrak{S}$ is iterative, we use the Level Omega Iteration Lemma. There is a least element in $\mathcal{F}$, namely the constant 0, and it is equal to its product in $\mathcal{F}$ with any element of $\mathcal{F}$. If $\{\theta_n\}_{n=0}^{\infty}$ is a monotonically increasing infinite sequence of elements of $\mathcal{F}$, then we define a real-valued function θ by means of the equality

$$\theta(u, E) = \lim_{n\to\infty} \theta_n(u, E) \tag{10}$$

and prove that $\theta \in \mathcal{F}$ (then it is obvious that θ is the least upper bound of the given sequence). The proof that θ belongs to $\mathcal{F}$ is based on the fact that the limit of a bounded monotonically increasing sequence of measures on $\mathcal{B}$ is again a measure on $\mathcal{B}$, and the limit of a bounded monotonically increasing sequence of real functions on M which are Borel measurable relative to $\mathcal{B}$ is again a Borel measurable function. The continuity of the mappings of the form $\lambda\tau.\,\Sigma(\varkappa, \tau, I_M^{\sim})$ is seen immediately. The continuity of the mappings of the form $\lambda\tau.\,\tau\varkappa$ is seen by application of the monotone convergence theorem. To show the continuity of the mappings of the form $\lambda\tau.\,\varkappa\tau$, it is sufficient to prove the following statement of Helly type: if $\{\theta_n\}_{n=0}^{\infty}$ is a monotonically increasing infinite sequence of elements of $\mathcal{F}$, the real-valued function θ is defined by means of the equality (10), and Z is a Borel measurable bounded non-negative function on M, then

$$\int Z(v)\theta(u, dv) = \lim_{n\to\infty} \int Z(v)\theta_n(u, dv)$$

for all u in M (one first verifies this equality in the case when Z is a function of the form $\lambda v.\,\| v \in E_0 \|$, where E_0 is some fixed set from $\mathcal{B}$, and then reduces the general case to this particular one by representing an arbitrary Borel measurable bounded non-negative function Z as the limit of a monotonically increasing sequence of linear combinations of functions of the above form).

6. The expression for the iteration can be obtained by supplementing an application of Proposition II.4.6 to the application of the Level Omega Iteration Lemma. The proof of the formula for $\rho_m(v_0, E)$ is by induction on m.

Remark 2. If $\theta \in \mathcal{F}$, and Z is a Borel measurable bounded non-negative function on M, then, for any u in M, the value of the integral $\int Z(v)\theta(u, dv)$ is some non-negative real number (in the particular case when $\theta(u, M) = 1$, the number in question is the expectation of the function Z with respect to the probability measure $\theta(u, \cdot)$). We shall denote this number by $\theta(u, Z)$, for short. The following equalities hold for all measurable elements f of $\mathcal{F}_p(M)$, all χ, φ, ψ in $\mathcal{F}$, all u in M and all Borel measurable bounded non-negative functions Z on M:

$$f^{\sim}(u, Z) = \begin{cases} Z(f(u)) & \text{if } u \in \operatorname{dom} f, \\ 0 & \text{if } u \notin \operatorname{dom} f, \end{cases} \tag{11}$$

(12) $$(\varphi\psi)(u, Z) = \psi(u, \lambda v.\, \varphi(v, Z)),$$

(13) $$\Pi(\varphi, \psi)(u, Z) = \varphi(u, \lambda s.\, \psi(u, \lambda t.\, Z(J(s, t)))),$$

(14) $$\Pi(\varphi, \psi)(u, Z) = \psi(u, \lambda t.\, \varphi(u, \lambda s.\, Z(J(s, t)))),$$

(15) $$\Sigma(\chi, \varphi, \psi)(u, Z) = \chi(u, \lambda s.\, H^{\sim}(s, \varphi(u, Z), \psi(u, Z))),$$

where

$$H^{\sim}(s, a, b) = \| s \in H^{-1}(\mathrm{true}) \| a + \| s \in H^{-1}(\mathrm{false}) \| b$$

for all s in M and all non-negative real numbers a, b. The equality (12) is, up to denotations, the equality (9) established in the proof of Proposition 3. The equality (15) follows from the definition of Σ, the representation of the values of the measures as integrals and the linear properties of the integral. For proving (13) and (14), it is appropriate to prove first the equality

(16) $$\Pi(\varphi, \psi)(u, Z) = \int_{M^2} Z(J)\, d(\varphi(u, \cdot) \times \psi(u, \cdot)),$$

and then to obtain them from it by means of Fubini's theorem. The proof of (16), as well as the proof of (11), can be reduced to proving the equality in the case of

$$Z = \lambda r.\, \| r \in E_0 \|,$$

where E_0 is some fixed set from $\mathfrak{B}$.

5. Combinatory spaces of functionals

Remark 2 in the previous section contains some formulas which can be used for the construction of combinatory spaces not necessarily connected with probability. Here we shall give such a construction, which is taken (with small modifications) from the paper Skordev [1980a]. A result will be presented (also from that paper) showing that each combinatory space is isomorphic to some combinatory space constructed in this way.

We suppose that an infinite set M is given together with an injection J of M^2 into M and mappings T and F of M into M such that $\mathrm{rng}\, T \cap \mathrm{rng}\, F = \emptyset$. Any quadruple $\langle M, J, T, F\rangle$ of this kind will be called a *coding structure*. Clearly, if $\langle M, J, L, R, T, F, H\rangle$ is a computational structure, then $\langle M, J, T, F\rangle$ is a coding structure, and each coding structure can be extended to a computational structure, since one could define L, R, H by the conditions

$$\mathrm{dom}\, L = \mathrm{dom}\, R = \mathrm{rng}\, J,\quad L(J(s, t)) = s,\quad R(J(s, t)) = t,$$

$$\mathrm{dom}\, H = \mathrm{rng}\, T \cup \mathrm{rng}\, F,\quad H(T(u)) = \mathrm{true},\quad H(F(u)) = \mathrm{false}.$$

A computational structure consists of mechanisms for coding

the ordered pairs and the logical values true and false, and of corresponding decoding mechanisms, whereas in a coding structure only the coding mechanisms are present. The reason to consider now coding structures instead of computational structures is that in some cases it is appropriate to allow using of slightly more general decoding mechanisms, namely ones with non-functional behaviour on the elements of M which are not codes.[99]

The following example of a coding structure will be used at the end of this section.

Example 1. Let $<\mathcal{F}_1, I_1, \mathcal{C}_1, \Pi_1, L_1, R_1, \Sigma_1, T_1, F_1>$ be an arbitrary combinatory space, M be the set $\mathcal{C}_1$, and J be the restriction of Π_1 to M^2. Let T, F be the mappings of M into M defined by means of the equalities

$$T(u)=T_1 u, \quad F(u)=F_1 u.$$

Then $<M, J, T, F>$ is a coding structure.

Coming back to the general case, we suppose that, besides the coding structure $<M, J, T, F>$, a partially ordered set V is also given, and our intention is to use its elements instead of the non-negative real numbers. For the time being, the set V is assumed to have at least two distinct elements (further assumptions will be needed later).

The set of all total mappings of M into V will be denoted by $\mathcal{Z}$. The partial ordering in V induces naturally a partial ordering in $\mathcal{Z}$. Namely the inequality $Z'' \geq Z'$, where Z', Z'' belong to $\mathcal{Z}$, means that $Z''(u) \geq Z'(u)$ for all u in M.

We shall denote by $\mathcal{F}$ the set of all mappings of $M \times \mathcal{Z}$ into V which are monotonically increasing with respect to their second argument.

Remark 1. The mappings in the above definition of $\mathcal{F}$ can be called functionals. A somewhat simpler definition of $\mathcal{F}$ can be given which is equivalent to the above one up to isomorphism and could make the construction more similar to a construction from Chapter 12 of Ivanov [1986]. Namely we could define $\mathcal{F}$ as the set of all monotonically increasing mappings of $\mathcal{Z}$ into itself. These mappings can be called operators, since they transform functions into functions. A natural one-to-one correspondence between the functionals θ and the operators Θ can be fixed by means of the equality

---

[99] We already came across some coding structures in Exercise II.4.22, but without naming them so.

$$\Theta(Z)(u) = \theta(u, Z).$$

The reader could try to translate the further definitions for the case of such a modified definition of $\mathcal{F}$.

The set $\mathcal{F}$ is considered with the natural partial ordering in it. Namely the inequality $\varphi \geq \psi$, where φ, ψ belong to $\mathcal{F}$, means that $\varphi(u, Z) \geq \psi(u, Z)$ for all u in M and all Z in $\mathcal{Z}$.

An operation of multiplication is defined in $\mathcal{F}$ by using the formula 4.(12). Namely, for any φ, ψ in $\mathcal{F}$, we define a mapping $\varphi\psi$ of $M \times \mathcal{Z}$ into V by means of the equality

(1) $$(\varphi\psi)(u, Z) = \psi(u, \lambda v.\ \varphi(v, Z)).$$

It is easily verified that $\varphi\psi$ belongs again to $\mathcal{F}$. The multiplication operation is monotonically increasing (as a consequence of the monotonicity of the elements of $\mathcal{F}$ with respect to their second arguments). This operation is also associative as seen from the following calculations

$$((\varphi\psi)\chi)(w, Z) = \chi(w, \lambda v.\ (\varphi\psi)(v, Z)) = \chi(w, \lambda v.\ \psi(v, \lambda u.\ \varphi(u, Z))),$$

$$(\varphi(\psi\chi))(w, Z) = (\psi\chi)(w, \lambda u.\ \varphi(u, Z)) = \chi(w, \lambda v.\ \psi(v, \lambda u.\ \varphi(u, Z))).$$

Thus $\mathcal{F}$ is a partially ordered semigroup.

Remark 2. In the variant of presentation mentioned in Remark 1, the multiplication in $\mathcal{F}$ must be defined as an ordinary composition of operators, and then its associativity becomes completely evident.

For each total mapping f of M into M, we define a mapping $f^{\sim}$ of $M \times \mathcal{Z}$ into V by using the formula 4.(11) i.e. we set

$$f^{\sim}(u, Z) = Z(f(u))$$

for all $u \in M$, $Z \in \mathcal{Z}$. Clearly, all such mappings $f^{\sim}$ belong to $\mathcal{F}$.

The element $I_M^{\sim}$ is an identity of the semigroup $\mathcal{F}$, as the following calculations show:

$$(I_M^{\sim}\psi)(u, Z) = \psi(u, \lambda v.\ I_M^{\sim}(v, Z)) = \psi(u, Z),$$

$$(\varphi I_M^{\sim})(u, Z) = I_M^{\sim}(u, \lambda v.\ \varphi(v, Z)) = \varphi(u, Z).$$

For each s in M, we, as usual, shall denote by $\bar{s}$ the constant total mapping of M into M with value s. Then $\bar{s}^{\sim} \in \mathcal{F}$ for all $s \in M$. The set of all $\bar{s}^{\sim}$, where $s \in M$, will be denoted by $\mathcal{C}$. Note that, for all s, u in $\mathcal{F}$ and all Z in $\mathcal{Z}$, we have

$$\bar{s}^{\sim}(u, Z) = Z(\bar{s}(u)) = Z(s).$$

For the definition of a binary operation Π in $\mathcal{F}$, we have two alternatives - to use either the formula 4.(13) or the formula 4.(14) (the right-hand expressions in both formulas make sense for all φ, ψ in $\mathcal{F}$, all u in M and all Z in $\mathcal{Z}$). It can be shown by means of examples that choosing the second of these alternatives would be not convenient for what follows. Therefore we choose the first one, i.e. we set

(2) $$\Pi(\varphi, \psi)(u, Z) = \varphi(u, \lambda s.\ \psi(u, \lambda t.\ Z(J(s, t))))$$

for all φ, ψ in $\mathcal{F}$, all u in M and all Z in $\mathcal{Z}$. It is easily seen that $\Pi(\varphi, \psi) \in \mathcal{F}$ for all φ, ψ in $\mathcal{F}$.

Now we are going to define a ternary operation Σ in $\mathcal{F}$. The formula 4.(15) is not directly usable for this in the general situation considered now, since the defining expression of $H^{\sim}(s, a, b)$ contains operations which are no longer present (in particular, no predicate H is given). However, this can be repaired. We make the assumption that a mapping H' of $M \times V^2$ into V is given such that H' is monotonically increasing with respect to its second and third arguments, and the equalities

(3) $$H'(T(u), a, b) = a, \quad H'(F(u), a, b) = b$$

hold for all u in M and all a, b in V. Such a mapping H' always exists. We can, for example, set

$$H'(s, a, b) = \begin{cases} a & \text{if } s \in \operatorname{rng} T, \\ b & \text{if } s \in \operatorname{rng} F, \\ o & \text{if } s \notin \operatorname{rng} T \cup \operatorname{rng} F, \end{cases}$$

where o is some fixed element of V. Having a mapping H' with the properties formulated above at our disposal, we set

(4) $$\Sigma(\chi, \varphi, \psi)(u, Z) = \chi(u, \lambda s.\ H'(s, \varphi(u, Z), \psi(u, Z)))$$

for all χ, φ, ψ in $\mathcal{F}$, all u in M and all Z in $\mathcal{Z}$, and we see that $\Sigma(\chi, \varphi, \psi) \in \mathcal{F}$ for all χ, φ, ψ in $\mathcal{F}$.

Proposition 1. Let L', R' be elements of $\mathcal{F}$ such that

(5) $$L'(J(s, t), Z) = Z(s), \quad R'(J(s, t), Z) = Z(t)$$

for all s, t in M and all Z in $\mathcal{Z}$.[100] Then the 9-tuple

---

[100] Such elements L', R' of $\mathcal{F}$ always exist. We can, for example, fix some element o of V and set

$$L'(u, Z) = \begin{cases} Z(s) & \text{if } u = J(s, t), \\ o & \text{if } u \notin \operatorname{rng} J, \end{cases}$$

$$\mathfrak{S} = \langle \mathcal{F}, I_M^{\sim}, \mathcal{C}, \Pi, L^{\bullet}, R^{\bullet}, \Sigma, T^{\sim}, F^{\sim} \rangle$$

is a combinatory space.

Proof. Some of the conditions from the definition of the notion of combinatory space are already verified. We shall present the verification of the remaining ones.

If $\theta \in \mathcal{F}$ and $s \in M$, then, for all u in M and all Z in $\mathcal{Z}$, we have

$$(\theta \bar{s}^{\sim})(u, Z) = \bar{s}^{\sim}(u, \lambda v.\ \theta(v, Z)) = \theta(\bar{s}^{\sim}(u), Z) = \theta(s, Z).$$

This immediately shows that condition II.1.(1) is satisfied.

If p and q are arbitrary elements of M, then, for all u in M and all Z in $\mathcal{Z}$, we have

$$\Pi(\bar{p}^{\sim}, \bar{q}^{\sim})(u, Z) = \bar{p}^{\sim}(u, \lambda s.\ \bar{q}^{\sim}(u, \lambda t.\ Z(J(s, t)))) =$$

$$\bar{p}^{\sim}(u, \lambda s.\ Z(J(s, q)))) = Z(J(p, q)) = \bar{r}^{\sim}(u, Z),$$

where $r = J(p, q)$, and hence

$$\Pi(\bar{p}^{\sim}, \bar{q}^{\sim}) = \bar{r}^{\sim},$$

$$(L^{\bullet}\ \Pi(\bar{p}^{\sim}, \bar{q}^{\sim}))(u, Z) = \bar{r}^{\sim}(u, \lambda v.\ L^{\bullet}(v, Z)) = L^{\bullet}(r, Z) =$$

$$Z(p) = \bar{p}^{\sim}(u, Z),$$

$$(R^{\bullet}\ \Pi(\bar{p}^{\sim}, \bar{q}^{\sim}))(u, Z) = \bar{r}^{\sim}(u, \lambda v.\ R^{\bullet}(v, Z)) = R^{\bullet}(r, Z) =$$

$$Z(q) = \bar{q}^{\sim}(u, Z).$$

Thus conditions II.1.(2)-II.1.(4) are verified. The monotonic increasing of the operation Σ is obvious from its definition, and hence condition II.1.(16) is also satisfied.

To show that $T^{\sim} \neq F^{\sim}$, we take two distinct elements a, b of $\mathcal{F}$ and set $Z = \lambda s.\ H^{\bullet}(s, a, b)$. Then $Z \in \mathcal{Z}$, and

$$T^{\sim}(u, Z) = Z(T(u)) = a, \quad F^{\sim}(u, Z) = Z(F(u)) = b,$$

hence $T^{\sim}(u, Z) \neq F^{\sim}(u, Z)$, for any u in M. Thus condition II.1.(8) is satisfied.

To check the conditions II.1.(5)-II.1.(7) and II.1.(9)-II.1.(15), suppose that some elements $\varphi, \psi, \chi, \theta$ of $\mathcal{F}$, some elements r, u of M and some Z from $\mathcal{Z}$ are given. The following calculations, where I denotes $I_M^{\sim}$, contain the verification of the listed conditions.

$$(\Pi(\varphi, \psi)\bar{r}^{\sim})(u, Z) = \Pi(\varphi, \psi)(r, Z) =$$

$$R^{\bullet}(u, Z) = \begin{cases} Z(t) & \text{if } u = J(s, t), \\ 0 & \text{if } u \notin \operatorname{rng} J \end{cases}$$

for all $u \in M$, $Z \in \mathcal{Z}$.

$$\varphi(r, \lambda s.\, \psi(r, \lambda t.\, Z(J(s, t)))) =$$
$$(\varphi \bar{r}^{\sim})(u, \lambda s.\, (\psi \bar{r}^{\sim})(u, \lambda t.\, Z(J(s, t)))) =$$
$$\Pi(\varphi \bar{r}^{\sim}, \psi \bar{r}^{\sim})(u, Z),$$

$$(\Pi(I, \psi \bar{r}^{\sim})\theta)(u, Z) = \theta(u, \lambda v.\, \Pi(I, \psi \bar{r}^{\sim})(v, Z) =$$
$$\theta(u, \lambda v.\, I(v, \lambda s.\, (\psi \bar{r}^{\sim})(v, \lambda t.\, Z(J(s, t)))) =$$
$$\theta(u, \lambda v.\, I(v, \lambda s.\, \psi(r, \lambda t.\, Z(J(s, t)))) =$$
$$\theta(u, \lambda v.\, \psi(r, \lambda t.\, Z(J(v, t)))) =$$
$$\theta(u, \lambda v.\, (\psi \bar{r}^{\sim})(u, \lambda t.\, Z(J(v, t)))) = \Pi(\theta, \psi \bar{r}^{\sim})(u, Z),$$

$$(\Pi(\bar{r}^{\sim}, I)\theta)(u, Z) = \theta(u, \lambda v.\, \Pi(\bar{r}^{\sim}, I)(v, Z)) =$$
$$\theta(u, \lambda v.\, \bar{r}^{\sim}(u, \lambda s.\, I(u, \lambda t.\, Z(J(s, t)))) =$$
$$\theta(u, \lambda v.\, \bar{r}^{\sim}(u, \lambda s.\, Z(J(s, u)))) =$$
$$\theta(u, \lambda v.\, Z(J(r, u)))) =$$
$$\bar{r}^{\sim}(u, \lambda s.\, \theta(u, \lambda v.\, Z(J(s, u)))) = \Pi(\bar{r}^{\sim}, \theta)(u, Z),$$

$$(T^{\sim} \bar{r}^{\sim})(u, Z) = T^{\sim}(r, Z) = Z(T(r)) = \overline{T(r)}^{\sim}(u, Z),$$

$$(F^{\sim} \bar{r}^{\sim})(u, Z) = F^{\sim}(r, Z) = Z(F(r)) = \overline{F(r)}^{\sim}(u, Z),$$

$$\Sigma(T^{\sim}, \varphi, \psi)(u, Z) = T^{\sim}(u, \lambda s.\, H'(s, \varphi(u, Z), \psi(u, Z))) =$$
$$H'(T(u), \varphi(u, Z), \psi(u, Z))) = \varphi(u, Z),$$

$$\Sigma(F^{\sim}, \varphi, \psi)(u, Z) = F^{\sim}(u, \lambda s.\, H'(s, \varphi(u, Z), \psi(u, Z))) =$$
$$H'(F(u), \varphi(u, Z), \psi(u, Z))) = \psi(u, Z),$$

$$(\theta\Sigma(\chi, \varphi, \psi))(u, Z) = \Sigma(\chi, \varphi, \psi))(u, \lambda v.\, \theta(v, Z) =$$
$$\chi(u, \lambda s.\, H'(s, \varphi(u, \lambda v.\, \theta(v, Z)), \psi(u, \lambda v.\, \theta(v, Z)))) =$$
$$\chi(u, \lambda s.\, H'(s, (\theta\varphi)(u, Z), (\theta\psi)(u, Z))) =$$
$$\Sigma(\chi, \theta\varphi, \theta\psi))(u, Z),$$

$$(\Sigma(\chi, \varphi, \psi)\bar{r}^{\sim})(u, Z) = \Sigma(\chi, \varphi, \psi)(r, Z) =$$
$$\chi(r, \lambda s.\, H'(s, \varphi(r, Z), \psi(r, Z))) =$$
$$(\chi \bar{r}^{\sim})(u, \lambda s.\, H'(s, (\varphi \bar{r}^{\sim})(u, Z), (\psi \bar{r}^{\sim})(u, Z))) =$$
$$\Sigma(\chi \bar{r}^{\sim}, \varphi \bar{r}^{\sim}, \psi \bar{r}^{\sim})(u, Z),$$

$$(\Sigma(I, \varphi \bar{r}^{\sim}, \psi \bar{r}^{\sim})\theta)(u, Z) = \theta(u, \lambda v.\, \Sigma(I, \varphi \bar{r}^{\sim}, \psi \bar{r}^{\sim})(v, Z) =$$
$$\theta(u, \lambda v.\, I(v, \lambda s.\, H'(s, (\varphi \bar{r}^{\sim})(v, Z), (\psi \bar{r}^{\sim})(v, Z))) =$$
$$\theta(u, \lambda v.\, I(v, \lambda s.\, H'(s, \varphi(r, Z), \psi(r, Z))) =$$

$$\theta(u, \lambda v.\, H'(v, \varphi(r, Z), \psi(r, Z))) =$$
$$\theta(u, \lambda v.\, H'(v, (\varphi\bar{r}^{\sim})(u, Z), (\psi\bar{r}^{\sim})(u, Z))) =$$
$$\Sigma(\theta, \varphi\bar{r}^{\sim}, \psi\bar{r}^{\sim})(u, Z). \blacksquare$$

Remark 3. Let $\mathcal{Z}_0$ be a subset of $\mathcal{Z}$ satisfying the condition that $\lambda s.\, H'(s, a, b) \in \mathcal{Z}_0$ for some two distinct elements a, b of V (where H' is a mapping with the same properties as above). For each total mapping f of M into M, let f° denote the restriction of $f^\sim$ to the set $M \times \mathcal{Z}_0$. Let $\mathcal{C}_0$ be the set of all $\bar{s}^\circ$, where $s \in M$. Suppose also that a set $\mathcal{F}_0$ of mappings of $M \times \mathcal{Z}_0$ into V is given which are monotonically increasing with respect to their second argument, and the following conditions are satisfied:

(a) $\{I_M{}^\circ, T^\circ, F^\circ\} \cup \mathcal{C}_0 \subseteq \mathcal{F}_0$;

(b) the expressions in the right-hand sides of the equalities (1), (2) and (4) make sense for all $u \in M$, $Z \in \mathcal{Z}_0$, and their values as functions of u, Z belong to $\mathcal{F}_0$ whenever φ, ψ, χ belong to $\mathcal{F}_0$.

Then we can define multiplication and operations Π and Σ in $\mathcal{F}_0$ by means of the same formulas (1), (2), (4). The multiplication in $\mathcal{F}_0$ will be again monotonically increasing and associative, and therefore $\mathcal{F}_0$ also can be considered as a partially ordered semigroup. Let L', R' be now elements of $\mathcal{F}_0$ satisfying the equalities (5) for all s, t in M and all Z in $\mathcal{Z}_0$. Using the proof of Proposition 1 with the needed obvious changes, we can prove that the 9-tuple

$$\mathfrak{S}_0 = \langle \mathcal{F}_0, I_M{}^\circ, \mathcal{C}_0, \Pi, L', R', \Sigma, T^\circ, F^\circ \rangle$$

is again a combinatory space.

Example 2. Following Ivanov [1986, Exercise 27.6], let us suppose that a partial ordering is given also in the set M, the mappings J, T, F are monotonically increasing, H' is monotonically increasing also with respect to its first argument, and L, R are monotonically increasing total mappings of M into M, satisfying the usual condition that

$$L(J(s, t)) = s, \quad R(J(s, t)) = t \tag{6}$$

for all s, t in M. Remark 3 can be applied for obtaining the statement of the mentioned exercise. Let $\mathcal{Z}_0$ consist of those mappings of M into V which are monotonically increasing, and let $\mathcal{F}_0$ consist of those mappings of $M\times\mathcal{Z}_0$ into V which are monotonically increasing with respect to both their arguments. Let $L' = L^\circ$, $R' = R^\circ$. It can be easily seen that all assumptions of Remark 3 are satisfied in this case, and hence the corresponding 9-tuple $\mathfrak{S}_0$ is a combinatory space.

Remark 4. The combinatory spaces constructed on the basis of Remark 3 are, in general, not necessarily symmetric. The book Skordev [1980] contains two examples of symmetric combinatory spaces of the considered kind. In fact they are probabilistic examples, but, roughly speaking, using expectations instead of probabilities. We shall formulate appropriate versions of these examples in the two theorems below. Unfortunately, the proofs of these theorems, and especially of the second one, use too much analysis and topology, and therefore we decided to omit the proofs this time.

Theorem 1 (Skordev [1980, Ch. II, Proposition 5.5.2]). Let the set M be a compact Hausdorff topological space, and the mappings J, T, F be continuous. Let V be the set of the non-negative real numbers, and $\mathcal{Z}_0$ be the set of all continuous non-negative functions on M. Let $\mathcal{F}_0$ be the set of all mappings θ of $M\times\mathcal{Z}_0$ into V which have the following properties:

(a) for any fixed u in M, the functional $\lambda Z.\,\theta(u, Z)$ is linear and has a norm not greater than 1;[101]

(b) for any fixed Z in $\mathcal{Z}_0$, the function $\lambda u.\,\theta(u, Z)$ is continuous.

Let L', R' be elements of $\mathcal{F}_0$ satisfying the equalities (5) for all s, t in M and all Z in $\mathcal{Z}_0$.[102] Let Z^t and Z^f be

---

[101] I.e. $\theta(u, a_1Z_1+a_2Z_2)=a_1\theta(u, Z_1)+a_2\theta(u, Z_2)$ for all Z_1, Z_2 in $\mathcal{Z}_0$ and all non-negative real numbers a_1, a_2, and the inequality $\theta(u, Z)\le \sup \operatorname{rng} Z$ holds for all $Z\in\mathcal{Z}_0$.

[102] Such L', R' are, for example, L° and R° (cf. Remark 3) if L and R are continuous total mappings of M into M satisfying the conditions (6) for all s, t in M. As an example of a compact Hausdorff topological space admitting such a continuous pairing mechanism we indicate the set of all infinite sequences of zeros and ones supplied with the Tychonoff topology. In this case we can find also continuous mappings T, F and functions Z^t, Z^f with the

functions from $\mathcal{Z}_0$ such that

$$Z^{t}(T(u)) = Z^{f}(F(u)) = 1, \quad Z^{t}(u) + Z^{f}(u) \leq 1$$

for all u in M, and let H' be the mapping of $M \times V^2$ into V defined by means of the equality

$$H'(s, a, b) = Z^{t}(s)a + Z^{f}(s)b.$$

Then the assumptions in Remark 3 are satisfied, and hence the corresponding 9-tuple $\mathfrak{S}_0$ is a combinatory space. Moreover, this combinatory space is symmetric, and the operation Π in it can be defined also by means of the formula 4.(14).

Theorem 2 (Skordev [1980, Ch. II, Proposition 5.5.3, and Ch. III, Section 3.2, Example 15]). Let the set M be a locally compact Hausdorff topological space, and the mappings J, T, F be continuous. Let V be the set of the non-negative real numbers, and $\mathcal{Z}_0$ be the set of all bounded lower semicontinuous non-negative functions on M. Let $\mathcal{F}_0$ be the set of all mappings θ of $M \times \mathcal{Z}_0$ into V which have the following properties:

(a) for any fixed u in M, the functional $\lambda Z.\, \theta(u, Z)$ is linear and has a norm not greater than 1;

(b) for any fixed Z in $\mathcal{Z}_0$, the function $\lambda u.\, \theta(u, Z)$ is lower semicontinuous;

(c) whenever $\mathcal{U}$ is a directed upwards subset of $\mathcal{Z}_0$, and $\sup \mathcal{U} \in \mathcal{Z}_0$,[103] then, for all u in M, the equality

$$\theta(u, \sup \mathcal{U}) = \sup\{\theta(u, Z) : Z \in \mathcal{U}\}$$

holds.

Let L', R' be elements of $\mathcal{F}_0$ satisfying the equalities (5) for all s, t in M and all Z in $\mathcal{Z}_0$.[104] Let H' be a map-

---

properties formulated next. For instance, we may set $T(u)$ and $F(u)$ to be always the sequence consisting only of ones and the sequence consisting only of zeros, respectively, $Z^{t}(u)$ to be always the first member of u, and $Z^{f}(u)$ to be always equal to $1 - Z^{t}(u)$.

[103] The values of $\sup \mathcal{U}$ are determined in a pointwise way.

[104] Such L', R' surely exist if there are continuous total mappings L and R of M into M satisfying the conditions (6) for all s, t in M. In this connection we note that any locally compact Hausdorff topological space B can

ping of $M\times V^2$ into V defined in the same way as in Theorem 1 (with Z^t, Z^f belonging to the new $\mathcal{Z}_0$). Then the assumptions in Remark 3 are satisfied, and hence the corresponding 9-tuple $\mathfrak{S}_0$ is a combinatory space. Moreover, this combinatory space is symmetric, and the operation Π in it can be defined also by means of the formula 4.(14). The combinatory space $\mathfrak{S}_0$ is iterative, and, for arbitrary σ, χ in $\mathcal{F}_0$ and arbitrary $u\in M$, $Z\in\mathcal{Z}_0$, the equality

$$[\sigma, \chi](u, Z) = \sum_{m=0}^{\infty} \rho_m(u, Z)$$

holds, where ρ_m is defined by means of the equality

$$\rho_m(v_0, Z) = \sigma(v_0, \lambda v_1 . \sigma(v_1, \lambda v_2 . \dots$$
$$\sigma(v_{m-1}, \lambda v_m . Z(v_m)\chi(v_m, Z^f)\prod_{j=0}^{m-1}\chi(v_j, Z^t))\dots)).$$

Proposition 1 and the remarks after it do not give sufficient conditions for assuring that the constructed combinatory spaces are iterative. We shall give now two such conditions.

Proposition 2 (cf. Exercise 27.7 in Ivanov [1986]). Let every chain in V (including the empty one) has a least upper bound. Then the combinatory space $\mathfrak{S}$ considered in Proposition 1 is iterative.

Proof. We shall apply the Unrestricted Iteration Lemma (Proposition II.4.5). The assumption (i) in it (each chain in $\mathcal{F}$ has a least upper bound) is obviously satisfied (the least upper bound can be constructed by a pointwise transition to least upper bounds in V). It remains to verify the assumption (ii), namely that the mappings $\lambda\tau.\varphi\tau$, with fixed φ in $\mathcal{F}$, and the mappings $\lambda\tau.\tau z$, with fixed z in $\mathcal{C}$, are continuous with respect to least upper bounds of arbitrary chains. But this continuity is clear from the

---

be embedded in a locally compact Hausdorff topological space with a continuous pairing mechanism. This can be done by means of an appropriate extension of the topological structure from the space B to the set B^* of the Moschovakis computational structure $\mathfrak{M}_B$. To satisfy all assumptions of the theorem, we can take also T, F from $\mathfrak{M}_B$ and set $Z^t(u)=0$, $Z^f(u)=1$ for all u in B°, $Z^t(u)=1$, $Z^f(u)=0$ for the other u.

equalities

$$(\varphi\tau)(u, Z) = \tau(u, \lambda v.\ \varphi(v, Z)),$$

$$(\tau \bar{r}^{\sim})(u, Z) = \tau(r, Z)$$

and the pointwise character of the least upper bounds of the chains in $\mathcal{F}$. ∎

Proposition 3. Let V have a least element, and let each monotonically increasing infinite sequence of elements of V has a least upper bound. Let $\mathcal{F}_0$ consists of all elements θ of $\mathcal{F}$ such that, for any fixed u in M, the mapping $\lambda Z.\ \theta(u, Z)$ of $\mathcal{Z}$ into V is continuous with respect to least upper bounds of monotonically increasing infinite sequences in $\mathcal{Z}$. Let L' and R' belong to $\mathcal{F}_0$, and H', considered as a function of its second or third argument, be continuous with respect to least upper bounds of monotonically increasing infinite sequences in V. Then $\mathcal{F}_0$ is a subsemigroup of the partially ordered semigroup $\mathcal{F}$, and if Π_0 and Σ_0 are the restrictions of Π and Σ to $\mathcal{F}_0^2$ and to $\mathcal{F}_0^3$, respectively, then the 9-tuple

$$\mathfrak{S}_0 = \langle \mathcal{F}_0, I_M^{\sim}, \mathcal{C}, \Pi_0, L', R', \Sigma_0, T^{\sim}, F^{\sim} \rangle$$

is an iterative combinatory space.

Proof. It is easily seen that $\mathcal{F}_0$ is closed under multiplication and under operations Π and Σ (when considering the case of Σ, one uses that the made continuity assumption about H' implies the continuity of H', considered as a function of its second and its third argument simultaneously). For each total mapping f of M into M, the corresponding $f^{\sim}$ belongs to $\mathcal{F}_0$. In particular, $I_M^{\sim}, T^{\sim}, F^{\sim}$ belong to $\mathcal{F}_0$, and $\mathcal{C} \subseteq \mathcal{F}_0$. Since L' and R' also belong to $\mathcal{F}_0$, and, by Proposition 1, $\mathfrak{S}$ is a combinatory space, it is clear that $\mathfrak{S}_0$ is also a combinatory space. So it remains only to show that $\mathfrak{S}_0$ is iterative. This will be done by application of the Level Omega Iteration Lemma.

Let o be the least element of V. Then $o = \lambda u Z.\ o$ is the least element of $\mathcal{F}_0$, and $\tau o = o$ for each τ in $\mathcal{F}_0$, since

$$(\tau o)(u, Z) = o(u, \lambda v.\ \tau(v, Z)) = o = o(u, Z)$$

for all $u \in M$, $Z \in \mathcal{Z}$.

To show that each monotonically increasing infinite sequence of elements of $\mathcal{F}_0$ has a least upper bound in $\mathcal{F}_0$, suppose that $\{\theta_k\}_{k=0}^{\infty}$ is such a sequence. This sequence has a least upper bound θ in $\mathcal{F}$, and the problem is to show that $\theta \in \mathcal{F}_0$. For that purpose, suppose some u in M is fixed, and a monotonically increasing sequence $\{Z_n\}_{n=0}^{\infty}$ of elements of $\mathcal{Z}$ is given. Let Z be the least upper bound of this sequence in $\mathcal{Z}$. Then

$$\theta(u, Z) = \sup\{\theta_k(u, Z)\}_{k=0}^{\infty} = \sup\{\sup\{\theta_k(u, Z_n)\}_{n=0}^{\infty}\}_{k=0}^{\infty} =$$
$$\sup\{\sup\{\theta_k(u, Z_n)\}_{k=0}^{\infty}\}_{n=0}^{\infty} = \sup\{\theta(u, Z_n)\}_{n=0}^{\infty}.$$

Thus for any fixed u in M, $\lambda Z.\, \theta(u, Z)$ is continuous with respect to least upper bounds of monotonically increasing infinite sequences in $\mathcal{Z}$, and hence $\theta \in \mathcal{F}_0$.

Now let $\varkappa$ be some fixed element of $\mathcal{F}_0$. As seen in the proof of Proposition 2, the mapping $\lambda\tau.\, \varkappa\tau$ is continuous with respect to least upper bounds of arbitrary chains in $\mathcal{F}$, hence, in particular, it will be continuous with respect to least upper bounds of monotonically increasing infinite sequences in $\mathcal{F}_0$. As to the mappings $\lambda\tau.\, \tau\varkappa$ and $\lambda\tau.\, \Sigma(\varkappa, \tau, I_M^{\sim})$, their continuity follows from the equalities

$$(\tau\varkappa)(u, Z) = \varkappa(u, \lambda v.\, \tau(v, Z)),$$
$$\Sigma(\varkappa, \tau, I_M^{\sim})(u, Z) = \varkappa(u, \lambda s.\, H'(s, \tau(u, Z), I_M^{\sim}(u, Z)))$$

and the continuity of $\varkappa$ and H'. ∎

We shall end this section by showing that the combinatory spaces described in Remark 3 are general enough in the sense that every given combinatory space is isomorphic to some of them. This will be seen from the following proposition.

Proposition 4. Let $\langle \mathcal{F}_1, I_1, \mathcal{C}_1, \Pi_1, L_1, R_1, \Sigma_1, T_1, F_1 \rangle$ be an arbitrary combinatory space, $\langle M, J, T, F \rangle$ be the corresponding coding structure from Example 1. Let V be the set of those elements α of $\mathcal{F}_1$ which satisfy the equality $\alpha u = \alpha$ for all u in M,[105] and let M and V be equipped with the partial orderings induced from $\mathcal{F}_1$. Let H' be the restriction of the operation Σ_1 to the set $M \times V^2$. For each element τ of $\mathcal{F}_1$, let $\bar{\tau}$ be the mapping of M into V defined by means of the equality

---

[105] Some information about the elements with this property can be found in Exercise II.1.26.

$$\bar{\tau}(u) = \tau u,$$

and let $\mathcal{Z}_0$ be the set of all mappings $\bar{\tau}$, where $\tau \in \mathcal{F}_1$.[106] For each element θ of $\mathcal{F}_1$, let $\theta^\vee$ be the mapping of $M \times \mathcal{Z}_0$ into V defined by means of the equality

$$\theta^\vee(u, \bar{\tau}) = \tau\theta u,$$

and let $\mathcal{F}_0$ be the set of all mappings $\theta^\vee$, where $\theta \in \mathcal{F}_1$. Let $\mathcal{Z}_0$ and $\mathcal{F}_0$ be equipped by the partial orderings induced by the partial ordering in V. Let the denotations f° and $\mathcal{C}_0$ have the same meaning as in Remark 3,[107] and let $L' = L_1^\vee$, $R' = R_1^\vee$. Then:

(i) the set V is infinite;

(ii) the mappings J, T, F, H' are monotonically increasing with respect to all their arguments, and the equalities (3) hold for all u in M and all a, b in V;

(iii) $\lambda s.\, H'(s, a, b) \in \mathcal{Z}_0$ for all a, b in V;

(iv) all mappings $\theta^\vee$, where $\theta \in \mathcal{F}_1$, are monotonically increasing with respect to both their arguments;

(v) the mapping $\lambda\theta.\, \theta^\vee$ is an isomorphism between the partially ordered sets $\mathcal{F}_1$ and $\mathcal{F}_0$;

(vi) for all φ, ψ, χ in $\mathcal{F}_1$, all u in M and all Z in $\mathcal{Z}_0$, we have the equalities

$$(\varphi\psi)^\vee(u, Z) = \psi^\vee(u, \lambda v.\, \varphi^\vee(v, Z)),$$

$$(\Pi_1(\varphi, \psi))^\vee(u, Z) = \varphi^\vee(u, \lambda s.\, \psi^\vee(u, \lambda t.\, Z(J(s, t)))),$$

$$(\Sigma_1(\chi, \varphi, \psi))^\vee(u, Z) = \chi^\vee(u, \lambda s.\, H'(s, \varphi^\vee(u, Z), \psi^\vee(u, Z)));$$

(vii) the equalities $I_1^\vee = I_M^\circ$, $T_1^\vee = T^\circ$, $F_1^\vee = F^\circ$ hold, and the image of $\mathcal{C}_1$ under $\lambda\theta.\, \theta^\vee$ is equal to $\mathcal{C}_0$;

(viii) for all s, t in M and all Z in $\mathcal{Z}_0$, the equalities (5) hold.

---

[106] In the case when $\tau \in M$, the above definition of $\bar{\tau}$ gives the result that $\bar{\tau}(u) = \tau$ for all u in M, hence there is no collision between this definition and the previously given definition of $\bar{s}$ for $s \in M$.

[107] I.e. $f^\circ(u, \bar{\tau}) = \bar{\tau}(f(u)) = \tau f(u)$ for all total mappings f of M into M and all $u \in M, \tau \in \mathcal{F}_1$, and $\mathcal{C}_0$ is the set of all $\bar{s}^\circ$, where $s \in M$.

Proof. We shall verify only the equalities in (vi), and the rest will be left to the reader. Let φ, ψ, χ belong to $\mathcal{F}_1$, u be an element of M, and Z be an element of $\mathcal{Z}_0$. By the definition of $\mathcal{Z}_0$, $Z=\overline{\tau}$ for some τ in $\mathcal{F}_1$. Then

$$\psi^{\vee}(u, \lambda v.\ \varphi^{\vee}(v, Z))=\psi^{\vee}(u, \lambda v.\ \tau\varphi v)=$$
$$\psi^{\vee}(u, \overline{\tau\varphi})=\tau\varphi\psi u=(\varphi\psi)^{\vee}(u, Z),$$

$$\varphi^{\vee}(u, \lambda s.\ \psi^{\vee}(u, \lambda t.\ Z(J(s, t))))=$$
$$\varphi^{\vee}(u, \lambda s.\ \psi^{\vee}(u, \lambda t.\ \tau\Pi_1(s, t)))=$$
$$\varphi^{\vee}(u, \lambda s.\ \psi^{\vee}(u, \lambda t.\ \tau\Pi_1(s, I_1)t))=$$
$$\varphi^{\vee}(u, \lambda s.\ \psi^{\vee}(u, \overline{\tau\Pi_1(s, I_1)}))=$$
$$\varphi^{\vee}(u, \lambda s.\ \tau\Pi_1(s, I_1)\psi u)=$$
$$\varphi^{\vee}(u, \lambda s.\ \tau\Pi_1(I_1, \psi u)s)=$$
$$\varphi^{\vee}(u, \overline{\tau\Pi_1(I_1, \psi u)})=\tau\Pi_1(I_1, \psi u)\varphi u=$$
$$\tau\Pi_1(\varphi, \psi)u=(\Pi_1(\varphi, \psi))^{\vee}(u, Z),$$

$$\chi^{\vee}(u, \lambda s.\ H'(s, \varphi^{\vee}(u, Z), \psi^{\vee}(u, Z)))=$$
$$\chi^{\vee}(u, \lambda s.\ \Sigma_1(s, \tau\varphi u, \tau\psi u))=$$
$$\chi^{\vee}(u, \lambda s.\ \tau\Sigma_1(I_1, \varphi u, \psi u)s)=$$
$$\chi^{\vee}(u, \overline{\tau\Sigma_1(I_1, \varphi u, \psi u)})=\tau\Sigma_1(I_1, \varphi u, \psi u)\chi u=$$
$$\tau\Sigma_1(\chi, \varphi, \psi)u=(\Sigma_1(\chi, \varphi, \psi))^{\vee}(u, Z). \blacksquare$$

In other words, Proposition 4 says that the given combinatory space $\langle\mathcal{F}_1, I_1, \mathcal{C}_1, \Pi_1, L_1, R_1, \Sigma_1, T_1, F_1\rangle$ is isomorphic to the combinatory space $\mathfrak{S}_0=\langle\mathcal{F}_0, I_M^{\circ}, \mathcal{C}_0, \Pi, L', R', \Sigma, T^{\circ}, F^{\circ}\rangle$ constructed as in Remark 3 on the basis of the coding structure $\langle M, J, T, F\rangle$, the sets V, $\mathcal{Z}_0$, $\mathcal{F}_0$ and the mappings H', L', R' specified in the proposition. The combinatory space $\mathfrak{S}_0$ constructed in this way would be also of the kind considered in Example 2 (with the mentioned partial ordering in M) if we generalize that example conveniently (by taking into account the remark after Exercise 27.6 in Ivanov [1986] and, in addition, allowing L' and R' to be not necessarily generated by monotonically increasing total mappings).

6. Combinatory spaces connected with complexity of data processing

In this section, a computational structure

$$\mathfrak{A}=\langle M, J, L, R, T, F, H\rangle$$

is supposed to be given. A construction will be presented which is based on the following intuitive idea about the complexity of data processing. We consider devices (possibly non-deterministic) which transform elements of M into elements of M, and we suppose that, whenever a concrete transformation of this sort is completed by some device, then some object is defined which characterizes the complexity of the concrete transformation. This object could be, for example, a number measuring the duration of the work of the device, or a number measuring the cost of the concrete data processing, or a vector consisting of both mentioned numbers. If the device sometimes uses external sources of information, the object in question could be also the number of times during the work when such external sources have been used.

When a data processing device of this kind is given, we could use as a mathematical description of it the set of all ordered triples $\langle u, k, v\rangle$, such that it is possible the work of the device with input data u to terminate with output data v and complexity k of the data processing.

A natural assumption about the objects measuring the complexity is that an associative operation of addition is defined for them, and this operation has a zero element. Therefore a semigroup E will be supposed to be given with the semigroup operation denoted as addition, and it will be assumed that there is an element 0 of E such that the equalities $k+0=0+k=k$ hold for all k in E. Since E is a semigroup, the associativity law

$$(k+l)+m=k+(l+m)$$

must hold for all k, l, m in E. However, we shall not assume that the semigroup operation in E is necessarily commutative (this enables, for example, E to consist of strings, and the concatenation operation to play the role of addition).

After all what has been said above, it is clear that the used mathematical descriptions of devices will be subsets of the Cartesian product $M\times E\times M$. We shall construct an iterative combinatory space whose semigroup will have all such subsets as its elements. The multiplication of such subsets will correspond to sequential composition of devices, and the other components of the combinatory space will also have a natural intuitive interpretation.

Proposition 1. Let $\mathcal{F}$ be the set of all subsets of the Cartesian product $M\times E\times M$, and let $\mathcal{F}$ be partially ordered by the inclusion relation. Let a multiplication in $\mathcal{F}$ be defined by the following equality, where φ, ψ denote arbitrary elements of $\mathcal{F}$:

$$\varphi\psi = \{\langle u, k, w\rangle : \exists v \exists i \exists j (\langle u, i, v\rangle \in \psi \ \& \ \langle v, j, w\rangle \in \varphi \ \& \ i + j = k)\}.$$

Then $\mathcal{F}$ is a partially ordered semigroup.

The proof of this proposition is left to the reader. We note only that, instead of a direct proof, one can use also an embedding Ξ of $\mathcal{F}$ into the partially ordered semigroup $\mathcal{F}_m(M \times E)$, namely

$$\Xi(\theta) = \{\langle\langle u, i\rangle, \langle v, j\rangle\rangle : i \in E \ \& \ \exists k (\langle u, k, v\rangle \in \theta \ \& \ i + k = j)\}. \tag{1}$$

For each binary relation f in M (in particular, for each partial function f in M), we set

$$f^{\sim} = \{\langle u, 0, v\rangle : \langle u, v\rangle \in f\}.$$

We shall denote by $\mathcal{C}$ the set of those $f^{\sim}$ which correspond to constant total functions f in M.

The promised construction of an iterative combinatory space is described in the following proposition.

Proposition 2. Let $\mathcal{F}$ be the partially ordered semigroup from Proposition 1, and let a binary operation Π and a ternary operation Σ be defined in $\mathcal{F}$ as follows:

$$\Pi(\varphi, \psi) = \{\langle u, k, w\rangle : \exists s \exists t \exists i \exists j (\langle u, i, s\rangle \in \varphi \ \& \ \langle u, j, t\rangle \in \psi \ \& \ J(s, t) = w \ \& \ i + j = k)\},$$

$$\Sigma(\chi, \varphi, \psi) = \{\langle u, k, w\rangle : \exists i \exists j ((\langle u, i, \text{true}\rangle \in H\chi \ \& \ \langle u, j, w\rangle \in \varphi \lor \langle u, i, \text{false}\rangle \in H\chi \ \& \ \langle u, j, w\rangle \in \psi) \ \& \ i + j = k)\},$$

where

$$H\chi = \{\langle u, i, p\rangle : \exists v (\langle u, i, v\rangle \in \chi \ \& \ H(v) = p)\}.$$

Then:

(i) the 9-tuple

$$\mathfrak{S} = \langle \mathcal{F}, I_M^{\sim}, \mathcal{C}, \Pi, L^{\sim}, R^{\sim}, \Sigma, T^{\sim}, F^{\sim}\rangle$$

is an iterative combinatory space;

(ii) the combinatory space $\mathfrak{S}$ is symmetric iff the addition operation in E is commutative;

(iii) for any σ, χ in $\mathcal{F}$, any u, w in M and any k in E, the condition

$$\langle u, k, w\rangle \in [\sigma, \chi]$$

is equivalent to the existence of a natural number m such that, for some sequence $v_0, \ldots, v_m$ of elements of M and some sequences $i_0, \ldots, i_m$ and $j_0, \ldots, j_{m-1}$ of elements of

E, the following conditions are satisfied:

$$v_0 = u, \quad v_m = w,$$
$$\langle v_l, i_l, \text{true}\rangle \in H\chi \ \& \ \langle v_l, j_l, v_{l+1}\rangle \in \sigma, \quad l = 0, \ldots, m-1,$$
$$\langle v_m, i_m, \text{false}\rangle \in H\chi,$$
$$(i_0 + j_0) + \ldots + (i_{m-1} + j_{m-1}) + i_m = k.$$

The proof of this proposition is again left to the reader, since it contains almost nothing essentially different from the proofs of other statements of a similar type which occur in the preceding chapters and sections. The only more specific moment is the proof that the symmetry of $\otimes$ implies commutativity of the addition in E. This can be done by taking arbitrary elements i, j of E and applying the condition II.1.(7*) to an arbitrary element x of $\mathscr{C}$ and to

$$\varphi = M \times \{i\} \times \{s\}, \quad \theta = M \times \{j\} \times \{s\},$$

where s is some fixed element of M.

Remark 1. If we are interested only in deterministic devices then we could use a smaller combinatory space, whose semigroup consists only of the subsets θ of $M \times E \times M$ satisfying the condition that

$$\langle u, i, v\rangle \in \theta \ \& \ \langle u, j, w\rangle \in \theta \Longrightarrow i = j \ \& \ v = w$$

for all u, v, w in M and all i, j in K. We shall call such subsets *functional*. Let $\mathscr{F}_0$ be the set of all functional subsets of $M \times E \times M$. The set $\mathscr{F}_0$ contains among its elements all $f^{\sim}$ corresponding to partial functions f in M, and it is closed under the multiplication defined in Proposition 1 and the operations Π, Σ defined in Proposition 2. Therefore we obtain a combinatory space by restriction of the mentioned operations to $\mathscr{F}_0$ and taking it as the semigroup of the space. It is easily seen that this combinatory space is also iterative, and the same expression for the iteration holds in it. Note that each element θ of $\mathscr{F}_0$ can be determined also by means of two functions having one and the same domain, namely the partial function from M to M, consisting of all pairs $\langle u, v\rangle$, where $\langle u, k, v\rangle \in \theta$ for some k, and the partial function from M to E, consisting of all pairs $\langle u, k\rangle$, where $\langle u, k, v\rangle \in \theta$ for some v. We shall call these partial functions *the output component* and *the complexity component* of θ. We may use this terminology also in the case when the whole combinatory space from Proposition 2 is considered, but θ is a functional element of $\mathscr{F}$.

An illustrative example follows. The intuitive idea in it (containing obvious idealizations) can be described as follows. One considers input-output behaviour and duration of the work of devices which transform natural numbers into natural numbers. The class of those devices is studied which can be constructed by the methods of combining mentioned in Section 2 from primitive ones corresponding to L, R, T, F and from three additional sorts of primitive devices: ones transforming momentarily any given number u into $u+1$, other ones transforming momentarily any given number u into $u \dot{-} 1$, and third ones which do not change the given number, but cause a delay equal to 1 (some unit of time being fixed). The example shows that any partial recursive function can be computed by some device from this class with a delay given by an arbitrarily chosen partial recursive function with the same domain (assuming momentary interactions between the components of the compound devices and momentary coding and decoding by means of J, L, R, T, F, H).

Example 1. Let $\mathfrak{A}$ be a standard computational structure on the natural numbers, E be the semigroup of the natural numbers with the usual addition operation, and $\mathfrak{S}$ be the combinatory space from Proposition 2 corresponding to these $\mathfrak{A}$ and E. Let $\mathcal{B}=\{S^{\sim}, P^{\sim}, \delta\}$, where

$$S=\lambda u.\, u+1, \quad P=\lambda u.\, u \dot{-} 1, \quad \delta=\{\langle u, 1, u\rangle : u\in\mathbb{N}\}.$$

We claim that $\mathrm{COMP}_{\mathfrak{S}}(\mathcal{B})$ consists of the recursively enumerable functional elements of $\mathcal{F}$, and obviously these are exactly the functional elements of $\mathcal{F}$ having partial recursive output and complexity components. The fact that all elements of $\mathrm{COMP}_{\mathfrak{S}}(\mathcal{B})$ are recursively enumerable and functional is seen by induction on the construction of these elements. For the proof of the converse statement, suppose θ is an arbitrary recursively enumerable functional element of $\mathcal{F}$. Let f and g be the output and the complexity component of θ, respectively. Using the partial recursiveness of f and g and applying Theorem I.3.1, one can prove that the elements $f^{\sim}$ and $g^{\sim}$ of $\mathcal{F}$ are $\mathfrak{S}$-computable in $\{S^{\sim}, P^{\sim}\}$. On the other hand, the following equality holds:

$$\theta=L[\delta\Pi(L, P^{\sim}R), R]\Pi(f^{\sim}, g^{\sim}).$$

The $\mathfrak{S}$-computability of θ is clear from this equality.

Remark 2. In the thesis Ignatov [1979], the combinatory space from Remark 1 is considered in the case when $\mathfrak{A}$ is the computational structure from Example I.1.8, and E is the same as in Example 1. Computability in this combinatory space is used for estimating the complexity of computation of concrete recursive functions. Special attention is paid to computability in the set $\mathcal{B}$ consisting, so to say, of the functions S and P with complexity 1. Intuitively, the computability of an element θ in this set $\mathcal{B}$ means that the value of the output component of θ can be effectively computed, starting from the value of the argument, by using as many additions and subtractions of 1 as the complexity component indicates. A number of concrete elements θ are shown to be computable in $\mathcal{B}$, and some other ones are shown to be not computable in $\mathcal{B}$ (for example, whatever the natural number n is, the element, whose both components are $\lambda u. nu$, is computable in $\mathcal{B}$, and whenever θ is an element computable in $\mathcal{B}$ and having output component $\lambda u. nu$, then the value of the complexity component of θ at u cannot be less than $(n-1)u$). Maybe a further study of the computability in this combinatory space could lead to some more profound results.

7. Combinatory spaces connected with side effects of data processing

Again a computational structure

$$\mathfrak{U} = \langle M, J, L, R, T, F, H \rangle$$

is supposed to be given. In addition, a non-empty set E is supposed also to be given, and its elements will be now regarded as states of the environment in which the data processing is carried out. The way of running of the processing by a given device can, in general, depend not only on the input data, but also on the state of the environment at the start of the process. On the other hand, the processing may sometimes have the side effect of changing this state, and therefore, the state at the end of the processing will be not necessarily the same as at the start. In the case of non-deterministic devices the mentioned dependencies are not necessarily functional, and the mathematical counterpart of the above intuitive ideas will be a binary relation in the Cartesian product $M \times E$. A pair $\langle\langle u, i\rangle, \langle v, j\rangle\rangle$ of elements of $M \times E$ will belong to this relation if output v in state j of the environment is a possible result of a processing started with input u in state i of the environment. In this sense such relations will be used as mathematical descriptions of devices, and a combinatory space will be constructed whose semigroup will consist of all such relations. The semigroup multiplication will be the ordinary composition of relations, since now it corresponds again to the sequential composition of devices.

The following proposition contains the construction in question.

Proposition 1. Let $\mathcal{F}$ be the partially ordered semigroup $\mathcal{F}_{\mathrm{m}}(M \times E)$ of all binary relations in $M \times E$. For each binary relation f in M, let

$$f^{\sim} = \{\langle\langle u, k\rangle, \langle v, k\rangle\rangle : \langle u, v\rangle \in f,\ k \in E\}.$$

Let $\mathcal{C}$ be the set of all $f^{\sim}$ corresponding to constant total functions in M. Let Π and Σ be, respectively, a binary and a ternary operation in $\mathcal{F}$, defined as follows:

$$\Pi(\varphi, \psi) = \{\langle\langle u, i\rangle, \langle w, k\rangle\rangle : \exists s \exists t \exists j(\langle\langle u, i\rangle, \langle s, j\rangle\rangle \in \varphi \ \& \ \langle\langle u, j\rangle, \langle t, k\rangle\rangle \in \psi \ \& \ J(s, t) = w)\},$$

$$\Sigma(\chi, \varphi, \psi) = \{\langle\langle u, i\rangle, \langle w, k\rangle\rangle : \exists j(\langle\langle u, i\rangle, \langle \mathrm{true}, j\rangle\rangle \in H\chi \ \& \ \langle\langle u, j\rangle, \langle w, k\rangle\rangle \in \varphi \ \vee \ \langle\langle u, i\rangle, \langle \mathrm{false}, j\rangle\rangle \in H\chi \ \& \ \langle\langle u, j\rangle, \langle w, k\rangle\rangle \in \psi)\},$$

where

$$H\chi = \{\langle\langle u, i\rangle, \langle p, j\rangle\rangle : \exists v(\langle\langle u, i\rangle, \langle v, j\rangle\rangle \in \chi \ \& \ H(v) = p)\}.$$

Then:

(i) the 9–tuple

$$\mathfrak{S} = \langle \mathcal{F}, I_M^{\sim}, \mathcal{C}, \Pi, L^{\sim}, R^{\sim}, \Sigma, T^{\sim}, F^{\sim}\rangle$$

is an iterative combinatory space;

(ii) the combinatory space $\mathfrak{S}$ is symmetric iff the set E consists of a single element;

(iii) for any σ, χ in $\mathcal{F}$, any u, w in M and any i, k in E, the condition

$$\langle\langle u, i\rangle, \langle w, k\rangle\rangle \in [\sigma, \chi]$$

is equivalent to the existence of a natural number m such that, for some sequence $v_0, \ldots, v_m$ of elements of M and some sequences $i_0, \ldots, i_m$ and $j_0, \ldots, j_m$ of elements of E, the following conditions are satisfied:

$$v_0 = u, \quad v_m = w, \quad i_0 = i, \quad j_m = k,$$
$$\langle\langle v_l, i_l\rangle, \langle \mathrm{true}, j_l\rangle\rangle \in H\chi, \quad l = 0, \ldots, m-1,$$
$$\langle\langle v_l, j_l\rangle, \langle v_{l+1}, i_{l+1}\rangle\rangle \in \sigma, \quad l = 0, \ldots, m-1,$$
$$\langle\langle v_m, i_m\rangle, \langle \mathrm{false}, j_m\rangle\rangle \in H\chi.$$

The proof again will be left to the reader. We give only the following hint in connection with the statement (ii): to prove the implication from the symmetry of $\mathfrak{S}$ to the statement that all elements of E are equal each other, suppose i, j are arbitrary elements of E and apply the condition II.1.(7*) to an arbitrary element x of $\mathcal{C}$ and to

$$\varphi = (M \times E) \times (\{\langle s, i\rangle\}, \quad \psi = (M \times E) \times (\{\langle s, j\rangle\},$$

where s is some fixed element of M.

The combinatory spaces connected with complexity of data processing (cf. the previous section) can be embedded in combinatory spaces of the type considered now, and for that purpose the mapping Ξ defined by the equality 6.(1) can be used. Here is the precise formulation of this fact.

Proposition 2. Let an associative operation of addition be defined in the set E, and let this operation has a zero element 0. Let

$$\mathfrak{S} = \langle \mathcal{F}, I_M^{\sim}, \mathcal{C}, \Pi, L^{\sim}, R^{\sim}, \Sigma, T^{\sim}, F^{\sim} \rangle$$

be the combinatory space from the above proposition, and let

$$\mathfrak{S}_0 = \langle \mathcal{F}_0, I_0, \mathcal{C}_0, \Pi_0, L_0, R_0, \Sigma_0, T_0, F_0 \rangle$$

be the combinatory space from Proposition 6.2. Let Ξ be the mapping of $\mathcal{F}_0$ into $\mathcal{F}$ defined by means of the equality 6.(1). Then Ξ is an isomorphic embedding of the partially ordered semigroup $\mathcal{F}_0$ into the partially ordered semigroup $\mathcal{F}$, the image of $\mathcal{C}_0$ under Ξ is equal to $\mathcal{C}$, and the following equalities hold (where $[\sigma, \chi]_0$ denotes iteration in $\mathfrak{S}_0$):

$$\Xi(I_0) = I_M^{\sim},\quad \Xi(L_0) = L^{\sim},\quad \Xi(R_0) = R^{\sim},\quad \Xi(T_0) = T^{\sim},\quad \Xi(F_0) = F^{\sim},$$
$$\Xi(\Pi_0(\varphi, \psi)) = \Pi(\Xi(\varphi), \Xi(\psi)),$$
$$\Xi(\Sigma_0(\chi, \varphi, \psi)) = \Sigma(\Xi(\chi), \Xi(\varphi), \Xi(\psi)),$$
$$\Xi([\sigma, \chi]_0) = [\Xi(\sigma), \Xi(\chi)]$$

(the last three hold for all $\varphi, \psi, \chi, \sigma$ in $\mathcal{F}_0$).

The verification of everything what is claimed in the above proposition contains no difficult moments, and we leave this verification to the reader. We note only that the last equality could be verified either by using the characterizations of the iteration in the both combinatory spaces or by using the Knaster-Tarski-Kleene representation of least fixed points and the fact that the image of $\mathcal{F}_0$ under Ξ is closed with respect to least upper bounds of monotonically increasing sequences.

Remark 1. If E has more than one element, then the image of $\mathcal{F}_0$ under Ξ in the above proposition is surely different from the whole $\mathcal{F}$. This can be seen, for example, by using the fact that, whenever θ is an element of $\mathcal{F}_0$, then the following condition is satisfied:

$$\forall u \forall v \forall i \forall j (\langle\langle u, i\rangle, \langle v, j\rangle\rangle \in \Xi(\theta) \Longrightarrow$$
$$\exists k (\langle\langle u, j\rangle, \langle v, k\rangle\rangle \in \Xi(\theta)).$$

Remark 2. A smaller combinatory space than the combinatory space from Proposition 1 can be constructed by using the partially ordered semigroup $\mathcal{F}_p(M \times E)$ instead of $\mathcal{F}_m(M \times E)$. We shall not enter into details, since they are similar to things mentioned in Remark 6.2 (instead of the name "complexity component" now the name "environment component" will be appropriate).

8. Some combinatory spaces of set-valued partial mappings

As usually, a computational structure

$$\mathfrak{A} = \langle M, J, L, R, T, F, H \rangle$$

is supposed to be given. For the sake of simplicity, the predicate H will be assumed to be total.

We shall consider partial mappings of M into the set of its subsets. The intuitive idea behind our considerations will be the following one. When describing the behaviour of a non-deterministic computational procedure, it is sometimes reasonable to proceed as follows:

(i) to specify the input data for which the execution of the procedure necessarily terminates,

(ii) to specify which are the possible output data corresponding to each concrete instance of the input data mentioned above, and

(iii) to pay no attention to what happens for the other input data.

The description obtained in this way can be represented by a function, which is defined only for the input data mentioned in (i) and transforms each instance of these data into the set of all possible output data corresponding to it. A combinatory space will be constructed now which is related to this intuitive idea.

We shall denote by $\mathcal{F}$ the set of all functions θ such that $\mathrm{dom}\,\theta \subseteq M$, $\mathrm{rng}\,\theta \subseteq \mathcal{P}M$, where $\mathcal{P}M$ is the set of all subsets of M. We introduce in $\mathcal{F}$ the partial ordering which is usual for sets of partial functions, namely, for any φ and ψ in $\mathcal{F}$, we adopt the convention

$$\varphi \geq \psi \iff \mathrm{dom}\,\varphi \supseteq \mathrm{dom}\,\psi \;\&\; \forall u \in \mathrm{dom}\,\psi\, (\varphi(u) = \psi(u))$$

(i.e. $\varphi \geq \psi$ is equivalent to $\varphi \supseteq \psi$, taking into account the interpretation of φ and ψ as subsets of $M \times \mathcal{P}M$). For any φ and ψ in $\mathcal{F}$, we shall denote by $\varphi\psi$ the element θ of $\mathcal{F}$ determined by means of the following conditions:

(a) $\mathrm{dom}\,\theta = \{u \in \mathrm{dom}\,\psi : \psi(u) \subseteq \mathrm{dom}\,\varphi\}$;

(b) for any u in $\mathrm{dom}\,\theta$, the equality

$$\theta(u) = \bigcup\{\varphi(v) : v \in \psi(u)\}$$

holds.

Proposition 1. The set $\mathcal{F}$, considered with the partial ordering and the multiplication introduced above, is a partially ordered semigroup.

Proof. A straightforward verification. ∎

Remark 1. In Exercise II.4.13, a combinatory space has been defined whose semigroup consists of all pairs $\langle f, A\rangle$ with $f \subseteq M\times M$, $A\subseteq M$. Let us denote now that partially ordered semigroup by $\mathcal{F}_0$. For each such pair $\langle f, A\rangle$ belonging to $\mathcal{F}_0$, let $\Xi(\langle f, A\rangle)$ be the function θ from A to $\mathcal{P}M$ defined by means of the equality

$$\theta(u) = \{v : \langle u, v\rangle \in f\}.$$

Then Ξ is a monotonically increasing mapping of $\mathcal{F}_0$ onto $\mathcal{F}$, and this mapping preserves the multiplication. If $\langle f, A\rangle$, describes some non-deterministic computational procedure in the way explained in the footnote to Exercise II.4.13, then the corresponding $\Xi(\langle f, A\rangle)$ will describe the same procedure in the way considered now.

For each partial function f in M, let $f^{\sim}$ be the element of $\mathcal{F}$ determined by the conditions that $\mathrm{dom}\, f^{\sim} = \mathrm{dom}\, f$ and $f^{\sim}(u) = \{f(u)\}$ for all u in $\mathrm{dom}\, f$. Let $\mathcal{C}$ be the set of all $f^{\sim}$ corresponding to constant total functions in M.

Remark 2. If f is a partial function in M, then its representation in the combinatory space from Exercise II.4.13 is $\langle f, \mathrm{dom}\, f\rangle$, and obviously $\Xi(\langle f, \mathrm{dom}\, f\rangle) = f^{\sim}$. Therefore the set $\mathcal{C}$ defined above is the image under Ξ of the set $\mathcal{C}$ from that exercise.

A binary operation Π will be defined in $\mathcal{F}$ as follows: if φ and ψ are arbitrary elements of $\mathcal{F}$, then $\Pi(\varphi, \psi)$ is the element θ of $\mathcal{F}$ such that

$$\mathrm{dom}\,\theta = \{u \in \mathrm{dom}\,\varphi : \varphi(u) \neq \emptyset \Longrightarrow u \in \mathrm{dom}\,\psi\},$$ [108]

and, for all u in $\mathrm{dom}\,\theta$ and all w in M, the equivalence

$$w \in \theta(u) \Longleftrightarrow \exists s \in \varphi(u)\, \exists t \in \psi(u)\, (w = J(s, t))$$

holds.

For each χ in $\mathcal{F}$, we define a partial mapping $H\chi$ of M into the set of the subsets of $\{\mathbf{true}, \mathbf{false}\}$ as follows: $\mathrm{dom}(H\chi) = \mathrm{dom}\,\chi$ and, for all u in $\mathrm{dom}\,\chi$, $(H\chi)(u)$ is the image of the set $\chi(u)$ under H.

For any χ, φ, ψ in $\mathcal{F}$, we shall denote by $\Sigma(\chi, \varphi, \psi)$

---

[108] Compare with the definition of the operation Π in Exercise I.8.3.

the element θ of $\mathcal{F}$ such that

$$\mathrm{dom}\,\theta=\{u\in\mathrm{dom}\,\chi:\ (\mathrm{true}\in(H\chi)(u)\Longrightarrow u\in\mathrm{dom}\,\varphi)\ \&\ (\mathrm{false}\in(H\chi)(u)\Longrightarrow u\in\mathrm{dom}\,\psi)\},$$ [109]

and, for all u in $\mathrm{dom}\,\theta$ and all w in M, the equivalence

$$w\in\theta(u)\Longleftrightarrow \mathrm{true}\in(H\chi)(u)\ \&\ w\in\varphi(u)\ \vee\ \mathrm{false}\in(H\chi)(u)\ \&\ w\in\psi(u)$$

holds.

Remark 3. It can be easily verified that the mapping Ξ of $\mathcal{F}_0$ onto $\mathcal{F}$ preserves the operations Π and Σ.

Two further notions will be introduced, and they will be similar to notions introduced in Exercise II.4.17. Suppose σ and χ are some elements of $\mathcal{F}$. An element u of M will be called *σ,χ-regular* iff the following condition is satisfied

$$u\in\mathrm{dom}\,\chi\ \&\ (\mathrm{true}\in(H\chi)(u)\Longrightarrow u\in\mathrm{dom}\,\sigma).$$

An element w of M will be called a *σ,χ-successor of* the element u iff

$$u\in\mathrm{dom}\,\chi\ \&\ u\in\mathrm{dom}\,\sigma\ \&\ \mathrm{true}\in(H\chi)(u)\ \&\ w\in\sigma(u).$$

It is appropriate also to introduce the notion of a σ,χ-path. As in Exercise II.4.18, a sequence (finite or infinite) of elements of M will be called a *σ,χ-path* iff each term of this sequence except for the first one is a σ,χ-successor of the previous term of the sequence. A σ,χ-path is called to *begin at* (to *end at*) a given element v of M iff v is the first (the last) term of the given σ,χ-path.

Proposition 2. The 9-tuple

$$\mathfrak{S}=\langle\mathcal{F},\ I_M^{\sim},\ \mathcal{C},\ \Pi,\ L^{\sim},\ R^{\sim},\ \Sigma,\ T^{\sim},\ F^{\sim}\rangle$$

is an iterative combinatory space. If σ and χ are some elements of $\mathcal{F}$ then $[\sigma,\chi]$ is the element θ of $\mathcal{F}$ determined by means of the following conditions:

(a) $\mathrm{dom}\,\theta$ is the intersection of all subsets Q of M such that, whenever an element of M is σ,χ-regular, and all its σ,χ-successors belong to M, then this element also belongs to M;

(b) for any $u\in\mathrm{dom}\,\theta$, $w\in M$, the condition $w\in\theta(u)$ is equivalent to the condition that $w\in\mathrm{dom}\,\chi$, $\mathrm{false}\in(H\chi)(w)$, and a finite σ,χ-path exists which begins at u and ends at w.

---

[109] Compare with the definition of the operation Σ in Exercise I.8.3.

The proof of this proposition consists of simpler variants of the solutions of Exercises II.4.13 and II.4.17. We leave this proof to the reader, whom we give the advice to study those exercises and the hint to the second of them.

Remark 4. The domain of $[\sigma, \chi]$ can be characterized also by using the notion of σ, χ-path in the same way as in Exercise II.4.18. Namely $\mathrm{dom}[\sigma, \chi]$ consists of the elements of M such that all σ, χ-paths beginning at them are finite and consist only of σ, χ-regular elements.

Remark 5. Using the characterizations of the corresponding iterations in Exercise II.4.17 and in the above proposition, one can verify that the mapping Ξ preserves also the iteration.

Remark 6. It can be easily seen that the combinatory space $\mathfrak{S}$ from Proposition 2 is not symmetric. A symmetric and again iterative combinatory space can be obtained if we replace $\mathcal{F}$ by the smaller set consisting of the partial mappings of M into the set of the non-empty subsets of M (compare with Exercise II.4.19). Note that in this case $\mathrm{dom}\Pi(\varphi, \psi)$ is simply the intersection of $\mathrm{dom}\varphi$ and $\mathrm{dom}\psi$.

Remark 7. Another modification of the combinatory space from Proposition 2 can be obtained if we replace $\mathcal{F}$ by its subset consisting of the partial mappings of M into the set of the finite subsets of M. Again an iterative combinatory space is obtained, and this time the characterization of $\mathrm{dom}[\sigma, \chi]$ by using the notion of a σ, χ-path can be modified as in Exercise II.4.20. In other words, $[\sigma, \chi]$ can be characterized as the mapping θ of M into $\mathcal{P}M$ which is determined by the condition (b) from Proposition 2, and the condition that $\mathrm{dom}\theta$ consists of the elements u of M having the following properties: all σ, χ-paths beginning at u consist only of σ, χ-regular elements, and there is a finite upper bound for the lengths of these σ, χ-paths.

Remark 8. A topological generalization of the statement in the above remark is possible. Suppose the set M in the given computational structure $\mathfrak{U}$ is a Hausdorff topological space, the sets $\mathrm{dom}L$, $\mathrm{dom}R$, $H^{-1}(\mathrm{true})$, $H^{-1}(\mathrm{false})$ are open, and the mappings J, L, R, T, F are continuous (the statement from the above remark will correspond to the case when the topology is the discrete one). Then an iterative combinatory space arises also if we replace the set $\mathcal{F}$ by its subset consisting of those partial mappings of M into $\mathcal{P}M$ which have open domains and are upper semicontinuous in them in the sense of Berge [1966, Chapter VI, § 1] (the upper semicontinuity in this sense requires, in particular, the values of the mappings to be compact subsets of M). It can be shown that the iteration in this combinatory space can be characterized in the same way as in the previous remark. For the proofs, cf. Skordev [1980, Chapter II, Proposition 5.9.7, and Chapter III, Section 3.2, Example 20]. In the same sections of that book also the case is considered when the additional restriction is imposed on the mappings their values to be connected subsets of the topological space M. In the case of such mappings, it turns out that, for any fixed u in the domain of $[\sigma, \chi]$, all σ, χ-paths beginning at u and ending at elements of the set $[\sigma, \chi](u)$ have one and the same length. These topological considerations have some relation to the interval analysis in the sense of Moore [1966].

Remark 9. Propositions 1 and 2 remain valid if we replace the partial ordering in $\mathcal{F}$ by another one which is defined in the following way:

$$\varphi \geq \psi \iff \mathrm{dom}\varphi \supseteq \mathrm{dom}\psi \ \& \ \forall u \in \mathrm{dom}\psi\,(\varphi(u) \subseteq \psi(u))$$

(a motivation for using such a partial ordering can be derived from the above-mentioned relation to the interval analysis). To prove that the new combinatory space obtained in this way will be iterative, and the iteration in it will be the same, we may use the part (b) of the Unrestricted Iteration Lemma (Proposition II.4.5). The same change of the partial ordering can be made also in the modifications of $\mathfrak{S}$ indicated in Remarks 6-8 (in the case of the spaces from Remarks 7 and 8, part (b) of the Level Omega Iteration Lemma is also applicable).

In the case when the given computational structure is a standard computational structure on the natural numbers, some results about $\mathfrak{S}$-computability will be formulated which are counterparts of the statements of Exercises III.2.13, III.2.16, III.2.17 and III.2.18 (and even can be deduced from them by using the mapping Ξ). These results are listed in the next proposition. We shall omit their proofs, and our advice to the reader is to prove them by simplifying the proposed way of solution for the mentioned exercises.

Proposition 3. Let $\mathfrak{A}$ be a standard computational structure on the natural numbers, and let $S=\lambda u.\, u+1$, $P=\lambda u.\, u \dot{-} 1$. Then:

(i) $\mathrm{COMP}_{\mathfrak{S}}(\{S^{\sim}, P^{\sim}, \lambda u.\, \mathbb{N}\})$ consists of the elements θ of $\mathcal{F}$ such that all values of θ are non-empty, $\mathrm{dom}\theta$ is a Π^1_1-set, and the set $\{\langle u, v\rangle : u \in \mathrm{dom}\theta \ \& \ v \in \theta(u)\}$ is the intersection of the Cartesian product $(\mathrm{dom}\theta)\times\mathbb{N}$ with some recursively enumerable binary relation;

(ii) $\mathrm{COMP}_{\otimes}(\{S^{\sim}, P^{\sim}, \lambda u.\{0, 1\}\})$ consists of the elements θ of $\mathcal{F}$ such that all values of θ are finite and non-empty, the set $\{\langle u, v\rangle : u \in \mathrm{dom}\,\theta \ \&\ v \in \theta(u)\}$ is recursively enumerable, and there is a partial recursive function which transforms each u from $\mathrm{dom}\,\theta$ into the cardinality of the corresponding set $\theta(u)$;[110]

(iii) $\mathrm{COMP}_{\otimes}(\{S^{\sim}, P^{\sim}, \lambda u.\mathbb{N}, \lambda u.\emptyset\})$ consists of the elements θ of $\mathcal{F}$ such that $\mathrm{dom}\,\theta$ is a Π^1_1-set, and the set $\{\langle u, v\rangle : u \in \mathrm{dom}\,\theta \ \&\ v \in \theta(u)\}$ is the intersection of the Cartesian product $(\mathrm{dom}\,\theta)\times\mathbb{N}$ with some recursively enumerable binary relation;

(iv) $\mathrm{COMP}_{\otimes}(\{S^{\sim}, P^{\sim}, \lambda u.\{0, 1\}, \lambda u.\emptyset\})$ consists of the elements θ of $\mathcal{F}$ such that all values of θ are finite, the set $\{\langle u, v\rangle : u \in \mathrm{dom}\,\theta \ \&\ v \in \theta(u)\}$ is recursively enumerable, and there is a partial recursive function which transforms each u from $\mathrm{dom}\,\theta$ into the cardinality of the corresponding set $\theta(u)$.

Set-valued partial mappings can be used for the description of non-deterministic computational procedures also in another way, which is in the spirit of the ideas from Section I.8 (especially of S. Nikolova's ideas which are embodied in Exercise I.8.3). The change in the kind of the description can be expressed by a change in the clause (i) at the beginning of the present section. Namely the words "the execution of the procedure necessarily terminates" must be replaced now by the words "unproductive termination of the execution of the procedure is impossible" (or, in a variant closer to Nikolova's ideas, by the words "failures during the execution of the procedure are impossible"). The corresponding combinatory space will be the same as in Proposition 1 up to the partial ordering in $\mathcal{F}$ and to the iteration, which will be quite different and will look as follows (compare with the partial ordering and the iteration defined in Exercise I.8.3). The partial ordering will be the following one:

$$\varphi \geq \psi \iff \mathrm{dom}\,\varphi \subseteq \mathrm{dom}\,\psi \ \&\ \forall u \in \mathrm{dom}\,\varphi\,(\varphi(u) \supseteq \psi(u))$$

(this is the inverse partial ordering of that one which has been mentioned in Remark 9). The iteration of σ controlled by χ will be now the mapping θ of M into $\mathcal{P}M$ which is determined by the condition (b) from Proposition 2, and the condition that $\mathrm{dom}\,\theta$ consists of the elements of M such

---

[110] A statement equivalent to this is established in Skordev [1980, Chapter IV, Section 1.2, Example 8].

that all σ, χ-paths beginning at them consist only of σ, χ-regular elements.

9. Some combinatory spaces of hybrid nature

Many of the examples of combinatory spaces considered until now have an intuitive interpretation connected with certain kinds of descriptions of data processing devices or procedures. In the different examples, the corresponding descriptions reflect different aspects of the behaviour of the devices or procedures. E.g., we have probabilistic examples of combinatory spaces and examples of combinatory spaces connected with the complexity of data processing. Besides such examples of combinatory spaces, it is possible to construct also combinatory spaces of descriptions which reflect simultaneously several aspects of the behaviour in question. We shall indicate now some combinatory spaces of such a hybrid nature.

We shall show first how to construct combinatory spaces connected simultaneously with probability and with complexity of data processing.

We suppose that a computational structure

$$\mathfrak{A} = \langle M, J, L, R, T, F, H \rangle$$

and a semigroup E are given. The semigroup operation in E will be denoted as addition, and we assume the existence of an element 0 of E such that $k+0=0+k=k$ for all k in E. The intuitive idea about the data processing devices, which are the object of study, is the following one. The devices proceed in a probabilistic manner such that, given any element u of M as input data, for each v in M and each k in E, there is a definite probability that v will be produced as output data, and the complexity of data processing will be equal to k. Mathematically, this state of affairs can be described by a function θ from $M \times E \times M$ into the interval $[0, 1]$ such that, for any u in M, the equality

$$\sum_{k} \sum_{v} \theta(u, k, v) \leq 1$$

holds. Let $\mathcal{F}$ be the set of all such functions θ. We introduce a partial ordering in $\mathcal{F}$ in the natural way, namely $\varphi \geq \psi$ means that $\varphi(u, k, v) \geq \psi(u, k, v)$ for all u, v in M and all k in E. For any two elements φ and ψ of $\mathcal{F}$, we define a real-valued function $\varphi\psi$ by means of the equality

$$\varphi\psi = \lambda u k w. \sum_i \sum_j \sum_v \| i + j = k \| \psi(v, j, w) \varphi(u, i, v).$$

One can prove that $\varphi\psi$ belongs to $\mathcal{F}$ again, and the multiplication in $\mathcal{F}$ defined in this way is an associative operation. We shall not give the proof, but we shall mention that it is possible to reduce the proof of the associativity to the case of multiplication of mappings of $(M \times E)^2$ into $[0, 1]$, and for that purpose it is appropriate to set

$$\Xi(\theta)(\langle u, i\rangle, \langle v, j\rangle) = \sum_k \| i + k = j \| \theta(u, k, v)$$

for each θ in $\mathcal{F}$ and all $\langle u, i\rangle$, $\langle v, j\rangle$ in $M \times E$ (compare with the equality 6.(1)). Since the multiplication in $\mathcal{F}$ is obviously monotonically increasing, the set $\mathcal{F}$ becomes a partially ordered semigroup.

For each partial mapping f of M into M, we define an element $f^{\sim}$ of $\mathcal{F}$ by means of the equality

$$f^{\sim}(u, k, v) = \| \langle u, v\rangle \in f \;\&\; k = 0 \|.$$

We denote by $\mathcal{C}$ the set of those $f^{\sim}$ which correspond to constant total functions in M.

For any two elements φ and ψ in $\mathcal{F}$, we define a real-valued function $\Pi(\varphi, \psi)$ on the set $M \times E \times M$ in the following way: for any u, v in M and any k in E, we set

$$\Pi(\varphi, \psi)(u, k, v) = \sum_i \sum_j \| i + j = k \| \varphi(u, i, L(v)) \psi(u, j, R(v))$$

if $v \in \operatorname{rng} J$, and we set $\Pi(\varphi, \psi)(u, k, v) = 0$ otherwise. It can be proved that $\Pi(\varphi, \psi)$ belongs to $\mathcal{F}$ again.

For any χ in $\mathcal{F}$, we define a real valued function $H\chi$ on the set $M \times E \times \{\mathbf{true}, \mathbf{false}\}$ by setting

$$(H\chi)(u, i, p) = \sum_s \chi(u, i, s) \| s \in H^{-1}(p) \|.$$

Then we set

$$\Sigma(\chi, \varphi, \psi)(u, k, v) =$$

$$\sum_i \sum_j \| i + j = k \| ((H\chi)(u, i, \mathbf{true}) \varphi(u, j, v)) + ((H\chi)(u, i, \mathbf{false}) \psi(u, j, v))$$

for all χ, φ, ψ in $\mathcal{F}$, all u, v in M and all k in E. It can be proved that $\Sigma(\chi, \varphi, \psi) \in \mathcal{F}$ for all χ, φ, ψ in $\mathcal{F}$.

The definitions we gave can be motivated by means of intuitive reasons concerning descriptions of data processing devices and of some combinations of such devices. The following proposition (compare with Propositions 4.1 and 6.2) gives a logical justification of these definitions.

Proposition 1. The 9-tuple

$$\mathfrak{S} = \langle \mathcal{F},\ I_M^{\sim},\ \mathcal{C},\ \Pi,\ L^{\sim},\ R^{\sim},\ \Sigma,\ T^{\sim},\ F^{\sim} \rangle$$

is an iterative combinatory space. For arbitrary σ, χ in $\mathcal{F}$, arbitrary u, w in M and any k in E, the equality

$$[\sigma, \chi](u, k, w) = \sum_{m=0}^{\infty} \rho_m(u, k, w)$$

holds, where

$$\rho_m(u, k, w) = \sum_{v_0} \cdots \sum_{v_m} \sum_{i_0} \cdots \sum_{i_m} \sum_{j_0} \cdots \sum_{j_{m-1}} \| v_0 = u \ \&\ v_m = w \| \times$$

$$(H\chi)(v_m, i_m, \text{false}) \prod_{l=0}^{m-1} ((H\chi)(v_l, i_l, \text{true})\sigma(v_l, j_l, v_{l+1})) \times$$

$$\| (i_0 + j_0) + \ldots + (i_{m-1} + j_{m-1}) + i_m = k \|.$$

The combinatory space $\mathfrak{S}$ is symmetric iff the addition operation in E is commutative.

We shall not present the proof, but we hope that the readers, who have carried out the proofs of the two other above-mentioned propositions, will be able to carry out also this one.

We shall mention quite briefly another example of combinatory spaces of a hybrid nature. The aspects described by the elements of the space now will be the following one: input-output relation, including complexity of the processing, and set of the input data for which the processing necessarily terminates. Thus features of the combinatory spaces from Exercise II.4.13 and from Proposition 6.2 will be put together.

We assume that $\mathfrak{U}$ and E are the same as before, and, for the sake of simplicity, it is appropriate to suppose that the predicate H is total. We shall denote by $\mathcal{F}$ the set of all ordered pairs $\langle \theta, A \rangle$, where $\theta \subseteq M \times E \times M$, $A \subseteq M$. A partial ordering is introduced in $\mathcal{F}$ by means of the following convention:

$$\langle \varphi, A \rangle \geq \langle \psi, B \rangle \Longleftrightarrow \varphi \geq \psi \ \&\ A \geq B \ \&$$
$$\forall u \in B\, \forall k\, \forall v\, (\langle u, k, v \rangle \in \varphi \Longrightarrow \langle u, k, v \rangle \in \psi).$$

A multiplication in $\mathcal{F}$ is defined by means of the equality

$$\langle\varphi, A\rangle\langle\psi, B\rangle = \langle\varphi\psi, \{u\in B : \forall k \forall v (\langle u, k, v\rangle \in\psi \Rightarrow v\in A)\}\rangle,$$

where $\varphi\psi$ is the product of φ and ψ as elements of the combinatory space from Proposition 6.2 (i.e. as elements of the semigroup from Proposition 6.1). We think the rest of the construction of the combinatory space can be left to the reader. The combinatory space constructed in this way is again iterative, and the characterization of the iteration in it is a certain hybrid of the characterizations of the iterations in the above-mentioned two kinds of combinatory spaces (i.e. a hybrid of the characterizations from Exercise II.4.17 and Proposition 6.2).

10. Products of combinatory spaces

In this section, we shall generalize a construction from Exercise II.1.40, namely the construction of the power-space $\mathfrak{S}^K$, where $\mathfrak{S}$ is an arbitrary combinatory space, and K is an arbitrary non-empty set. We shall generalize also the statement of Exercise II.3.9 that $\mathfrak{S}^K$ is iterative, whenever $\mathfrak{S}$ is iterative.

We suppose that K is some non-empty set, and a combinatory space

$$\mathfrak{S}_k = \langle\mathcal{F}_k, I_k, \mathcal{C}_k, \Pi_k, L_k, R_k, \Sigma_k, T_k, F_k\rangle$$

is assigned to each k from K. We shall denote by $\mathcal{F}$ the set of all functions θ such that $\mathrm{dom}\,\theta = K$, and $\theta(k)\in\mathcal{F}_k$ for all k in K. We make the set $\mathcal{F}$ to be a partially ordered semigroup by the conventions that, for any φ, ψ in $\mathcal{F}$, the inequality $\varphi\geq\psi$ holds iff, for all k in K, the inequality $\varphi(k)\geq\psi(k)$ holds in $\mathfrak{S}_k$, and $\varphi\psi$ is the element θ of $\mathcal{F}$ such that $\theta(k) = \varphi(k)\psi(k)$ for all k in K. We set

$$I = \lambda k.\, I_k,\quad L = \lambda k.\, L_k,\quad R = \lambda k.\, R_k,\quad T = \lambda k.\, T_k,\quad F = \lambda k.\, F_k,$$
$$\mathcal{C} = \{\theta\in\mathcal{F} : \forall k\in K\, (\theta(k)\in\mathcal{C}_k)\}.$$

A binary operation Π and a ternary operation Σ are defined in $\mathcal{F}$ by means of the equalities

$$\Pi(\varphi, \psi) = \lambda k.\, \Pi_k(\varphi_k, \psi_k),\quad \Sigma(\chi, \varphi, \psi) = \lambda k.\, \Sigma_k(\chi_k, \varphi_k, \psi_k).$$

The following two propositions can be verified immediately (however, the verification of the first of them makes use of the Axiom of Choice in the general case).

Proposition 1. For any k in K and any x in $\mathcal{C}_k$, there is an element θ of $\mathcal{C}$ such that $\theta(k) = x$.

Proposition 2. For all φ, ψ in $\mathscr{C}$, the elements $\Pi(\varphi, \psi)$, $T\varphi$, $F\varphi$ of $\mathscr{F}$ also belong to $\mathscr{C}$.

The main result in this section reads as follows.

Theorem 1. Let

$$\mathfrak{S} = \langle \mathscr{F}, I, \mathscr{C}', \Pi, L, R, \Sigma, T, F \rangle,$$

where $\mathscr{C}'$ is a subset of $\mathscr{C}$ such that:

(i) for any k in K and any x in $\mathscr{C}_k$, there is an element θ of $\mathscr{C}'$ such that $\theta(k) = x$;

(ii) for all φ, ψ in $\mathscr{C}'$, the elements $\Pi(\varphi, \psi)$, $T\varphi$, $F\varphi$ of $\mathscr{F}$ also belong to $\mathscr{C}'$.

Then:

(a) $\mathfrak{S}$ is a combinatory space;

(b) if the combinatory spaces $\mathfrak{S}_k$ are iterative for all k in K, then $\mathfrak{S}$ is also iterative, and, for all σ, χ in $\mathscr{F}$ and all k in K, the equality

$$[\sigma, \chi](k) = [\sigma(k), \chi(k)]_k$$

holds, where $[\cdot, \cdot]_k$ means iteration in $\mathfrak{S}_k$.

Proof. The conditions II.1.(2), II.1.(9) and II.1.(10) from the definition of the notion of combinatory space are satisfied due to the assumption that $\mathscr{C}'$ satisfies condition (ii) above. To verify condition II.1.(1), suppose that φ and ψ are such elements of $\mathscr{F}$ that $\varphi\theta \geq \psi\theta$ for all θ in $\mathscr{C}'$. Then, taking arbitrary k from K and arbitrary x in $\mathscr{C}_k$, we, by condition (i) above, can find an element θ of $\mathscr{C}'$ such that $\theta(k) = x$. Using this θ, we get

$$\varphi(k)x = \varphi(k)\theta(k) = (\varphi\theta)(k) \geq (\psi\theta)(k) = \psi(k)\theta(k) = \psi(k)x.$$

Since x was chosen arbitrary in $\mathscr{C}_k$, we conclude that $\varphi(k) \geq \psi(k)$. But k was also arbitrary, hence $\varphi \geq \psi$. The verification of all other conditions in the definition of the notion of combinatory space is straightforward.

For the proof of (b), suppose that $\mathfrak{S}_k$ is iterative for all k in K. Let σ, χ be arbitrary elements of $\mathscr{F}$, and let $\iota = \lambda k.\,[\sigma(k), \chi(k)]_k$. We shall show that ι is the iteration of σ controlled by χ in the combinatory space $\mathfrak{S}$. Clearly, $\iota = \Sigma(\chi, \iota\sigma, I)$. To check the second condition in the definition of iteration, suppose that $\mathscr{A}$ is some subset of $\mathscr{C}$ invariant with respect to σ, and τ, ρ are elements of $\mathscr{F}$ satisfying the inequality

$$\tau \underset{\mathscr{A}}{\geq} \Sigma(\chi, \tau\sigma, \rho). \tag{1}$$

We shall prove that $\tau \underset{\mathcal{A}}{\geq} \rho\iota$. For that purpose, we suppose that θ is an arbitrary element of $\mathcal{A}$. We have to prove the inequality $\tau\theta \geq \rho\iota\theta$. Let k be an arbitrary element of K. We must prove that $\tau(k)\theta(k) \geq \rho(k)\iota(k)\theta(k)$. Let us denote by $\mathcal{A}_k$ the set of the values at k of the elements of $\mathcal{A}$. By this definition, $\theta(k) \in \mathcal{A}_k$. Obviously, $\mathcal{A}_k \subseteq \mathcal{C}_k$. It is easily verified that, whenever some elements φ and ψ of $\mathcal{F}$ satisfy the inequality $\varphi \underset{\mathcal{A}}{\geq} \psi$, then the inequality $\varphi(k) \underset{\mathcal{A}_k}{\geq} \psi(k)$ also holds. Hence the inequality (1) implies the inequality

(2) $$\tau(k) \underset{\mathcal{A}_k}{\geq} \Sigma_k(\chi(k), \tau(k)\sigma(k), \rho(k)).$$

Now we shall show that $\mathcal{A}_k$ is invariant with respect to $\sigma(k)$. Let α and β be elements of $\mathcal{F}_k$ satisfying the inequality $\alpha \underset{\mathcal{A}_k}{\geq} \beta$. We have to prove that $\alpha\sigma(k) \underset{\mathcal{A}_k}{\geq} \beta\sigma(k)$. To do this we consider elements φ and ψ of $\mathcal{F}$ such that $\varphi(k) = \alpha$, $\psi(k) = \beta$, and $\varphi(k') = \psi(k')$ for all k' in $K \setminus \{k\}$. It is easily seen that $\varphi \underset{\mathcal{A}}{\geq} \psi$. Since $\mathcal{A}$ is invariant with respect to σ, this implies the inequality $\varphi\sigma \underset{\mathcal{A}}{\geq} \psi\sigma$. Hence the inequality $\varphi(k)\sigma(k) \underset{\mathcal{A}_k}{\geq} \psi(k)\sigma(k)$ holds, i.e. $\alpha\sigma(k) \underset{\mathcal{A}_k}{\geq} \beta\sigma(k)$. Thus we proved the invariance of $\mathcal{A}_k$ with respect to $\sigma(k)$. This invariance, together with the inequality (2), implies the inequality

$$\tau(k) \underset{\mathcal{A}_k}{\geq} \rho(k)\iota(k).$$

Taking into account the already mentioned fact that $\theta(k)$ belongs to $\mathcal{A}_k$, we get the needed inequality

$$\tau(k)\theta(k) \geq \rho(k)\iota(k)\theta(k). \blacksquare$$

By Propositions 1 and 2, the set $\mathcal{C}$ is an example of a set $\mathcal{C}'$ satisfying the assumptions of the above theorem. In general, smaller sets $\mathcal{C}'$ could also happen to satisfy these assumptions. For instance, to obtain the statements concerning $\mathfrak{S}^K$ as corollaries of the theorem, we can take $\mathcal{C}'$ to be the set of all constant mappings of K into $\mathcal{C}$.

REFERENCES

Abian, S., Brown, A. B.

1961. A theorem on partially ordered sets, with applications to fixed points theorems. Canad. J. Math., 13, 78-82.

Backus, J.

1978. Can programming be liberated from the von Neumann style? A functional style and its algebra of programs. Comm. of the ACM, 21, 613-641.

Bekić, H.

1969. Definable operations in general algebras, and the theory of automata and flowcharts. IBM Laboratory, Vienna (manuscript).

Berge, C.

1966. Espaces topologiques. Fonctions multivoques. Deuxième édition. Dunod, Paris.

Bird, R.

1975. Non recursive functionals. Z. Math. Logik Grundlag. Math., 21, 41-46.

Birkhoff, G.

1948. Lattice Theory. A. M. S. Coll. Publ., 25, rev. ed. New York.

Blikle, A.

1971. Nets; complete lattices with a composition. Bull. Acad. Polon. Sci., Sér. Sci. Math. Astr. Phys., 19, 1123-1127.

1972. Iterative systems; an algebraic approach. Bull. Acad. Polon. Sci., Sér. Sci. Math. Astr. Phys., 20, 51-55.

1972a. Complex iterative systems. Bull. Acad. Polon. Sci., Sér. Sci. Math. Astr. Phys., 20, 57-61.

1973. An algebraic approach to mathematical theory of programs. Prace CO PAN, 119, Warszawa.

1974. An extended approach to mathematical analysis of programs. Prace CO PAN, 169, Warsaw.

Böhm, C., Jacopini, G.

1966. Flow diagrams; Turing machines and languages with only two formation rules. Comm. of the ACM, 9, 366-371.

Bourbaki, N.

1949-50. Sur le théorème de Zorn. Archiv der Mathematik, 2, 434-437.

Buchberger, B.

1974. On certain decompositions of Gödel numberings. Z. Math. Logik Grundlag. Math., 16, 85-96.

Chen, T. Y.

1984. On the fixpoints of nondeterministic recursive definitions. J. Computer and System Sciences, 29, 58-79.

Cooper, D. C.

1967. Böhm and Jacopini's reduction of flow charts. Comm. of the ACM, 10, p. 463, p. 473.

de Bakker, J. W.

1971. Recursive Procedures. Math. Centre Tracts, 24, Amsterdam.

de Bakker, J. W., Scott, D.

1969. A theory of programs. IBM Seminar, Vienna (manuscript).

Ditchev, A.

1981. Computability in the Moschovakis' sense and its relation to partial recursiveness via enumerations. Serdica Bulg. Math. Publ., 7, 117-130 (in Russian).

1983. Non-reducibility of certain Cartesian degrees to other ones and the interrelation between two kinds of computability. Sofia Univ., Fac. of Math. and Mech., Sofia (Ph. D. Dissertation, in Bulgarian).

1984. On the computability in the Moschovakis' sense and its relation to partial recursiveness via enumerations. In: Mathematical Logic, Proc. of the Conference on Mathematical Logic Dedicated to the Memory of A. A. Markov (1903-1979), Sofia, September 22-23, 1980, Publ. House of the Bulg. Acad. Sci., Sofia, 34-46 (in Russian).

1987. Search computability and computability with numberings are equivalent in the case of finite set of objects. In: Mathematical Logic and its Applications (ed. D. Skordev), Plenum Press, New York - London, 233-242.

Egli, H.

1975. A mathematical model for nondeterministic computations. Zurich, ETH.

Ershov, A. P.

1958. On operator algorithms. Doklady Acad. Sci. USSR, 122, no. 6, 967-970 (in Russian).

1981. Abstract computability on algebraic structures. Lect. Notes in Computer Science, 122, 397-420.

Fenstad, J. E.

1980. General Recursion Theory, an Axiomatic Approach. Springer-Verlag, New York-Heidelberg-Berlin.

Fitting, M. C.

1981. Fundamentals of Generalized Recursion Theory. North-Holland, Amsterdam-New York-Oxford.

Fraïssé, R.

1961. Une notion de récursivité relative. In: Infinitistic Methods, Proc. Symp. Foundations of Math. (Warsaw, 1959), Pergamon Press, Oxford, 323-328.

Friedman, H.

1971. Algorithmic procedures, generalized Turing algorithms, and elementary recursion theory. In: Logic Colloquium '69 (eds. R. O. Gandy, C. M. E. Yates), North-Holland, Amsterdam, 361-389.

Georgieva, N. V.

1979. On some equivalences of McCarthy. C. R. Acad. Bulgare Sci., 32, no. 6, 721-724 (in Russian).

1980. Normal form theorems for some recursive elements and mappings. C. R. Acad. Bulgare Sci., 33, no. 12, 1577-1580 (in Russian).

1983. On some properties of the conditional expressions. C. R. Acad. Bulgare Sci., 36, no. 6, 725-728 (in Russian).

1984. A theorem on the representability of general recursive functions. In: Mathematical Logic, Proc. of the Conference on Mathematical Logic Dedicated to the Memory of A. A. Markov (1903-1979), Sofia, September 22-23, 1980, Publ. House of the Bulg. Acad. Sci., Sofia, 27-33.

Goguen, J. A.

1967. *L*-fuzzy sets. J. Math. Anal. Appl., 18, 145-147.

Hitchcock, P., Park, D.

1973. Induction rules and termination proofs. In: Automata, Languages and Programming (ed. M. Nivat), North-Holland, Amsterdam, 225-251.

Ignatov, O.

1979. Application of the combinatory spaces to the study of the complexity of computations. Sofia Univ., Fac. of Math. and Mech., Sofia (Master Thesis, in Bulgarian).

Ivanov, L. L.

1977. Recursiveness in natural combinatory spaces. Sofia Univ., Fac. of Math. and Mech., Sofia (Master Thesis, in Bulgarian).

1978. Natural combinatory spaces. Serdica Bulg. Math. Publ., **4**, 296-310 (in Russian).

1980. Iterative operative spaces. Sofia Univ., Fac. of Math. and Mech., Sofia (Ph. D. Dissertation, in Bulgarian).

1980 a. Iterative operative spaces. C. R. Acad. Bulgare Sci., **33**, no. 6, 735-738 (in Russian).

1980 b. Some examples of iterative operative spaces. C. R. Acad. Bulgare Sci., **33**, no. 7, 877-879 (in Russian).

1981. *P*-recursiveness in iterative combinatory spaces. Serdica Bulg. Math. Publ., **7**, 281-297 (in Russian).

1983. Iterative operatory spaces and the system of Scott and de Bakker. Serdica Bulg. Math. Publ., **9**, 275-288.

1984. Kleene-recursiveness and iterative operatory spaces. In: Mathematical Logic, Proc. of the Conference on Mathematical Logic Dedicated to the Memory of A. A. Markov (1903-1979), Sofia, September 22-23, 1980, Publ. House of the Bulg. Acad. Sci., Sofia, 47-62.

1984 a. First order axioms for the foundations of Recursion Theory. In: Extended Abstracts of Short Talks of the 1982 Summer Institute on Recursion Theory (ed. I. Kalantari), Recursive Function Theory Newsletter, 55-60.

1986. Algebraic Recursion Theory. Ellis Horwood, Chichester.

1990. Operative vs. combinatory spaces. J. Symbolic Logic, **55**, 561-572.

19??. Skordev spaces. Annuaire Univ. Sofia, Fac. Math.

Méc., 80 (to appear).

Kelley, J. L.

1975. General Topology. Graduate Texts in Mathematics, 27, Springer-Verlag, New York-Heidelberg-Berlin.

Kleene, S. C.

1952. Introduction to Metamathematics. Van Nostrand, New York - Toronto.

1959. Recursive functionals and quantifiers of finite types I. Trans. Amer. Math. Soc., 91, 1-52.

Knaster, B.

1928. Un théorème sur les fonctions d'ensembles. Annales Soc. Pol. Math., 6, 133-134.

Kreisel, G.

1965. Model theoretic invariants: applications to recursive and hyperarithmetic operations. In: The Theory of Models, Proc. 1963 Internat. Symp. Berkeley, North-Holland, Amsterdam, 190-205.

Lacombe, D.

1964. Deux généralizations de la notion de récursivité. C. R. Acad. Sci. Paris, 258, 3141-3143.

1964a. Deux généralizations de la notion de récursivité relative. C. R. Acad. Sci. Paris, 258, 3410-3413.

Leszczyłowski, J.

1971. A theorem on resolving equations in the space of languages. Bull. Acad. Polon. Sci., Sér. Sci. Math. Astr. Phys., 19, 967-970.

Loève, M.

1977. Probability Theory I. Graduate Texts in Mathematics, 45, Springer-Verlag, New York-Heidelberg-Berlin.

Lukanova, R.

1978. Pseudocombinatory spaces and recursiveness in them. Sofia Univ., Dept. of Math. and Mech., Sofia (Master Thesis, in Bulgarian).

1986. Pseudocombinatory spaces and recursiveness in them. MTA SZTAKI Tanulmanyok, 182, 89-96.

Manna, Z.

1971. Mathematical theory of partial correctness. Lect. Notes in Math, 188, 252-269.

Markowsky, G.

1976. Chain-complete posets and directed sets with applications. Algebra Univ., 6, 53-68.

Mazurkiewicz, A.

1971. Proving algorithms by tail functions. Information and Control, 18, 220-226.

McCarthy, J.

1962. Towards a mathematical theory of computation. In: Information Processing 1962, Proc. IFIP Congress 62 (Munich, 1962), North-Holland, Amsterdam, 21-28.

1963. A basis for a mathematical theory of computation. In: Computer Programming and Formal Systems (ed. P. Braffort, D. Hirshberg), North-Holland, Amsterdam, 33-70.

Moore, R. E.

1966. Interval Analysis. Prentice-Hall, Englewood Cliffs.

Moschovakis, Y. N.

1969. Abstract first order computability. I. Trans. Amer. Math. Soc., 138, 427-464.

1974. Elementary Induction on Abstract Structures. North-Holland, Amsterdam.

Myhill, J.

1961. Note on degrees of partial functions. Proc. Amer. Math. Soc., 12, 519-521.

Pazova, E. G.

1978. Investigation of a concrete combinatory space. Sofia Univ., Fac. of Math. and Mech., Sofia (Master Thesis, in Bulgarian).

Petrov, V. P., Skordev, D. G.

1979. Combinatory structures. Serdica Bulg. Math. Publ., 5, 128-148 (in Russian).

Platek, R.

1966. Foundations of recursion theory. Dissertation, Stanford Univ.

Rogers, H., Jr.

1967. Theory of Recursive Functions and Effective Computability. McGraw-Hill, New York.

Santos, E. S.

1969. Probabilistic Turing machines and computability.

Proc. Amer. Math. Soc., 22, 704-710.

1971. Computability by probabilistic Turing machines. Trans. Amer. Math. Soc., 159, 165-184.

Sasso, L. P.

1975. A survey of partial degrees. J. Symbolic Logic, **40**, 130-140.

Scott, D.

1971. The lattice of flow diagrams. Lect. Notes in Math, **188**, 311-366.

1975. Lambda calculus and recursion theory. In: Proc. of the Third Scandinavian Logic Symposium, North-Holland, Amsterdam, 154-193.

Shepherdson, J. C.

1975. Computation over abstract structures. In: Logic Colloquium '73 (eds. H. E. Rose, J. C. Shepherdson), North-Holland, Amsterdam, 445-513.

Skordev, D. G.

1963. Computable and μ-recursive operators. Izv. Mat. Inst., Bulg. Acad. Sci., 7, 5-43 (in Bulgarian, with German and Russian summaries).

1973. Some examples of universal functions recursively definable by small equation systems. In: Studies in the Theory of Algorithms and Mathematical Logic, vol. I, Computational Centre of the Acad. Sci. USSR, Moscow, 134-177 (in Russian).

1975. Some topological examples of iterative combinatory spaces. C. R. Acad. Bulgare Sci., 28, no. 12, 1575-1578 (in Russian).

1976. On Turing computable operators. Annuaire Univ. Sofia, Fac. Math. Méc., 67, 1972/1973, 103-112.

1976a. Certain combinatory spaces that are connected with the complexity of data processing. C. R. Acad. Bulgare Sci., 29, no 1, 7-10 (in Russian).

1976b. Recursion theory on iterative combinatory spaces. Bull. Acad. Polon. Sci., Sér. Sci. Math. Astr. Phys., **24**, 23-31.

1976c. On the partial ordering of the set $\mathfrak{C}$ in combinatory spaces. C. R. Acad. Bulgare Sci., 29, no. 2, 151-154 (in Russian).

1976d. The concept of search computability from the point of view of the theory of combinatory spaces. Serdica Bulg. Math. Publ., 2, 343-349 (in Russian).

1976e. An axiomatic treatment of recursiveness for some kinds of multi-valued functions (abstract). J. Symbolic Logic, **41**, 555–556.

1977. Simplification of some definitions in the theory of combinatory spaces. C. R. Acad. Bulgare Sci., **30**, no. 7, 947–950.

1978. A normal form theorem for recursive operators in iterative combinatory spaces. Z. Math. Logik Grundlag. Math., **24**, 115–124.

1979. The first recursion theorem for iterative combinatory spaces. Z. Math. Logik Grundlag. Math., **25**, 69–77.

1980. Combinatory Spaces and Recursiveness in them. Publ. House of the Bulg. Acad. Sci., Sofia (in Russian, with English summary).

1980a. Semicombinatory spaces. C. R. Acad. Bulgare Sci., **33**, no. 6, 151–154 (in Russian).

1982. An algebraic treatment of flow diagrams and its application to Generalized Recursion Theory. In: Universal Algebra and Applications, Banach Center Publications, vol. 9, PWN – Polish Scientific Publishers, Warsaw, 277–287.

1982a. An application of Abstract Recursion Theory for investigating the capabilities of the Functional Programming Systems. In: Mathematical Theory and Practice of Software Systems, Computational Centre of the Sib. Div. Acad. Sci. USSR, Novosibirsk, 7–16 (in Russian).

1984. The First Recursion Theorem for iterative semicombinatory spaces. In: Mathematical Logic, Proc. of the Conference on Mathematical Logic Dedicated to the Memory of A. A. Markov (1903–1979), Sofia, September 22–23, 1980, Publ. House of the Bulg. Acad. Sci., Sofia, 89–111 (in Russian).

1984a. A formal system for proving some properties of programs in iterative combinatory spaces. Annales Societatis Mathematicae Polonae, Series IV: Fundamenta Informaticae, VII.3, 359–365.

1987. On the analog of the partial recursive functions for the case of non-deterministic computations. In: Proc. of the Sixteenth Spring Conference of the Union of Bulg. Mathematicians, Sunny Beach, April 6–10, 1987, Publ. House of the Bulg. Acad. Sci., Sofia, 266–272.

1989. On some formal systems for the theory of iterative semi-combinatory spaces. Annuaire Univ. Sofia, Fac. Math. Méc., 79, 1985, livre 1, 323–347 (in Russian, with English summary).

Soskov, I.

1979. Prime computable functions of finitely many arguments with argument and function values in the basic set. Sofia Univ., Fac. of Math. and Mech., Sofia (Master Thesis, in Bulgarian).

1983. Computability in partial algebraic systems. Sofia Univ., Fac. of Math. and Mech., Sofia (Ph. D. Dissertation, in Bulgarian).

1984. Prime computable functions in the basic set. In: Mathematical Logic, Proc. of the Conference on Mathematical Logic Dedicated to the Memory of A. A. Markov (1903-1979), Sofia, September 22-23, 1980, Publ. House of the Bulg. Acad. Sci., Sofia, 112-138 (in Russian).

1985. The connection between prime computability and recursiveness in functional combinatory spaces. In: Mathematical Theory of Programming, Computational Centre of the Sib. Div. Acad. Sci. USSR, Novosibirsk, 4-11 (in Russian).

1987. Prime computability on partial structures. In: Mathematical Logic and its Applications (ed. D. Skordev), Plenum Press, New York - London, 341-350.

Soskova, A.

1979. Some problems connected with the definition of the notion of prime computability. Sofia Univ., Fac. of Math. and Mech., Sofia (Master Thesis, in Bulgarian).

Tabakov, M.

1977. Primitive recursive probabilistic functions. Sofia Univ., Fac. of Math. and Mech., Sofia (Master Thesis, in Bulgarian).

Tarski, A.

1955. A lattice-theoretical fixpoint theorem and its applications. Pacific J. Math., 5, 1955, 285-309.

Uspensky, V. A.

1955. On computable operations. Dokl. Akad. Nauk SSSR, 103, 773-776.

Wand, M.

1973. A concrete approach to abstract recursive definitions. In: Automata, Languages and Programming (ed. M. Nivat), North-Holland, Amsterdam, 331-341.

Zashev, J.

1983. Recursion theory in partially ordered combinatory

models. Sofia Univ., Fac. of Math. and Mech., Sofia (Ph. D. Dissertation, in Bulgarian).

1984. Basic recursion theory in partially ordered models of some fragments of the combinatory logic. C. R. Acad. Bulgare Sci., 37, no. 5, 561-564.

1984a. ∇-combinatory algebras and abstract Plotkin's models. C. R. Acad. Bulgare Sci., 37, no. 6, 711-714.

1985. Abstract Plotkin models. In: Mathematical Theory of Programming, Computational Centre of the Sib. Div. Acad. Sci. USSR, Novosibirsk, 12-33 (in Russian).

1986. *B*-combinatory algebras. Serdica Bulg. Math. Publ., 12, 225-237.

1987. Recursion theory in *B*-combinatory algebras. Serdica Bulg. Math. Publ., 13, 210-223 (in Russian).

1990. Least fixed points in preassociative combinatory algebras. In: Mathematical Logic (ed. P. P. Petkov), Plenum Press, New York - London, 389-397.

ADDITIONAL BIBLIOGRAPHY

1. List of publications referred to in the survey part of the preface of the book Skordev [1980]

(the bibliographical data for the items containing square brackets can be found in the main list of references)

M. A. Arbib, E. G. Manes. Fuzzy machines in a category. Bull. Austral. Math. Soc., 13, 1975, 169-210.

Blikle [1971, 1972, 1972a, 1973, 1974]

de Bakker [1971]

J. W. de Bakker, W. P. de Roever. A calculus for recursive program schemes. In: Automata, Languages and Programming (ed. M. Nivat), North-Holland, Amsterdam, 1973, 167-196.

de Bakker and Scott [1969]

C. C. Elgot. Monadic computations and iterative algebraic theories. In: Logic Colloquium '73 (eds. H. E. Rose, J. C. Shepherdson), North-Holland, Amsterdam, 1975, 175-230.

Friedman [1971]

Hitchcock and Park [1973]

B. Jónsson, A. Tarski. Boolean algebras with operators. II. Amer. J. Math., 74, 1952, 127-162.

Kleene [1952][111]

A. I. Mal'cev, Symmetric grupoids. Matem. sb., 31, 1952, 136-151 (in Russian).

Z. Manna, J. Vuillemin. Fixpoint approach to the theory of computation. In: Automata, Languages and Programming (ed. M. Nivat), North-Holland, Amsterdam, 1973, 273-291.

Mazurkiewicz [1971]

Mc Carthy [1962, 1963]

---

[111] In fact, the Russian translation of this book is referred to in Skordev [1980].

K. Menger. Statistical metrics. Proc. Nat. Acad. Sci. Wash., **28**, no. 12, 1942, 535-537.

K. Menger. Algebra of Analysis. Notre Dame Math. Lect., **3**, 1944, Notre Dame, Indiana.

K. Menger. Tri-operational algebra. Rep. Math. Colloquium, 2nd ser., issue 5-6, 1945, p. 3, issue 7, 1946, p. 46.

K. Menger. Ensembles flous et fonctions aléatoires. C. R. Acad. Sci. Paris, **232**, no. 22, 1951, 2001-2003.

K. Menger. Probabilistic theories of relations. Proc. Nat. Acad. Sci. Wash., **37**, no. 3, 1951, 178-180.

K. Menger. Probabilistic geometry. Proc. Nat. Acad. Sci. Wash., **37**, no. 4, 1951, 226-229.

K. Menger. An axiomatic theory of functions and fluents. In: The Axiomatic Method (Proc. Intern. Symposium, Berkeley, 1958), North-Holland, Amsterdam, 1959, 454-473.

K. Menger. The algebra of functions: past, present, future. Rend. mat. e applic., **20**, no. 3-4, 1961, 409-430.

K. Menger. On substitutive algebra and its syntax. Z. Math. Logik Grundlag. Math., **10**, 1964, 81-104.

K. Menger. Superassociative systems and logical functors. Math. Ann., **157**, 1964, 81-104.

J. H. Morris, Jr. Another recursion induction principle. Comm. of the ACM, **14**, 1971, 351-354.

Moschovakis [1969, 1974]

Y. N. Moschovakis. Axioms for computation theories - first draft. In Logic Colloquium '69 (ed. R. O. Gandy, C. E. M. Yates), North-Holland, Amsterdam, 1971, 199-255.

Y. N. Moschovakis. On nonmonotone inductive definability. Fund. Math., **82**, 1974, 39-83.

D. Park. Fixpoint induction and proofs of program semantics. In: Machine Intelligence, vol. 5 (ed. B. Meltzer, D. Michie), Edinburgh Univ. Press, 1970, 59-78.

Platek [1966]

J. Robinson. Axioms for number theoretic functions. In: Selected Questions of Algebra and Logic, Nauka, Novosibirsk, 1973, 253-263.

E. S. Santos. Fuzzy and probabilistic programs. Information Sci., **10**, 1976, 331-345.

B. M. Schein. Restrictive bisemigroups. Izv. Vyssh. Uchebn. Zaved. Mat., 1965, no. 1, 168-179 (in Russian).

B. M. Schein. Restrictive-multiplicative algebras. Izv.

Vyssh. Uchebn. Zaved. Mat., 1970, no. 4, 91-102 (in Russian).

B. Schweizer, A. Sklar. The algebra of functions. Math. Ann., **139**, 1960, 366-382, **143**, 1961, 440-447, **161**, 1965, 171-196.

B. Schweizer, A. Sklar. The axiomatic characterization of functions. Z. Math. Logik Grundlag. Math., **23**, 1977, 373-382.

Scott [1971]

D. Scott. Continuous lattices. Lect. Notes in Math., **274**, 1972, 97-136.

D. Skordev. A generalization of the theory of recursive functions. Soviet Math. Doklady, **15**, no. 5, 1974, 1756-1760 (Translation from Dokl. Akad. Nauk SSSR, **219**, no. 5, 1974, 1079-1082)[112]

D. Skordev. On multiple-valued functions of several variables. C. R. Acad. Bulgare Sci., **28**, no. 7, 1975, 885-888 (in Russian).

V. V. Vagner. Restrictive semigroups. Izv. Vyssh. Uchebn. Zaved. Mat., 1962, no. 6, 19-27 (in Russian).

2. Some works having relationship to the theory of combinatory spaces, but not mentioned in the text of the present book

G. Gargov. Decidability of the basic combinatory propositional dynamic logic. In: Mathematical Theory of Programming, Computational Centre of the Sib. Div. Acad. Sci. USSR, Novosibirsk, 1985, 42-49 (in Russian).

N. Georgieva. Notes on the equivalence of some programs. In: Summer School on Mathematical Logic and its Applications (Primorsko, September 22-28, 1983), Abstracts of short communications), 1983, 46-48.

N. Georgieva. On the equivalence of Cooper's programs. In: Mathematical Theory of Programming, Computational Centre of the Sib. Div. Acad. Sci. USSR, Novosibirsk, 1985, 119-131 (in Russian).

L. Ivanov. An axiomatization of counters. Serdica Bulg. Math. Publ., **12**, 1986, 358-364.

---

[112] In fact, only the Russian original of this paper is referred to in Skordev [1980].

L. Ivanov. Distributive spaces. In: Mathematical Logic and its Applications (ed. D. Skordev), Plenum Press, New York - London, 1988, 265-272.

S. Nikolova. Computability in sense of Moschovakis over multiple-valued abstract structures. C. R. Acad. Bulgare Sci., **43**, no. 1, 1990, 33-35.

S. Nikolova. Admissibility in abstract structures of arbitrary power. Sofia Univ., Fac. of Math. and Computer Sci., Sofia, 1991 (Ph. D. Dissertation, in Bulgarian).

S. Passy. Combinatory Dynamic Logic. Sofia, Sofia Univ., Fac. of Math. and Mech., Sofia, 1985, (Ph. D. Dissertation).

S. Passy, T. Tinchev. PDL with Data Constants. Information Processing Letters, 20, 1985, 35-41.

S. Passy, T. Tinchev. Quantifiers in Combinatory PDL: Completeness, Definability, Incompleteness. Lecture Notes in Computer Science, **199**, 1985, 512-519.

A. Radensky. An infinite expansion in iterative combinatorial spaces and in functional programming systems. C. R. Acad. Bulgare Sci., **35**, no. 5, 1982, 569-571.

A. Radensky. Functional programming: FP-systems, lazy evaluation and nondeterminism. Sofia Univ., Fac. of Math. and Computer Sci., Sofia, 1988 (Doctoral Dissertation, in Bulgarian).

D. Skordev. A method for computing the values of recursively defined functions. In: Proc. Sci. Conf. in honour of the 1300th Anniversary of the Bulg. State and the 10th Anniversary of the Higher Pedagogigal School in Shoumen (Shoumen, 1981), 1982, 176-187 (in Russian).

D. Skordev. A reduction of polyadic recursive programs to monadic ones. In: Symp. on Math. Found. of Computer Sci. (Diedrichshagen, 1982), Seminarbericht Nr. 52, Sektion Mathematik der Humboldt-Univ. zu Berlin, Berlin, 1983, 124-132.

A. Soskova. Effective algebraic systems. Sofia Univ., Fac. of Math. and Computer Sci., Sofia, 1990 (Ph. D. Dissertation, in Bulgarian).

T. Tinchev. Extensions of the propositional dynamic logic. Sofia Univ., Fac. of Math. and Mech., Sofia, 1986 (Ph. D. Dissertation, in Bulgarian).

T. Tinchev, D. Vakarelov. Propositional Dynamic Logic with least fixed points which are programs. In: Summer School on Math. Logic and its Applications (Primorsko, September 22-28, 1983), Abstracts of short communications), 1983, 64-67.

INDEX OF NAMES

INDEX OF DEFINITIONS

GPSR Compliance

The European Union's (EU) General Product Safety Regulation (GPSR) is a set of rules that requires consumer products to be safe and our obligations to ensure this.

If you have any concerns about our products, you can contact us on ProductSafety@springernature.com

In case Publisher is established outside the EU, the EU authorized representative is:

Springer Nature Customer Service Center GmbH
Europaplatz 3
69115 Heidelberg, Germany

Batch number: 10607320

Printed by Printforce, the Netherlands